# The Handbook of Environmental Chemistry

Editors-in-Chief: O. Hutzinger · D. Barceló · A. Kostianoy

Volume 5 Water Pollution
Part S/1

# The Handbook of Environmental Chemistry
Recently Published and Forthcoming Volumes

**Environmental Specimen Banking**
Volume Editors: S. A. Wise and P. P. R. Becker
Vol. 3/S, 2009

**Polymers: Chances and Risks**
Volume Editors: P. Eyerer, M. Weller
and C. Hübner
Vol. 3/V, 2009

**The Black Sea Environment**
Volume Editors: A. Kostianoy and A. Kosarev
Vol. 5/Q, 2008

**Emerging Contaminants from Industrial and Municipal Waste**
Removal Technologies
Volume Editors: D. Barceló and M. Petrovic
Vol. 5/S/2, 2008

**Emerging Contaminants from Industrial and Municipal Waste**
Occurrence, Analysis and Effects
Volume Editors: D. Barceló and M. Petrovic
Vol. 5/S/1, 2008

**Fuel Oxygenates**
Volume Editor: D. Barceló
Vol. 5/R, 2007

**The Rhine**
Volume Editor: T. P. Knepper
Vol. 5/L, 2006

**Persistent Organic Pollutants in the Great Lakes**
Volume Editor: R. A. Hites
Vol. 5/N, 2006

**Antifouling Paint Biocides**
Volume Editor: I. Konstantinou
Vol. 5/O, 2006

**Estuaries**
Volume Editor: P. J. Wangersky
Vol. 5/H, 2006

**The Caspian Sea Environment**
Volume Editors: A. Kostianoy and A. Kosarev
Vol. 5/P, 2005

**Marine Organic Matter: Biomarkers, Isotopes and DNA**
Volume Editor: J. K. Volkman
Vol. 2/N, 2005

**Environmental Photochemistry Part II**
Volume Editors: P. Boule, D. Bahnemann
and P. Robertson
Vol. 2/M, 2005

**Air Quality in Airplane Cabins and Similar Enclosed Spaces**
Volume Editor: M. B. Hocking
Vol. 4/H, 2005

**Environmental Effects of Marine Finfish Aquaculture**
Volume Editor: B. T. Hargrave
Vol. 5/M, 2005

**The Mediterranean Sea**
Volume Editor: A. Saliot
Vol. 5/K, 2005

**Environmental Impact Assessment of Recycled Wastes on Surface and Ground Waters**
Engineering Modeling and Sustainability
Volume Editor: T. A. Kassim
Vol. 5/F (3 Vols.), 2005

**Oxidants and Antioxidant Defense Systems**
Volume Editor: T. Grune
Vol. 2/O, 2005

# Emerging Contaminants from Industrial and Municipal Waste

## Occurrence, Analysis and Effects

Volume Editors: Damià Barceló · Mira Petrovic

With contributions by
M. L. de Alda · D. Barceló · J. Blasco · A. DelValls · M. Farre
M. Gros · M. Huerta-Fontela · M. Kuster · M. Petrovic · C. Postigo
J. Radjenovic · T. Smital · F. Ventura

Springer

Environmental chemistry is a rather young and interdisciplinary field of science. Its aim is a complete description of the environment and of transformations occurring on a local or global scale. Environmental chemistry also gives an account of the impact of man's activities on the natural environment by describing observed changes.
The Handbook of Environmental Chemistry provides the compilation of today's knowledge. Contributions are written by leading experts with practical experience in their fields. The Handbook will grow with the increase in our scientific understanding and should provide a valuable source not only for scientists, but also for environmental managers and decision-makers.
The Handbook of Environmental Chemistry is published in a series of five volumes:

Volume 1: The Natural Environment and the Biogeochemical Cycles
Volume 2: Reactions and Processes
Volume 3: Anthropogenic Compounds
Volume 4: Air Pollution
Volume 5: Water Pollution

The series Volume 1 The Natural Environment and the Biogeochemical Cycles describes the natural environment and gives an account of the global cycles for elements and classes of natural compounds.
The series Volume 2 Reactions and Processes is an account of physical transport, and chemical and biological transformations of chemicals in the environment.
The series Volume 3 Anthropogenic Compounds describes synthetic compounds, and compound classes as well as elements and naturally occurring chemical entities which are mobilized by man's activities.
The series Volume 4 Air Pollution and Volume 5 Water Pollution deal with the description of civilization's effects on the atmosphere and hydrosphere.
Within the individual series articles do not appear in a predetermined sequence. Instead, we invite contributors as our knowledge matures enough to warrant a handbook article.
Suggestions for new topics from the scientific community to members of the Advisory Board or to the Publisher are very welcome.

ISBN 978-3-540-74793-2 e-ISBN 978-3-540-74795-6
DOI 10.1007/978-3-540-74795-6

The Handbook of Environmental Chemistry ISSN 1433-6863

Library of Congress Control Number: 2008936837



Cover design: WMXDesign GmbH, Heidelberg
Typesetting and Production: le-tex publishing services oHG, Leipzig

Printed on acid-free paper

9 8 7 6 5 4 3 2 1 0

springer.com

## Editors-in-Chief



## Volume Editors

Prof. Dr. Damià Barceló
Dept. of Environmental Chemistry
IIQAB - CSIC
Jordi Girona, 18–26
08034 Barcelona, Spain
*dbcqam@iiqab.csic.es*

Mira Petrovic
Dept. of Environmental Chemistry
IIQAB - CSIC
Jordi Girona, 18–26
08034 Barcelona, Spain
*mpeqam@cid.csic.es*

## Advisory Board

Prof. Dr. M. A. K. Khalil
Department of Physics
Portland State University
Science Building II, Room 410
P.O. Box 751
Portland, OR 97207-0751, USA
*aslam@global.phy.pdx.edu*

Prof. Dr. D. Mackay
Department of Chemical Engineering
and Applied Chemistry
University of Toronto
Toronto, ON, M5S 1A4, Canada

Prof. Dr. A. H. Neilson
Swedish Environmental Research Institute
P.O. Box 21060
10031 Stockholm, Sweden
*ahsdair@ivl.se*

Prof. Dr. J. Paasivirta
Department of Chemistry
University of Jyväskylä
Survontie 9
P.O. Box 35
40351 Jyväskylä, Finland

Prof. Dr. Dr. H. Parlar
Institut für Lebensmitteltechnologie
und Analytische Chemie
Technische Universität München
85350 Freising-Weihenstephan, Germany

Prof. Dr. S. H. Safe
Department of Veterinary
Physiology and Pharmacology
College of Veterinary Medicine
Texas A & M University
College Station, TX 77843-4466, USA
*ssafe@cvm.tamu.edu*

Prof. P. J. Wangersky
University of Victoria
Centre for Earth and Ocean Research
P.O. Box 1700
Victoria, BC, V8W 3P6, Canada
*wangers@telus. net*

## The Handbook of Environmental Chemistry Also Available Electronically

For all customers who have a standing order to The Handbook of Environmental Chemistry, we offer the electronic version via SpringerLink free of charge. Please contact your librarian who can receive a password or free access to the full articles by registering at:

springerlink.com

If you do not have a subscription, you can still view the tables of contents of the volumes and the abstract of each article by going to the SpringerLink Homepage, clicking on "Browse by Online Libraries", then "Chemical Sciences", and finally choose The Handbook of Environmental Chemistry.

You will find information about the

- Editorial Board
- Aims and Scope
- Instructions for Authors
- Sample Contribution

at springer.com using the search function.

*Color figures* are published in full color within the electronic version on SpringerLink.

# Preface

This book on "Emerging Contaminants from Industrial and Municipal Waste" is based on the scientific developments and results achieved within the European Union (EU)-funded project EMCO (reduction of environmental risks posed by emerging contaminants, through advanced treatment of municipal and industrial wastes). One of the key elements of the EMCO project was to provide support to the various Western Balkans countries involved in the project as regards the implementation of the Water Framework Directive (WFD) (2000/60/EC). A regional network, as proposed by the EMCO project, aiming to ensure the comparability (and reliability) of measurement data obtained by screening methodologies for water quality management, would support the EU Water Initiative, which aims to promote co-operation between countries in order to better manage their water resources.

The EMCO project addressed basically two directives: Directive 91/271/EEC to reduce the pollution in Community surface waters caused by municipal waste and the IPPC Directive (Directive 96/61/EC). This Directive expands the range of pollutants that should be monitored in industrial effluent discharges like those from the paper and pulp industry, refineries, textiles and many other sectors. The EMCO project has devoted its attention to the wastewater treatment technologies, especially in the Western Balkan countries. It is obvious that building up and improving wastewater treatment plant performance in the public and private sectors will avoid direct pollution of receiving waters by urban and industrial activities.

The book is divided into two volumes: Vol. I—Occurrence, Analysis and Effects, and Vol. II—Removal Technologies.

Volume I is structured in several chapters covering advanced chemical analytical methods, the occurrence of emerging contaminants in wastewaters, environmental toxicology and environmental risk assessment. Advanced monitoring analytical methods for emerging contaminants cover the use of liquid chromatography combined with tandem mass spectrometric detection or hybrid mass spectrometric techniques. It is certainly known that without these advanced mass spectrometric tools it would not be possible to investigate the fate and behaviour of emerging pollutants at the wastewater treatment plants and receiving waters at the nanogram per litre level. Ecotoxicology is also a very relevant aspect that should be taken into consideration for emerging contaminants, and it is also covered in this book. Risk assessment methodologies will allow us to critically establish the good performance of an appropriate wastewater treatment technology for the removal of urban, agricultural and industrial wastewaters.

Volume II covers different treatment options for the removal of emerging contaminants and includes membrane bioreactors (MBR), ozonization and photocatalysis, and advanced sorbent materials together with more conventional natural systems, such as artificial recharge and constructed wetlands. The MBR is an emerging technology based on the use of membranes in combination with traditional biological treatment. It is considered as a promising technology able to achieve more efficient removal of micro-pollutants in comparison to conventional wastewater treatment plants. Other examples reported in the book are advances in nanomaterials, also an emerging field in wastewater treatment, which are providing great opportunities in the development of more effective wastewater treatment technologies.

Overall, this book is certainly timely since the interest in emerging contaminants and wastewater treatment has been growing considerably during the last few years, related to the availability of novel treatment options together with the advanced and highly sensitive analytical techniques. This book can also be considered, in a way, the follow-up of two previous books in this series entitled *Emerging Organic Pollutants in Waste Waters and Sludge*, Vols. 1 and 2 (5 1 and 5 0), published in 2004 and 2005. The present book is complementary to these volumes since here much more attention has been devoted to wastewater treatment systems, which are a key part of this book.

The book will be of interest to a broad audience of analytical chemists, environmental chemists, water management operators and technologists working in the field of wastewater treatment, or newcomers who want to learn more about the topic. Finally, we would like to thank all the contributing authors of this book for their time and effort in preparing this comprehensive compilation of research papers.

Barcelona, September 2008

D. Barceló
M. Petrovic

# Contents

## Contents of Volume 5, Part S/2

## Emerging Contaminants from Industrial and Municipal Waste

Removal Technologies

**Volume Editors: Barceló, D., Petrovic, M.**
ISBN: 978-3-540-79209-3

Hdb Env Chem Vol. 5, Part S/1 (2008): 1–35
DOI 10.1007/698_5_106

Published online: 18 April 2008

# Emerging Contaminants in Waste Waters: Sources and Occurrence

Mira Petrovic[1,2] (✉) · Jelena Radjenovic[1] · Cristina Postigo[1] · Marina Kuster[1] · Marinella Farre[1] · Maria López de Alda[1] · Damià Barceló[1]

[1]Department of Environmental Chemistry, IIQAB-CSIC, c/ Jordi Girona 18–26, 08034 Barcelona, Spain
*mpeqam@iiqab.csic.es*

[2]Institució Catalana de Reserca i Estudis Avanzats (ICREA), Barcelona, Spain



**Abstract** There is a growing concern about possible ecotoxicological importance of various classes of emerging contaminants in the environment. Numerous field studies designed to provide basic scientific information related to the occurrence and potential transport of specific classes of emerging contaminants in the environment are being conducted with the aim to identify the sources and points of entry of these contaminants into the environment, and to determine their concentrations in both input streams (i.e., urban and industrial wastewaters) and receiving environment. This chapter summarizes the data regarding the occurrence of emerging contaminants in urban and industrial wastewaters, including some prominent classes such as pharmaceuticals, hormones, illicit drugs, surfactants and their degradation products, plasticizers, and perfluorinated compounds.

**Keywords** Emerging contaminants · Municipal waste waters · Occurrence · Sources

**Abbreviations**

| | |
|---|---|
| AP | Alkylphenol |
| APEC | Alkylphenoxy carboxylates |

| | |
|---|---|
| APEO | Alkylphenol ethoxylate |
| BBP | Butylbenzyl phthalate |
| BE | Benzoylecgonine |
| BPA | Bisphenol A |
| CAFO | Concentrated animal feeding operation |
| CE | Cocaethylene |
| DA | Drug of abuse |
| DBP | Dibutyl phthalate |
| DEHP | Di(2-ethylhexyl) phthalate |
| DEP | Diethyl phthalate |
| DMP | Dimethyl phthalate |
| DnOP | Di-*n*-octyl phthalate |
| E1 | Estrone |
| E2 | Estradiol |
| E3 | Estriol |
| EDC | Endocrine disrupting compound |
| EDDP | 2-ethylidine-1,5-dimethyl-3,3-diphenylpyrrolidine perchlorate |
| EE2 | Ethinylestradiol |
| FTOH | Fluorotelomer alcohol |
| LC-MS/MS | Liquid chromatography and tandem mass spectrometry |
| LSD | Lysergic acid diethylamide |
| MDE or MDEA | Methylenedioxyethylamphetamine |
| MDMA | 3,4-Methylenedioxymetamphetamine hydrochloride |
| NP | Nonylphenol |
| NSAID | Non-steroidal anti-inflammatory drug |
| O-H-LSD | 2-Oxo-3-hydroxy-LSD |
| OPEO | Octylphenol ethoxylate |
| OTC | Over-the-counter (drug) |
| PAEs | Phthalate acid ester |
| PEC | Predicted environmental concentration |
| PFBS | Perfluorobutane sulfonate |
| PFCA | Perfluoro carboxylic acid |
| PFCs | Perfluorinated compound |
| PFNA | Perfluorononanoic acid |
| PFOS | Perfluorooctane sulphonate |
| PhAC | Pharmaceutically active compound |
| POP | Persistent organic pollutant |
| THC | $\Delta^9$-Tetrahydrocannabinol |
| WWTP | Wastewater treatment plant |

## 1 Introduction

Until the beginning of the 1990s, non-polar hazardous compounds, i.e., persistent organic pollutants (POP) and heavy metals, were the focus of interest and awareness as priority pollutants and consequently were part of intensive monitoring programs. Today, these compounds are less relevant for the industrialized countries since a drastic reduction of emission has been achieved due to the adoption of appropriate measures and elimination of the dominant pollution sources.

However, the emission of so-called "emerging" or "new" unregulated contaminants has emerged as an environmental problem and there is a widespread consensus that this kind of contamination may require legislative intervention.

A wide range of man-made chemicals, designed for use in industry, agriculture, and as consumer goods and chemicals unintentionally formed or produced as by-products of industrial processes or combustion, are potentially of environmental concern. The term "emerging contaminants" does not necessarily correspond to "new substances", i.e., newly introduced chemicals and their degradation products/metabolites or by-products, but also refers to compounds with previously unrecognized adverse effects on the ecosystems, including naturally occurring compounds. Therefore, "emerging contaminants" can be defined as contaminants that are currently not included in routine monitoring programmes and which may be candidates for future regulation, depending on research on their (eco)toxicity, potential health effects, public perception and on monitoring data regarding their occurrence in the various environmental compartments [1].

Today, there are several groups of compounds that emerged as particularly relevant:

- Algal and cyanobacterial toxins
- Brominated flame retardants
- Disinfection by-products
- Gasoline additives
- Hormones and other endocrine disrupting compounds
- Organometallics
- Organophosphate flame retardants and plasticisers
- Perfluorinated compounds
- Pharmaceuticals and personal care products
- Polar pesticides and their degradation/transformation products
- Surfactants and their metabolites

For most emerging contaminants, occurrence, risk assessment, and ecotoxicological data are not available, and therefore it is difficult to predict what health effects they may have on humans and aquatic organisms. Numerous field studies designed to provide basic scientific information related to the occurrence and potential transport of specific classes of emerging contaminants in the environment are being conducted with the aim to identify the sources and points of entry of these contaminants into the environment, and to determine their concentrations in both input streams (i.e., urban and industrial wastewaters) and receiving environment.

The objective of this chapter is to give an overview of recent monitoring data, focusing on urban and industrial wastewaters. It reports the levels detected for some prominent classes such as pharmaceuticals, hormones, illicit drugs, surfactants and their degradation products, plasticizers and perfluorinated compounds. Possible sources and routes of entry of selected emerging contaminants into the environment are also discussed.

## 2 Pharmaceutical Residues

### 2.1 Sources

Pharmaceutically active compounds (PhACs) are an important group of emerging environmental contaminants that has been an issue of increasing interest in the international scientific community. In the European Union (EU), around 3000 different PhACs are used in human medicine (i.e., analgesics and anti-inflammatory drugs, $\beta$-blockers, lipid regulators, antibiotics, etc), thus their main route into the aquatic environment is ingestion following excretion and disposal via wastewater. After administration, pharmaceutical can be excreted as an unchanged parent compound, in the form of metabolites or as conjugates of glucuronic and sulphuric acid, primarily via urine and faeces. By analyzing the excretion pathways of 212 PhAC, equaling 1409 products, Lienert et al. [2] concluded that on average, 64% ($\pm$27%) of each PhAC was excreted via urine, and 35% ($\pm$26%) via faeces. In urine, 42% ($\pm$28%) of each PhAC was excreted as metabolites. Figure 1 shows the average total fraction excreted via urine and the fraction of the non-metabolized parent compound for selected therapeutic groups.

Metabolites of drugs can be expected to be bioactive and even more persistent, due to their increased polarity. Also, conjugates of parent compounds can be cleaved back into the original drug during the sewage treatment in wastewater treatment plants (WWTPs) [3]. Besides these WWTP discharges that are usually a consequence of their incomplete removal, other environmental exposure pathways of PhACs are manufacturing and hospital effluents, land applications (e.g., biosolids and water reuse), concentrated animal feeding operations (CAFOs), and direct disposal/introduction into the environment. For example, a survey conducted in the USA reported that the vast majority of people were disposing of expired medications via municipal garbage or domestic sewage [4].

In comparison to conventional priority pollutants, PhACs are designed to have specific pharmacologic and physiologic effects at low doses and thus are inherently potent, often with unintended outcomes in wildlife. They can undergo different chemical, photolytic, and biological reactions that mod-

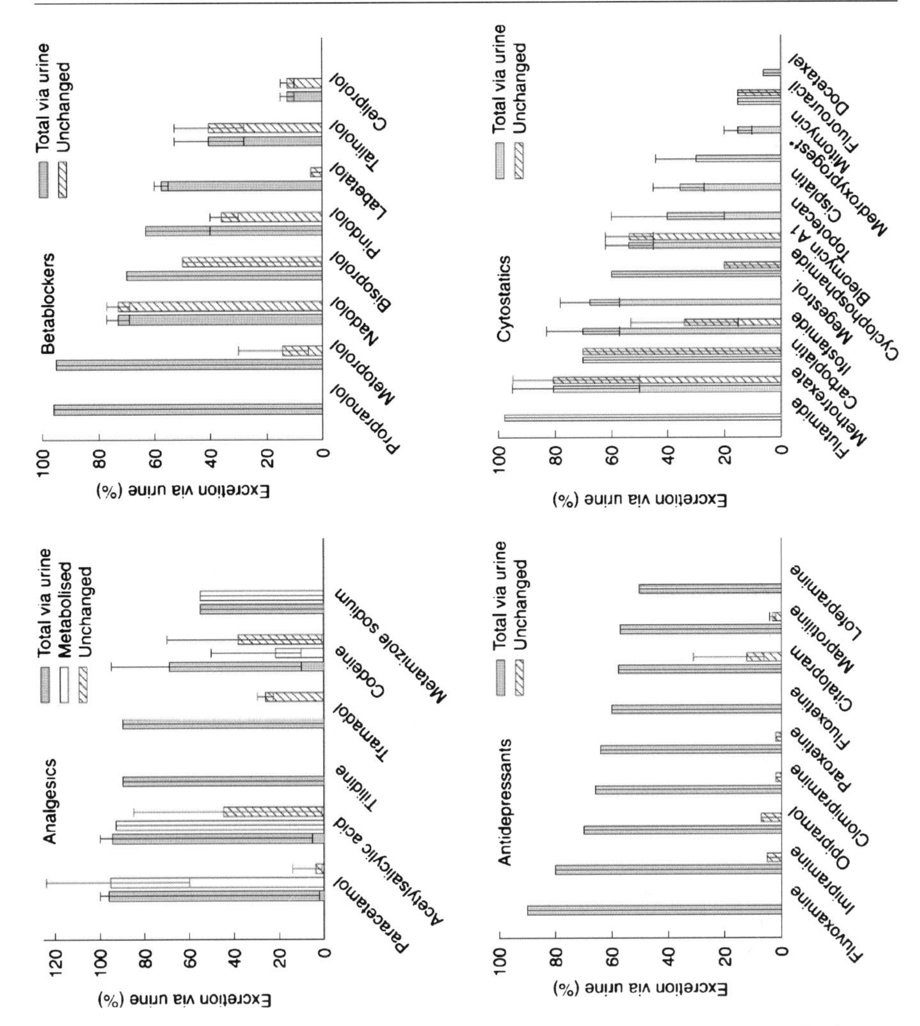


**Fig. 1** Excretion via urine of selected therapeutic groups. The average for each PhAC is shown. *Error bars* denote the minimal and maximal value detected for each PhAC. The total fraction excreted via urine and the fraction of the non-metabolised parent compound (unchanged) is shown. For clarity, excretion via feces is not included. If bars are missing, then respective data were missing (e.g., no data on metabolism for the analgesic tilidine). For antidepressants, β-blockers, and cytostatics, metabolism data were missing for most PhAC. Cytostatics: cyclophosphamide includes cyclophosphane; *p*, medroxyprogesteronacetate. Reprinted with permission from [2]. © IWA Publishing 2007

ify the structure and physical transport of a compound in the environmental media. Furthermore, many PhACs do not exhibit acute toxicity but have a significant cumulative effect on the metabolism of non-target organisms [5] and ecosystem as a whole [6]. Some pharmaceuticals such as antidepressants, β-blockers or lipid regulators, can be prone to biococentration/bioaccumulation in aquatic organisms [7–9]. These results have led to concerns about the ongoing exposure to PhACs, as a result of constant patient use. Also, little is known about their fate and transport in the natural aquatic environment [5, 10], especially when soil/sediment media is in question. There are only a few studies that have dealt with distribution of pharmaceuticals in a natural porous system [11–13]. Therefore, the occurrence of these emerging contaminants in different environmental compartments (e.g., natural waters, waste waters, soil, sludge, sediment) has become a serious issue for the scientific community.

## 2.2 Occurrence in Wastewaters

Due to their continuous input into the aquatic media through wastewater as a main point-source, PhACs are considered to be "pseudo-persistent". In a proper evaluation of persistency of a certain compound, both transformation of a compound in the environment and its supply rate should be taken into consideration [6]. Factors of environmental concern are production volume, ecotoxicity, and persistence. To the extent of feasibility, predicted environmental concentration (*PEC*) can be calculated, based on the excretion rates and portions of pharmaceutical production. Bendz et al. [14] estimated loads of several pharmaceuticals in the influent of a WWTP in Sweden, based on a per-capita consumption rate, number of inhabitants, and the percentage of excretion of drugs as parent compounds. In this attempt they used the following formula published by Alder et al. [15]:

$$PEC_{\mathrm{STPin}} = \frac{F_{\mathrm{API}}E}{PopAWW} \times \frac{10^{12}}{365} ,$$

where $PEC_{\mathrm{WWTPin}}$ is predicted concentration in the WWTP influent ($\mathrm{ng\,L^{-1}}$), $F_{\mathrm{API}}$ consumption of $\beta$-blockers per year ($\mathrm{kg\,yr^{-1}}$), $E$ fraction excreted as active substance without metabolization in urine and/or not absorbed (dimensionless), *Pop* population of Switzerland: 7.3 million inhabitants (cap) and *AWW* is amount of wastewater per capita and day ($400\,\mathrm{L\,cap^{-1}\,d^{-1}}$). The measured concentrations of some of them were of the same order of magnitude as the predicted ones (i.e., diclofenac, naproxen, and metoprolol). However, significantly lower concentrations of gemfibrozil, trimethoprim and atenolol, and significantly higher concentrations of carbamazepine were measured compared to the theoretical values. These discrepancies may be explained with seasonal variations in consumption rates and differences in excretion rates for humans depending on their age, sex, thyroid function, nutrition, etc [14]. In another study [16], predictions made out of excretion rates of atenolol (90%), sotalol (70%), metoprolol (5%) and propranolol (10%) and the data on their consumption in Switzerland gave $PEC_{\mathrm{WWTPin}}$ very similar to their measured concentrations in the influents of two Swiss WWTPs.

Estimations of pharmaceutical concentration in sewage have been usually performed by back-calculating the total prescribed mass from prescription rate data (number of defined daily doses) and excretion rates, partitioning, biodegradation, and the potential hydrolysis of conjugates [17, 18]. However, predictions based on annual sales of drugs are likely to be underestimating the loads of PhACs in the influents of WWTPs. This is because sales figures refer only to prescription drugs, and do not include over-the-counter drugs and Internet sales. Nevertheless, although these predictions have a high degree of uncertainty, they can focus attention on drugs that are candidates for further analytical studies.

The data on measured environmental loads of pharmaceutical residues is still scarce. The inputs of PhACs are generally considered to be constant and widely distributed. However, for some of them (e.g., antibiotics), differences between winter and summer influent loads were noted, probably because of higher attenuation in summer, and also less use of pharmaceuticals [19, 20]. On the other side, for other drugs (e.g., $\beta$-blockers, diuretics and anti-ulcer drugs) this seasonal variability was absent, which was consistent with the data on their occurrence [19].

Over the last 10 years, scattered data all over the world has demonstrated an increasing frequency of appearance in wastewater. The most ubiquitous drugs in WWTP influents are summarized in Table 1, together with their concentration ranges reported in literature.

The ubiquity of drugs is related to specific sales and practices in each country. For example, antihistamines, analgesics, and antidepressants are the families of drugs with major consumption in Spain, according to the National Health System. Indeed, in a study by Gros et al. [21] of the Ebro river basin, the highest influent loads from seven WWTPs were found for non-steroidal anti-inflammatory drugs (NSAIDs), lipid regulators, $\beta$-blockers and histamine $H_1$- and $H_2$-receptor antagonists. The total load of 29 monitored pharmaceuticals ranged from 1 to 5 g/day/1000 inhabitants for influent wastewater (Fig. 2). The results of a study in six WWTPs conducted in Italy [19] indicated high inputs of antibiotics sulfamethoxazole, ofloxacin, and ciprofloxacin, $\beta$-blocker atenolol, anti-histaminic ranitidine, diuretics furosemide and hydrochlorothiazide, and NSAID ibuprofen. A recent comprehensive reconnaissance of more than 70 individual wastewater contaminants in the region of Western Balkan (Bosnia and Herzegovina, Croatia, and Serbia) revealed the presence of 31 out of 44 analyzed pharmaceutical compounds at a concentration above the detection limit (typically 1–10 ng $L^{-1}$) [22]. The most abundant drug groups included analgesics and antiinflammatories, antimicrobials, $\beta$-blockers and lipid regulators, as shown in Fig. 3.

Generally, the most abundant loads are commonly reported for NSAIDs, which could be attributed to their wide consumption because they can be purchased without medical prescription (i.e., over-the counter (OTC) drugs).

**Table 1** Occurrence of pharmaceutical residues in WWTP influents

| Compound | Influent concentration ($\mu g\,L^{-1}$) | Refs. |
|---|---|---|
| ***Analgesics and anti-inflammatory drugs*** | | |
| Ibuprofen | 53.48–373.11; 150.73 $^a$ | [23] |
| | 0.381–1.13; 0.672 $^b$ | [25] |
| | 2.6–5.7 | [134] |
| | 8.45 $^a$; 16.5 $^c$ | [38] |
| | 23.4 $^a$ | [39] |
| | 34–168; 84 $^a$ | [37] |
| Ketoprofen | 0.108–0.369; 0.208 $^b$ | [25] |
| | 0.146 $^a$; 0.289 $^c$ | [38] |
| | 2.9 $^a$ | [39] |
| | 0.57 $^c$ | [40] |
| | 0.16–0.97; 0.451 $^a$ | [28] |
| Naproxen | 0.038–0.23; 0.1 $^b$ | [25] |
| | 1.8–4.6 | [134] |
| | 8.6 $^a$ | [39] |
| | 5.58 $^a$; 17.1 $^c$ | [38] |
| Diclofenac | 0.204 $^a$; 1.01 $^c$ | [38] |
| | 0.46 $^a$ | [39] |
| | 3.25 $^a$; 4.114 $^a$; 3.19 $^a$; 1.4 $^a$; 0.905 $^a$ | [33] |
| | 0.05–0.54; 0.25 $^a$ | [28] |
| | 2.94 $^c$ | [40] |
| Indomethacin | 0.23 $^a$; 0.64 $^c$ | [38] |
| | nd | [28] |
| Acetyl-salicylic acid | 0.47–19.4; 5.49 $^b$ | [25] |
| Salicylic acid | 13.7 $^a$; 27.8 $^c$ | [38] |
| Acetaminophen | 0.13–26.09; 10.194 $^a$ | [28] |
| | 29–246; 134 $^a$ | [37] |
| ***Lipid regulator and cholesterol lowering statin drugs*** | | |
| Gemfibrozil | 0.453 $^a$; 0.965 $^c$ | [38] |
| | nd–0.36; 0.155 $^a$ | [28] |
| Bezafibrate | 2.2 $^a$ | [39] |
| | 1.96 $^a$; 2.014 $^a$; 6.84 $^a$; 7.6 $^a$; 1.55 $^a$ | [33] |
| | nd–0.05; 0.023 $^a$ | [28] |
| Clofibric acid | nd–0.11; 0.072 $^a$ | [28] |
| | 0.36 $^c$ | [40] |
| ***Psychiatric drugs*** | | |
| Carbamazepine | 0.015–0.27; 0.054 $^b$ | [25] |
| | 1.85 $^a$; 1.2 $^a$; 0.704 $^a$; 0.67 $^a$; 0.325 $^a$ | [33] |
| | nd–0.95; 0.42 $^a$ | [28] |
| | 0.12–0.31; 0.15 $^a$ | [37] |
| Caffeine | 52–192; 118 $^a$ | [37] |

**Table 1** (continued)

| Compound | Influent concentration ($\mu g L^{-1}$) | Refs. |
|---|---|---|
| ***Antibiotics*** | | |
| Sulfamethoxazole | nd–0.87; 0.59 [a] | [28] |
| Ofloxacin | nd | [28] |
| Ciprofloxacin | 3.8 [b]; 4.6 [c] | [32] |
| Norfloxacin | 0.17 [b]; 0.21 [c] | [32] |
| Trimethoprim | 0.34 [b]; 0.93 [c] | [32] |
| | nd–4.22; 1.172 [a] | [28] |
| ***Antihistamines*** | | |
| Ranitidine | nd–0.29; 0.188 [a] | [28] |
| ***β-blockers*** | | |
| Atenolol | nd–0.74; 0.395 [a] | [28] |
| | $(0.971 \pm 0.03)$ [a] | [135] |
| Metoprolol | $(0.411 \pm 0.015)$ [a] | [135] |
| Sotalol | 0.12–0.2; 0.167 [a] | [28] |
| | $(0.529 \pm 0.01)$ [a] | [135] |
| Propranolol | 0.08–0.29; 0.168 [a] | [28] |
| | $(0.01 \pm 0.001)$ [a] | [135] |
| ***X-ray contrast media*** | | |
| Iopromide | 6.0–7.0 | [134] |
| | $(7.5 \pm 1.5)$ [a] | [136] |
| Diatrizoate | $(3.3 \pm 0.7)$ [a] | [136] |
| Iopamidol | $(4.3 \pm 0.9)$ [a] | [136] |

[a] mean,
[b] median,
[c] maximum concentrations.

For example, ibuprofen is usually detected at very high concentrations (in $\mu g L^{-1}$) [23–25]. Although the percentage of elimination of this drug is very high [21], it is still detected in rivers downstream WWTPs due to a very high usage in human medicine. Other very popular pain killers are acetaminophen (paracetamol) and aspirin (acetyl-salicylic acid). Acetyl-scalycilic acid is deacetylated in human organism into its more active form, salicylic acid, and two other metabolites, ortho-hydroxyhippuric acid and gentisic acid [26]. Ternes et al. [27] detected all three metabolites in sewage influent samples at very high $\mu g L^{-1}$ concentrations. Gros et al. [28] encountered an average concentration of 10.2 $\mu g L^{-1}$ in WWTP influents. The environmental loads of these drugs are expected to be substantially higher than the values predicted from their sales figures, as their use is often abused.

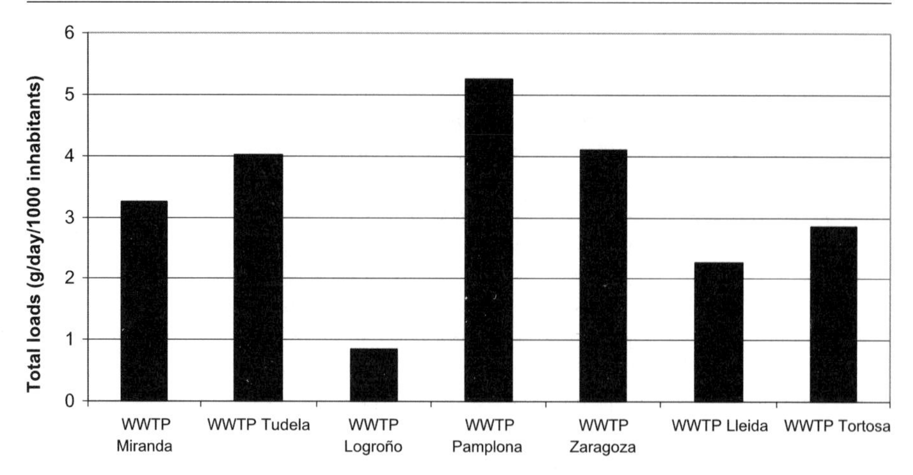


**Fig. 2** Total loads of 29 multi-class pharmaceuticals, expressed as g/day/1000 inhabitants, measured in the raw wastewater entering seven major WWTP in the Ebro River basin. Modified from [21]

Besides these OTC drugs, pharmaceuticals ubiquitous in raw sewage are also prescription drugs $\beta$-blockers [21, 24, 29]. Atenolol seems to be the most frequently found $\beta$-blocker worldwide in WWTP influents [19, 30]. Atenolol, metoprolol, and propranolol were detected at high influent concentrations in a study by Nikolai et al. [30] (i.e., 110–1200, 170–520 and 20–92 ng $L^{-1}$, respectively). As far as their toxicity is concerned, it is suspected that mixtures of $\beta$-blockers are concentration-additive, since they all have the same mode of toxic action in the aquatic environment [31]. These drugs are also used in high quantities and are not efficiently eliminated in WWTPs, thus they are frequently encountered in surface waters [21].

Antibiotic losses to the environment are considered to be substantial due to their widespread consumption in human and veterinary medicine. Sulfamethoxazole, trimethoprim, ciprofloxacin, norfloxacin, and cephalexin had the highest median influent concentrations in a WWTP in Brisbane, Australia (360, 340, 3800, 170, and 4600 ng $L^{-1}$, respectively) [32]. Other studies confirmed high ubiquity of several antibiotics (i.e., ofloxacin, trimethoprim, roxyhtromycin and sulfamethoxazole) in sewage influent, though at low ng $L^{-1}$ level [28, 33]. However, even at very low concentrations they can have significant ecotoxicological effects in the aquatic and terrestrial compartment [34, 35]. Indiscriminate or excessive use of antibiotics has been widely blamed for the appearance of so-called "super-bugs" that are antibiotic resistant. It is of crucial importance to control their emissions into the

**Fig. 3** Frequency of detection for individual pharmaceuticals (%) in the Croatian wastewaters (modified from [22]) ▶

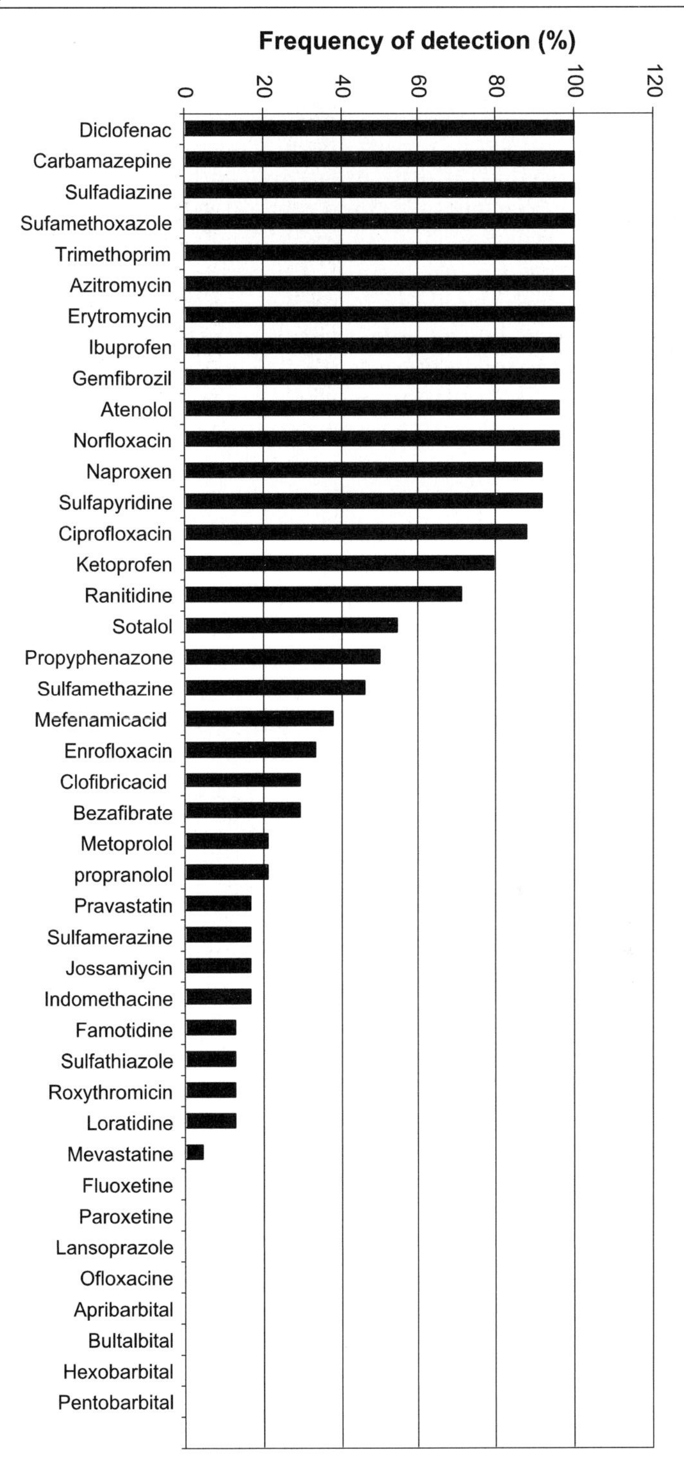
Frequency of detection (%)
0
20
40
60
80
100
120
Diclofenac
Carbamazepine
Sulfadiazine
Sufamethoxazole
Trimethoprim
Azitromycin
Erytromycin
Ibuprofen
Gemfibrozil
Atenolol
Norfloxacin
Naproxen
Sulfapyridine
Ciprofloxacin
Ketoprofen
Ranitidine
Sotalol
Propyphenazone
Sulfamethazine
Mefenamicacid
Enrofloxacin
Clofibricacid
Bezafibrate
Metoprolol
propranolol
Pravastatin
Sulfamerazine
Jossamiycin
Indomethacine
Famotidine
Sulfathiazole
Roxythromicin
Loratidine
Mevastatine
Fluoxetine
Paroxetine
Lansoprazole
Ofloxacine
Apribarbital
Bultalbital
Hexobarbital
Pentobarbital

environment through more cautious utilization and monitoring of outbreaks of drug-resistant infections.

The anti-epileptic drug carbamazepine is one of the most prominent drugs with a long history of clinical usage and it is frequently found in the environment [21, 24, 29, 36]. This drug has proven to be very recalcitrant since it by-passes sewage treatment [24, 36]. Common WWTP influent concentrations are in the order of magnitude of several hundreds $\mathrm{ng\,L^{-1}}$ [25, 28, 33, 37].

Lipid regulators are ordinarily applied drugs in clinical practice used to lower the level of cholesterol and regulate the metabolism of lipids. Clara et al. [33] detected a lipid regulator bezafibrate at concentrations up to $7.6\,\mu\mathrm{g\,L^{-1}}$, although normally they are found at lower $\mathrm{ng\,L^{-1}}$ range [28, 33, 38–40].

In all countries with developed medical care, X-ray contrast media can be expected to be present at appreciable quantities in sewage water. Clara et al. [33] detected iopromide at a mean concentration of $3.84\,\mu\mathrm{g\,L^{-1}}$ in the influent of a WWTP receiving hospital wastewater, while in WWTPs without a hospital within their drainage area this contrast media was not present. Iodinated X-ray contrast media are proved to contribute significantly total absorbable organic iodine in clinical wastewaters; up to $130\,\mu\mathrm{g\,L^{-1}}$ of iodine in the influent of municipal WWTP in Berlin, and $10\,\mathrm{mg\,L^{-1}}$ in hospital sewage was detected [41].

We could assume that a drug that is highly metabolized in humans will be subjected to extensive degradations in the environment, however, a high metabolic rate in humans does not necessarily mean that the lifetime of the pharmaceutical in the environment will be short. For some compounds, this assumption is correct (e.g., ibuprofen, diclofenac, propranolol, metoprolol, and carbamazepine), and they were found to be easily dissipated in the environment [42]. On the other side, atenolol, trimethoprim, and naproxen are substances with a low metabolic rate in humans, and they are excreted mainly unchanged or as acyl-glucuronide (naproxen), whereas their half-lives range from 10 days to 1 year [43]. Furthermore, monitoring of metabolic products should be included in risk-assessment analysis. Commonly, glucuronide and sulphate conjugates are the major Phase II metabolites that leave the biologically active group of the parent drug intact [44]. Some evidence suggests that these metabolites can be cleaved back into the original compound [45, 46]. Moreover, Bendz et al. [14] reported very high influent concentrations of metabolites of ibuprofen, carboxy-ibuprofen and hydroxyl-ibuprofen (10.75 and $0.99\,\mu\mathrm{g\,L^{-1}}$, respectively). Although more polar metabolites are presumed to be less hazardous to aquatic organisms, the European Medicines Agency (EMEA) guideline suggests environmental risk assessment of all human metabolites that constitute more than 10% of the total excretion of drug [47].

Due to their beneficial health effects and economic importance, the reduction of drug inputs into the environment through restricting or banning their use is not possible. Moreover, the use of pharmaceutical compounds is expected to grow with the increasing age of the population. The only possible way is to regulate their environmental pathways, perhaps on the source through labelling of medicinal products and/or developing disposal and awareness campaigns. Another option is to add sewage-treatment facilities in hospitals, and to enhance current wastewater-treatment techniques in order to eliminate more efficiently such polar pollutants.

## 3 Natural and Synthetic Estrogens

Estrogens are female steroid sex hormones based on a cholesterol structure. They are produced naturally in vertebrates in the gonads and adrenal cortex of both sexes and are responsible for the development of secondary sexual characteristics in the body. Their presence in the environment can cause negative effects to the endocrine functions of wildlife (e.g., aquatic organisms), posing an environmental risk. Estrogens reach the aquatic environment mainly due to incomplete removal in WWTP [48]. Other sources, such as livestock wastes will not be discussed in this section since these residues follow other pathways and do not end up in WWTPs.

### 3.1 Metabolism and Sources of Estrogens

In terms of binding to the human estrogen receptor, estradiol is the principal endogenous phenolic steroid estrogen. Estradiol is both metabolized reversibly and irreversibly. In the reversible metabolism, estradiol is transformed to estrone and estrone sulphate, meanwhile in the irreversible metabolism, estradiol is transformed to cathecol estrogens or estriol. These metabolites are mostly conjugated with glucuronides and, to a smaller extent, sulfates and excreted in the urine. A minor amount of the estrogens are excreted via feces as un-conjugated metabolites [49, 50].

Blocking the oxidation to estrone by, for instance, introducing an ethinyl group in position $17\alpha$ or $17\beta$ of estradiol leads to much more stable products, which remain longer in the body. The consequence of this increased stability is that the so-formed synthetic steroid ethinylestradiol is excreted up to 80% unchanged in its conjugated form [51].

The human daily excretion of estradiol, estrone, and estriol vary from men (1.6, 3.9, 1.5 μg) to women (3.5, 8, 4.8 μg) maintaining similar proportions with estrone being the most abundant estrogen [5]. Pregnant women show a different profile with higher levels of estradiol and estrone by a factor of ten, and estriol daily excretion at 6000 μg. Women taking contraceptives based on ethinylestradiol excrete 35 μg of this synthetic estrogen daily [52].

In addition to the natural endogenous estrogens discussed above, other estrogens have to be taken into account, such as natural and/or synthetic estrogens administered as medicine. One of the main applications of estrogens is in contraceptives. The estrogen content in birth control pills is usually in the range of 20 to 50 μg daily [53]. Besides contraception, the uses of estrogens can largely be put into three main groups: the management of the menopausal and postmenopausal syndrome (its widest use); physiological replacement therapy in deficiency states; and the treatment of prostatic cancer in men and of breast cancer in postmenopausal women.

The main sources of estrogens to WWTPs are therefore from the natural production of estrogens by humans, from hormone and estrogen replacement therapies and the intake of hormone contraceptives containing ethinylestradiol.

## 3.2
## Occurrence in Wastewater

The occurrence and environmental fate of estrogens have been reviewed in several articles [52, 54, 55]. The analysis of estrogens in wastewater has been discussed by Lopez de Alda et al. [56].

Estrogens are mainly excreted as their less active sulfate, glucuronide and sulfo-glucuronide conjugates [57]. However, in raw sewage and sewage-treatment plants (WWTPs), as well as in the environment, these conjugates may suffer deconjugation and act as precursors of the corresponding free steroids [58–61]. Thus, an appropriate evaluation of their occurrence and impact requires the analysis of both free and conjugated estrogens.

Most of the studies dealing with the investigation of estrogens in wastewaters have been performed in WWTPs receiving urban/domestic discharges and concentrations reported have been most usually in the ng/L range. Estradiol (E2) and estrone (E1) have been the free estrogens most frequently found, whereas estriol (E3) has been studied and detected only sporadically. However, E3 concentrations, when detected, have been usually higher than those of E2 and E1. In general, estrogens concentrations decrease in the order E3 > E1 > E2 (see Table 2 for examples). Thorough revision of all data available situates mean and median concentrations in the range of 9 to 20 ng/L for E2, 20 to 55 ng/L for E1 and 45 to 75 ng/L for E3 [58, 62–79].

The most studied synthetic estrogen, ethinylestradiol (EE2), has been either not detected [65, 67, 68] or detected at concentrations in general much lower than the other estrogens [58, 66, 77] (see Table 2). Levels higher than 100 ng/L have been only occasionally reported (e.g., 155 ng/L [63] and 138 ng/L [75]).

High levels of E1, E2, and E3 have also been reported by a few authors, e.g., 2100 ng/L of E2 [62], 200, 400, and 670 ng/L of E1 [62, 70, 79, 80] and 250 and 660 ng/L of E3 [79, 80].

**Table 2** Levels of free estrogens in wastewater reported in some selected studies. Values are given as minimum–maximum (average or median) concentrations in ng $L^{-1}$

| Estradiol | Estrone | Estriol | Ethinylestradiol | Refs. |
|---|---|---|---|---|
| 3–22 (9) | 8–52 (16) | n.a. | n.a. | [69] |
| 10–31 (25) | 16–60 (35) | 23–48 (31) | n.d. | [68] |
| 4.7–25 (12) | 25–132 (52) | 24–188 (80) | 0.4–13 (3) | [58] |
| n.d.–21 (5.7) | 10–57 (24) | 27–220 (110) | n.d. | [67] |
| n.d.–234 (89) | 9.4–232 (108) | n.d.–108 (23) | 2.4–138 (57) | [75] |

n.d. not detected;
n.a. not analysed

In general, it appears that the concentration of the un-conjugated estrogens in wastewater reflects roughly their excretion by the human body, where the high levels of estriol originate from pregnant women. This relation, however, is not found in influent wastewaters from WWTPs receiving industrial, or mainly industrial, wastes. In these cases, either estrone is the only estrogen detected [65] or the estrone concentration is significantly higher than that of estradiol and estriol [75].

The concentration of estrogens in wastewater entering WWTPs, together with other relevant data form the WWTP, such as influent flow-rate and the population served, has been used by some authors to calculate the loads of compounds (g/day) entering WWTPs. In a study dealing with the removal of pharmaceuticals, the calculated loads (mg/day/100 inhabitants) of estradiol (from not detected to 4), estrone (from not detected to 28) and ethinylestradiol (not detected) in six WWTPs were far below those of most of the other pharmaceuticals investigated [81]. Small loads of estrogens were also calculated by Ternes et al. [82] in a study performed in Germany (1 g/day E1, 0.5 g/day E2), and Brazil (5 g/day E1, 2.5 g/day E2).

In contrast to free estrogens, conjugated estrogen derivatives have been included only in a few studies [64, 65, 67, 74]. Mostly sulphates and glucuronides of E1, E2, and E3 have been included as target analytes and detected at similar levels as the free estrogens (see Table 3). Derivatives from the chemically more stable synthetic estrogen EE2 were studied by Gomes et al. [65], but no positive samples were found. Komori et al. [67] studied the presence of di-conjugated E2 derivatives and found high levels of the disulfate and moderately high levels of the sulfate-glucuronide derivative (see Table 3).

Although most estrogens are excreted as glucuronides the concentrations found at the entrance of WWTPs do not reflect this fact. Glucuronides levels are usually low; sulfates dominate the load of estrogens [74]. D'Azcenzo et al. [64] compared the amount of glucuronides and sulfates detected in female urine, a septic tank from a condominium and the entrance of a WWTP and found a higher percentage of sulfates (60%) at the entrance of the WWTP

**Table 3** Levels (ng $L^{-1}$) of conjugated estrogen derivatives detected in waste water

| Refs. | E1-3S | E2-S | E3-S | EE2-S | E1-G | E2-G | E2-2G | E3-G | EE2-G | E2-SG | E2-SS |
|---|---|---|---|---|---|---|---|---|---|---|---|
| [65] | 10–14 | n.a. | n.a. | n.d. | n.d. | n.a. | n.a. | n.d. | n.d. | n.a. | n.a. |
| [74] | 34 | 3.2 | n.a. | n.a. | 0.4 | 0.3 | n.a. | n.a. | n.a. | n.a. | n.a. |
| [64] | 27 | 9 | 47 | n.a. | 10 | n.d. | 9 | 39 | n.a. | n.a. | n.a. |
| [67] | 42 | 110 | 22 | n.a. | 11 | 18 | n.a. | 22 | n.a. | 5.5 | 77 |

S, sulphate;
G, glucuronide;
n.a., not analysed;
n.d., not detected

than in the septic tank (55%) and the female urine (22%), suggesting that glucuronides might be de-conjugated in the sewer moiety and reach the WWTP at lower levels. In contrast, sulfates appear to be more stable than glucuronides, probably because bacterial sulfatases are present at lower concentrations than glucuronidases and/or because they have low affinity towards steroid sulfates. One example presented by Huang et al. [83] showed that sulfatases enzymes convert only 30% of E2 sulfate into E2.

In conclusion, the levels of estrogens in wastewater are occasionally very high (>100 ng/L), although in average values are usually below 100 ng/L. The calculated loads of estrogens entering the WWTPs are relatively low compared to those of pharmaceutical residues. However, there is no sufficient data on the concentration of the conjugated derivatives and their loads. Their de-conjugation can pose a problem if elimination is not complete.

## 4 Drugs of Abuse

According to the World Drug Report 2007, about 200 million people use illicit drugs each year globally. Drugs of abuse (DAs) consumption seems now to be stabilized after the increasing trends observed over a decade [84, 85]. Similar to PhACs, these substances are considered to be "pseudo-persistent" in the environment, thus they have become a group of emerging environmental contaminants of interest. DAs reach aquatic systems mainly through sewage water. After drug ingestion, diverse proportions of the parent compound, conjugated forms and metabolites are excreted via urine and flushed towards municipal WWTPs. Some of them may not be efficiently or completely removed at WWTPs and therefore they will be released into the environment via WWTP effluents. In addition to WWTPs discharges, direct disposal into the environment is to a lesser extent another pathway to the aquatic media.

The toxicological or cumulative effect of these substances on the ecosystem has not yet been studied. These compounds have specific physiologic and psychological effects in humans at low-concentration doses (mg or even μg in the case of lysergic acid diethylamide), thus the evaluation of the exposure of the wildlife to the bioactive molecules may be of interest, according to their occurrence in the environment. Fate and transport in aquatic environments is also not known. Most of them are polar compounds that will be concentrated in aqueous environmental matrices; however, some of them, such as cannabinoids, are likely to bioaccumulate in organisms or concentrate in sediments due to their physico-chemical properties (octanol–water partition coefficient, solubility...). A study of the distribution of these compounds in the different environmental compartments may also be a matter of scientific interest.

Since 2004, several authors have developed analytical methodologies based on liquid chromatography and tandem mass spectrometry (LC-MS/MS) detection to evaluate the occurrence of drugs of abuse in sewage and natural waters [86–92]. The target drugs of abuse and metabolites studied so far belong to five different classes: cocainics, amphetamine-like compounds, opiates, cannabinoids, and lysergics. Although a lack of data on drugs of abuse residues in environmental waters is still remarkable, mean values of these substances reported so far in the peer-reviewed literature are summarized in Table 1. The table gathers levels of common drugs of abuse and their metabolites detected in influent waters collected at different European WWTPs located in Spain [86, 92], Ireland [88], Italy [87, 89], Switzerland [87] and Germany [90].

The ubiquity of the different target compounds is directly related to local patterns of drug abuse. The highest loads, thus the highest consumption, are usually reported for two cocainic compounds, namely, cocaine and its main metabolite benzoylecgonine (BE), that are commonly detected at the high $ng L^{-1}$ or even the $\mu g L^{-1}$ level. The highest concentrations have been found in influent waters collected at a WWTP located in Barcelona, where BE, an inactive metabolite of cocaine with a relatively long half-life, was present at a mean concentration of $4226 ng L^{-1}$ [92]. Cocaethylene (CE), which is a transesterification product of cocaine formed when cocaine is consumed together with ethanol, has not been detected at high levels; thus either this practice is rather limited or, what is more likely, CE transforms rapidly into metabolites not studied yet in WWTPs, such as norcocaethylene and ecgonine ethyl ester. Other cocaine metabolites, norcocaine and norbenzoylecgonine, have been studied at two WWTPs in Italy but their levels did not surpass $40 ng L^{-1}$.

From the studied opiates, only morphine has been found in some WWTPs at high $ng L^{-1}$ levels, resulting probably from its medical applications. Although morphine is excreted in urine mainly as glucuronide metabolites, cleavage of the conjugated molecules in wastewater is likely to occur in the light of the low levels found for morphine-3$\beta$-d-glucuronide (the only conjugated compound studied) in comparison with those usually detected for morphine [87]. Heroine has been either not detected or detected at very low concentrations due to its low consumption and its also rapid hydrolysis to morphine and 6-acetylmorphine (heroine is quite unstable in blood serum) [93]. The results of the study done in WWTPs located in Italy and Switzerland [87] indicate that methadone, that is a long-acting opioid agonist used for treating acute and chronic pain and for preventing opiate withdrawal, is commonly present at lower levels than its pharmacologic inactive metabolite 2-ethylidine-1,5-dimethyl-3,3-diphenylpyrrolidine perchlorate (EDDP); both compounds were found in both areas at ng $L^{-1}$.

Concerning lysergic acid diethylamide (LSD) and its metabolites nor-LSD and nor-iso LSD (nor-LSD) and 2-oxo-3-hydroxy-LSD (O-H-LSD), absence or very low concentrations have been reported in influent samples. These results are in line with the very low doses of LSD needed to produce an effect compared to those needed in the case of other drugs (μg vs. mg), as LSD is the most potent psychoactive drug known so far [93].

The most abundant amphetamine-like compound detected in influent sewage waters is the phenylethylamine ephedrine. Besides a recreational and illicit use, this drug presents medical applications as topical decongestant and bronchodilator in the treatment of asthma and in the reversal of hypotension states. The so-called "designer drugs" 3,4-methylenedioxymetamphetamine hydrochloride (MDMA or "ecstasy"), methylenedioxyethylamphetamine (MDE, MDEA or "Eve") and 3,4-methylenedioxyamphetamine (MDA or "Love pills", and metabolite of both MDE and MDMA), have been detected frequently at the ng $L^{-1}$ level in the different studied WWTPs. As shown in Table 4, amphetamine and methamphetamine are usually present in this type of matrix at lower concentrations than MDMA.

The presence of $\Delta^9$-tetrahydrocannabinol (THC), which is the most psychologically active constituent of Cannabis (the most widely used illicit drug), in influent sewage waters has been observed insignificant as compared to that of its metabolites since THC is extensively metabolized before excretion. 11-nor-9 carboxy THC (nor-THC) is the major THC urinary metabolite and 11-hydroxy-THC (OH-THC) is the main psychoactive metabolite in the body. Thus, monitoring of these metabolites seems to be more appropriate to study the occurrence of cannabinoids in waters.

Measured values of DAs in sewage waters provide real-time data to estimate drug abuse at the community level. This strategy was first proposed by Daughton in 2001 [94] and implemented 4 years later by Zucatto et al. [89] to estimate cocaine abuse in the north of Italy. Such estimations, obtained in a fairly cheap and anonymous way (avoiding potential privacy conflicts), allow the immediate adoption of appropriate measures by the responsible authorities to fight drug abuse by the population. Efficiency of removal of DAs in WWTPs is largely unknown and should be addressed in order to control their

**Table 4** Occurrence of drugs of abuse residues in WWTPs influents

| Compound | Concentration (ng $L^{-1}$) | Refs. |
|---|---|---|
| ***Cocainics*** | | |
| Cocaine | 225 $^{a}$, 79 $^{b}$ | [86] |
| | (421.4±83.3) $^{b}$, (218.4±58.4) $^{b}$ | [87] |
| | (489±117) $^{b}$ | [88] |
| | 42–120; 80.25 $^{b}$ | [89] |
| | (860.9±213.6) $^{b}$; 502.3 $^{b}$ | [92] |
| Norcocaine | (13.7±5.3) $^{b}$; (4.3±0.9) $^{b}$ | [87] |
| Benzoylecgonine | 2307 $^{a}$, 810 $^{b}$ | [86] |
| | (1132.1±197.2) $^{b}$, (547.4±169.4) $^{b}$ | [87] |
| | (290±11) $^{b}$ | [88] |
| | 390–750; 550 $^{b}$ | [89] |
| | 78 | [90] |
| | (4225.7±1142.8) $^{b}$; 1456.7 $^{b}$ | [92] |
| Norbenzoylecgonine | (36.6±7.8) $^{b}$, (18.8±5.6) $^{b}$ | [87] |
| Cocaethylene | (11.5±5.1) $^{b}$, (5.9±2.6) $^{b}$ | [87] |
| | (77.5±33.2) $^{b}$, (78.5) $^{b}$ | [92] |
| | n.d. | [88] |
| ***Opiates*** | | |
| Heroine | n.d., 2.4 $^{b}$ | [92] |
| Morphine | (83.3±11.8) $^{b}$, (204.4±49.9) $^{b}$ | [87] |
| | n.d | [88] |
| | 820 $^{a}$; 310 $^{c}$ | [90] |
| | (162.9±20.0) $^{b}$, 68.1 $^{b}$ | [92] |
| 6 Acetyl morphine | (11.8±8.5) $^{b}$, (10.4±4.8) $^{b}$ | [87] |
| | (12.8±3.1) $^{b}$, 8.4 $^{b}$ | [92] |
| Morphine-3$\beta$-d-glucuronide | (2.5±7.1) $^{b}$, (18.1±30) $^{b}$ | [87] |
| Methadone | (11.6±1.7) $^{b}$, (49.7±9.6) $^{b}$ | [87] |
| | n.d. | [88] |
| EDDP | (19.8±3.1) $^{b}$, (91.3±19.2) $^{b}$ | [87] |
| | n.d. | [88] |
| ***Amphetamine-like compounds*** | | |
| Amphetamine | 15 $^{a}$; 15 $^{b}$ | [86] |
| | (14.7±10.6) $^{b}$; < LOQ | [87] |
| | (41.1±9.1) $^{b}$; 20.8 $^{b}$ | [92] |
| Methamphetamine | (16.2±7.1) $^{b}$; < LOQ | [87] |
| | (18.2±5.8) $^{b}$; 4.8 $^{b}$ | [92] |
| | n.d. | [86] |
| MDMA | 91 $^{a}$; 49 $^{b}$ | [86] |
| | (14.2±14.5) $^{b}$, (13.6±12.6) $^{b}$ | [87] |
| | (133.6±29.8) $^{b}$, (135.13) $^{b}$ | [92] |
| | n.d. | [88] |

**Table 4** (continued)

| Compound | Concentration (ng $L^{-1}$) | Refs. |
|---|---|---|
| MDEA | $27^{a}$; $28^{b}$ | [86] |
| | $(1.5 \pm 3.8)^{b}$, < LOQ | [87] |
| MDA | $(4.6 \pm 7.3)^{b}$, < LOQ | [87] |
| Ephedrine | $(591.9 \pm 124.5)^{b}$, $399.3^{b}$ | [92] |
| ***LSD and its metabolites*** | | |
| LSD | $(2.8 \pm 1.2)^{b}$, $2.9^{b}$ | [92] |
| | n.d. | [86] |
| | n.d. | [88] |
| 2-oxo-3-hydroxy-LSD | $(5.6 \pm 12.1)^{b}$, $3.4^{b}$ | [92] |
| Nor-LSD & nor-iso LSD | $(4.3 \pm 1.8)^{b}$, $13.5^{b}$ | [92] |
| ***Cannabinoids*** | | |
| THC | nd; $14.24^{b}$ | [92] |
| 11-nor-9-carboxy-THC | $(62.7 \pm 5)^{b}$; $(91.2 \pm 24.7)^{b}$ | [87] |
| | $(4.3 \pm 7.8)^{b}$; $21.03^{b}$ | [92] |
| 11-hydroxy-THC | $(8.4 \pm 2.1)^{b}$; $46.3^{b}$ | [92] |

[a] maximum concentration,
[b] mean,
[c] median

release to the environment and avoid potential adverse effects in the aquatic ecosystem.

## 5 Surfactants (Alkylphenol Ethoxylates and Related Compounds)

Surfactants are produced in huge amounts and used in households as well as in industrial cleansing processes and as such they make up one of the most relevant organic pollutants of anthropogenic origin with the high potential to enter the environment. After use, detergents are usually discarded down the drain into sewer systems and afterwards treated in WWTP where they are completely or partially removed by a combination of sorption and biodegradation.

Among various classes of non-ionic, anionic, and cationic surfactants, alkylphenol ethoxylates (APEOs) are the group that raised the most concern. APEOs are effective nonionic surfactants, widely used as industrial cleaning agents and wherever their interfacial effects of detergency, (de)foaming, (de)emulsification, dispersion or solubilization can enhance products or process performance. Although parent APEOs are not classified as highly toxic substances ($EC_{50}$, 48 h, *Daphnia magna* 1.5 mg $L^{-1}$) their environmental acceptability is strongly disputed because of estrogenic metabolic products (alkylphenols (APs) and carboxylic derivatives (APECs)) generated during wastewater treatment. Because of these findings, APEOs are banned or restricted in Europe. Throughout northern Europe (Scandinavia, UK, and Germany) a voluntary ban on APEO use in household cleaning products began in 1995 and restrictions on industrial cleaning applications in 2000 [95]. This resulted in a significant reduction of APEO concentrations found. For example, in five Norwegian WWTP nonylphenol (NP) was found in the range of 0.2–7 μg $L^{-1}$ in the effluent samples in 2002, while concentrations below the detection limit (2 ng $L^{-1}$) were found in the 2004 samples [96], which is attributed to new restrictions implemented in 2002. Similarly, the NP concentrations in digested sewage sludge in Switzerland were around 1.3 g/kg dry sludge before the ban of NP surfactants in laundry detergents in 1986. In the 1990s, the NP concentrations in sludge ranged from 0.1 to 0.2 mg/kg dry sludge [97]. In Catalonia (Spain), typical levels of NP measured in WWTPs in 1998 and 1999 ranged from 100 to 200 μg $L^{-1}$ in influents, while 2002–2003 data show almost a 10-fold decrease (Fig. 4), which suggests a gradual withdraw and replacement of NPEOs by Spanish tanneries and textile industry [98].

However, mainly because of lower production costs, APEOs are still being used in substantial amounts in institutional and industrial applications. Hence information about the total concentrations of APEOs and their degradation products in environmental matrices is essential in assessing the environmental impact of these compounds.

Several extensive monitoring programs were conducted with the objective of determining the concentrations of APEO and their degradation products in raw and treated wastewaters. The concentrations of NPEOs (Table 5) in WWTP influents varies from less than 30 to 1035 μg $L^{-1}$. In industrial wastewaters (especially from tannery, textile, pulp, and paper industry) much higher values, up to 22 500 μg $L^{-1}$, are detected. Octylphenol ethoxylates (OPEOs) typically comprised 5–15% of total APEOs in WWTP influents, which is congruent with their lower commercial use. Concentrations found in WWTP effluents rarely exceeded 100 μg $L^{-1}$, corresponding to an elimination of the parent compound ranging from 80–98%.

However, their removal led to the formation of transformation products that are much more resistant to further microbial degradation. Acidic and neutral degradation products of NPEOs have been found to be rather resistant to further degradation, being NP the most recalcitrant intermediate. NPEO metabolites, NP and NPECs are already detected in WWTP influents, due to in-sewer degradation, in concentrations up to 40 μg $L^{-1}$. Recently, a comprehensive study in the region of Western Balkan (Bosnia and Herzegovina, Croatia, and Serbia) [22] showed widespread occurrence of surfactant-derived alkylphenolic compounds, although the concentration levels were

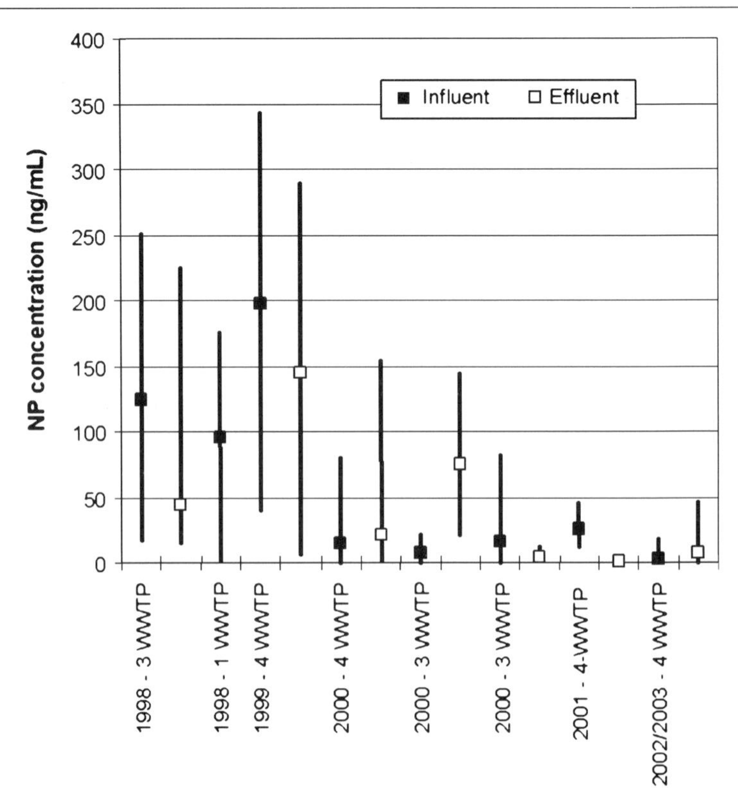


**Fig. 4** Concentration of NP in influents and effluents of WWTP in Catalonia (NE Spain) in the period from 1998 to 2003 (Adapted from [98])

relatively low and suggest a decreasing trend in comparison to some previous campaigns conducted in early 1990s [99]. The concentration of NP, as the most toxic and most potent estrogen disrupting compound derived from NPEO surfactants [100], was present in concentrations up to 4.4 $\mu g L^{-1}$ with an average value of 1.7 $\mu g L^{-1}$. It is interesting to mention that Croatia was one of the first countries that introduced water-quality criteria for NP with a maximum permissible concentration in ambient water of 1 $\mu g L^{-1}$ [101], 15 years before it was accepted as a priority pollutant in the EU Water Framework Directive. Besides NP, all municipal wastewaters contained measurable levels of other metabolites derived from NPEO surfactants, in particular NPEC. The composition of alkylphenolic compounds was highly variable and revealed a strong impact of various biotransformation and physico-chemical processes on the distribution of individual alkylphenolic compounds in various types of wastewater samples. The most abundant alkylphenolic species in non-treated wastewaters was NP, while NPEC were the dominant species in biologically treated effluents, which is in agreement with earlier reports on this subject [102].

**Table 5** Concentration ranges of alkylphenolic surfactants and their metabolites in raw wastewater entering WWTP

| Compounds | Country | Concentration ($\mu g\,L^{-1}$) | Refs. |
|---|---|---|---|
| NPEO | Germany | 120–270 | [137] |
| | Austria | 2.6–35 ($NP_1EO$) | [138] |
| | | 1.2–5.8 ($NP_2EO$) | |
| | Italy | 29–145 | [139] |
| | | 127–221 | [140] |
| | Spain | 27–880 (2120) [a] | [141–143] |
| | Switzerland | 96–430 | [144, 145] |
| | The Netherlands | < 0.1–125 | [146] |
| | | 50–22500 [a] | |
| | Croatia | 5–392 | [22] |
| NPEC | Spain | < 0.2–14 [a] | [147] |
| | | < 0.4–219 | [141, 143] |
| | Croatia | < 0.001–3.20 ($NPE_1C$) | [22] |
| | | < 0.001–4.37 ($NPE_2C$) | |
| NP | Belgium | < 0.4–219 [b] | [148] |
| | Italy | 2–40 | [149] |
| | Spain | < 0.5–22 | [141, 147] |
| | | 17–251 [a] | [143] |
| | The Netherlands | < lod-19 (40) [a] | [146] |
| | Croatia | 0.460–4.40 | [22] |
| | Norway | < 0.002–5.2 | [96] |
| | Austria | 1.05–8.6 | [138] |

[a] WWTP receiving high percentage of industrial wastewater
[b] effluent of a textile plant

Octylphenolic analogues of NPEOs and their metabolites represented only a small percentage of the total alkylphenolic compounds in all analyzed samples, typically less than 10%. This is important for the assessment of the endocrine disrupting potential associated with APEO surfactants and their metabolites, because OP is an endocrine disrupting compound (EDC) four times more potent than NP [100].

## 6 Perfluorinated Compounds

Perfluorinated compounds (PFCs) have been manufactured for more than 50 years, and released into the environment following production and use. As a result, PFCs are now acknowledged to be widespread environmental contaminants. PFCs repel both water and oil and these compounds are therefore ideal chemicals for surface treatments. These compounds have been used for many industrial applications such as stain repellents (such as Teflon), textile, paints, waxes, polishes, electronics, adhesives, and food packaging.

PFCs are both hydrophobic and lipophobic, and are highly stable in the environment. Many of the degradation products of PFCs have been found in the environment throughout the world, because of the strong carbon–fluorine (C–F) bond associated with FASs. In addition, the most important PFC: perfluorooctane sulphonate (PFOS) and perfluoro carboxylic acids (PFCAs) are also stable degradation products/metabolites of neutral PFC. These precursor compounds are more volatile and therefore more likely to undergo long-range atmospheric transport, with sufficient atmospheric lifetimes to reach remote locations, where they can break down.

Possible precursor compounds for PFCAs and PFOS are fluorotelomer alcohols (FTOHs). Fluorotelomeralcohols are manufactured as a raw material used in the synthesis of fluorotelomer-based surfactants and polymeric products. The manufacture of FTOHs usually results in a mixture containing six to 12 fluorinated carbon congeners, the 8 : 2 FTOH being the dominant one. Release of the volatile FTOH may occur all along the supply chain from production application.

PFOS and PFOA are environmentally persistent substances that have been detected worldwide in human blood, water, soils, sediments, air, and biota samples [103].

PFCs are currently receiving great attention because of their persistence [104, 105]], bioaccumulation [106], and potential health concerns including toxicity [107] and cancer promotion [108], and they are now included in different health programs in EEUU to provide a better assessment of the distribution, toxicity, and persistence of these compounds in humans [109]. Research questions include understanding the sources of perfluorinated compounds and their environmental fate and transport.

In the EU, there is currently no legislation on the use of PFCs associated with their (potential) environmental and/or human health effects. It should, however, be noted that some legislation which generally applies to the release of substances to the environment may be relevant to the release of PFOS. This is the case with the IPPC Directive 96/61/EC concerning integrated pollution prevention and control, which includes fluorine and its compounds in the "indicative list of the main polluting substances to be taken into account if they are relevant for fixing emission limit values" (Annex III to the Directive) [110].

Recent studies have attempted to explain the occurrence of PFOA in the Arctic environment by oceanic transport as a result of the manufacture and use of PFOA [104, 111, 112]. Armitage et al. assumed emissions via waste water treatment plants effluents and their predictions have indicated PFOA concentrations in the Northern Polar Zone (equivalent to the Arctic Ocean) would increase until about 2030 and then gradually decline as ocean concentrations adjust to lower emission rates.

**Table 6** Concentrations (ng $L^{-1}$) of perfluorinated compounds found in wastewaters and different environmental waters

| Type of water | Country and site | PFOS | PFOA | PFHpA | PFNA | PFDA | Refs. |
|---|---|---|---|---|---|---|---|
| Wastewater | | | | | | | |
| Effluent | Austria | 4.5–20 | 10–21 | 2.5–4.6 | 0–2 | 0–2 | [150] |
| Effluent | EEUU (New York) | 3–68 | 58–1050 | | 0–376 | 0–47 | [114] |
| Effluent | EEUU (Kentucky) | 8–993 | 8.3–334 | – | 0–15.7 | 0–201 | [115] |
| Effluent | EEUU (Georgia) | 0–70 | 7–227 | – | 0–54 | 0–86 | [115] |
| River | | | | | | | |
| Dalälven | Sweden | – | < 0.97 | 0.36 | < 0.14 | – | [151] |
| Vindelälven | Sweden | – | < 0.65 | 0.2 | 0.22 | – | [151] |
| Elbe | Germany | – | 7.6 | 2.7 | 0.27 | – | [151] |
| Oder | Poland | – | 3.8 | 0.73 | 0.73 | – | [151] |
| Vistula | Poland | – | 3.0 | 0.48 | 0.36 | – | [151] |
| Po | Italy | – | 200 | 6.6 | 1.46 | – | [151] |
| Danuve | Romania/ Ucrania | – | 16.4 | 0.95 | 0.27 | – | [151] |
| Daugava | Letonia | – | < 2.2 | 0.86 | 0.36 | – | [151] |
| Seine | France | – | 8.9 | 3.7 | 1.26 | – | [151] |
| Loire | France | – | 3.4 | 0.90 | 0.43 | – | [151] |
| Thames | UK | – | 23 | 4.1 | 0.79 | – | [151] |
| Rhine | Germany | – | 12.3 | 3.3 | 1.50 | – | [151] |
| Guadalquivir | Spain | – | 4.6 | 1.58 | 1.02 | – | [151] |
| Rhine | Germany (Breisach) | 26 | 2 | – | – | – | [120] |
| Rhine | Germany (Mainz) | 12 | 3 | – | – | – | [120] |
| Rhine | Germany (Ludwigshafen) | 5 | 2 | – | – | – | [120] |
| Ruhr | Germany (Duisburg) | 5 | 48 | – | – | – | [120] |
| Ruhr | Germany (Schwerte) | 14 | 177 | – | – | – | [120] |
| Elpe | Germany (Bestwig) | – | 1168 | – | – | – | [120] |
| Moehne | Germany (Heidelberg) | 193 | 3640 | 148 | – | – | [120] |
| Tenjin | Japan | 4.7 | 39 | – | – | – | [152] |
| Katsura | Japan | < 5.2 | 7.9 | – | – | – | [152] |
| Lake | | | | | | | |
| Shihwa | Korea | 89.11 | 19.22 | 2.50 | 3.26 | 1.98 | [153] |
| Maggiore | Italy | 7.8 | 2.4 | 2.4 | 0.6 | 3.7 | [153] |

**Table 6** (continued)

| Type of water | Country and site | PFOS | PFOA | PFHpA | PFNA | PFDA | Refs. |
|---|---|---|---|---|---|---|---|
| Huron | Canada | 4.2 | 3.6 | – | 3.6 | 3.7 | [154] |
| Ontario | Canada | 3.9 | 2.6 | – | 3.1 | – | [154] |
| Michigan | Canada | 3.8 | 3.4 | – | – | – | [154] |
| Sea | | | | | | | |
| Harbor | Norway | 71–749 | | 3–30 | Nd | 3–30 | [155] |
| Harbor | Iceland | 26–67 | | 6–14 | Nd | 6–14 | [155] |
| Harbor | Denmark | 129–650 | | 5–36 | Nd | 5–36 | [155] |
| Baltic Sea | | 232–1149 | | 18–59 | Nd | 18–59 | [155] |
| North Sea | | 12–395 | | | Nd | Nd | [156] |
| Black sea | | 33–1790 | | 1.0–19 | 1.4–7.2 | 1.9–19 | [157] |

PFCs reach the aquatic environment either through their release into rivers or via wastewater discharge into receiving waters. In Table 6 are summarized occurrence of PFCs reported in different aquatic environments reported in Europe during recent years. Different studies on EEUU reported high concentrations in wastewater, in a recent study by Logannathan et al. [113], PFCs including perfluoroalkyl sulfonates (PFASs; PFOS, PFOSA, PFHxS) and perfluoroalkyl carboxylates (PFACs; PFOA, PFNA, PFDA, PFDoDA, PFUnDA) were investigated in two wastewater treatment plats (WWTPs). The first plant was located in Kentucky and it was representative of a rural area. The second plant was located in Georgia and it was representative of an urban area. PFOS was a major contaminant in samples from Kentucky (8.2–990 ng $g^{-1}$ dry wt. in solid samples and 7.0–149 ng $L^{-1}$ in aqueous samples), followed by PFOA (8.3–219 ng $g^{-1}$ dry wt. in solid samples and 22–334 ng $L^{-1}$ in aqueous samples). PFOA was the predominant contaminant in samples from the urban WWTP (7.0–130 ng $g^{-1}$ dry wt. in solid samples and 1–227 ng $L^{-1}$ in aqueous samples), followed by PFOS (<2.5–77 ng $g^{-1}$ dry wt. in solid samples and 1.8–22 ng $L^{-1}$ in aqueous samples). PFHxS, PFNA, PFDA, and PFOSA were detected in most of the samples, whereas PFUnDA and PFDoDA were detected in very few samples. Concentrations of some PFCs, particularly PFOA, were slightly higher in effluent than in influent, suggesting that biodegradation of some precursors contributes to the increase in PFOA concentrations in wastewater treatment processes. These mass loading values were similar to the values reported by Sinclair and Kannan [114] for New York plants and slightly higher than values reported for a Pacific Northwestern WWTP [115].

In Europe these quantities were even higher. Fifteen effluents from representative industry sectors (printing, electronics, leather, metals, paper, photographic and textiles) from Austria were analysed for PFOS. The PFOS levels ranged from 0–2.5 μg/L (2.5 $\mu g L^{-1}$ for leather, 0.120 $\mu g L^{-1}$ for metal, 0.140–1.2 $\mu g L^{-1}$ at four paper sites, 1.2 $\mu g L^{-1}$ for photographic, not found in textiles or electronics) [116]. Concentrations from 0.05 to 8.2 $\mu g L^{-1}$ were quantifies in the effluents of urban wastewater in Spain [117]. Predominantly, however, they are adsorbed to sewage sludge [118]. The use of sludge for land treatment or its disposal on dump sites leads to a remobilization of these recalcitrant compounds. Also, their polarity and mobility in water and soil allow them to reach the sea or groundwater unaffected.

Several studies have reported the presence of PFCs in surface waters. The occurrence of PFOA and PFOS in several surface waters in Germany was described in 2004 [119]. In summer 2006, the discovery of perfluorinated compounds in waters of the Arnsberg district in the North Rhein Westfalian Sauerland region caused a stir [120]. In this study, 12 different perfluorinated surfactants in German rivers (the Rhine River and its main tributaries, as well as the Moehne River), canals and drinking waters of the Ruhr catchments area are presented. Furthermore, the main contamination source was identified as an agricultural area on the upper reaches of the Moehne River, which is an important tributary of the Ruhr River. PFOA was the compound quantified in higher concentrations, it was found at 519 $ng L^{-1}$ in drinking water and at 4385 $ng L^{-1}$ in surface waters. In this case, the concentrations were higher than the highly polluted Tokyo Bay. In addition, the Möhne Reservoir is a source of drinking water.

In a survey study of contamination of surface and drinking waters around Lake Maggiore in northern Italy, PFCs were investigated in conjunction with other polar anthropogenic environmental pollutants [121].

PFOS and PFOA were identified as major PFCs being PFOS the most abundant one. PFOS was detected in two river water samples (Creek Vevera and River Strona) at concentrations >20 $ng L^{-1}$, and in the Lake Maggiore at concentrations around 8 $ng L^{-1}$. In addition, detection of some compounds such as PFOS and PFOA at high concentrations in rain water suggested that atmospheric deposition contributes to the contamination of the lake by these substances.

In this sense, different studies are examining precipitation (rainwater) to test for the atmospheric transformation of FTOHs as a source of PFOA and other perfluorocarboxylic acids (PFCAs) [122, 123].

A number of studies have been carried out in recent years in order to measure the occurrence of PFCs in marine environments. Sea water is a particularly challenging matrix because of the lower levels ($pg L^{-1}$, part-per-quadrillion) of PFCs in sea water. Yamashita used LC/ESI-MS/MS to carry out a global survey of PFOS, PFOA, PFHS, perfluorobutane sulfonate(PFBS), perfluorononanoic acid (PFNA), and perfluoro octane sulphonamide in sea water samples [124]. This paper also provides a nice summary of PFOS and PFOA measurements in the livers of various marine animals.

## 7
## Industrial Chemicals (Corrosion Inhibitors and Plasticizers)

2-substituted benzothiazoles are a class of high-production-volume chemicals used as anticorrosion additives and biocides as well as vulcanization accelerators and antifungal agents in the paper and tanning industry. Owing to the wide application, they are regularly detected in the municipal wastewaters, being benzothiazole-2-sulfonate, benzothiazole and 2-hydroxybenzothiazole the most abundant, as shown by Kloepfer et al. [125, 126] (Fig. 5). The total concentration of six benzothiazoles in the wastewater of Berlin summed up to 3.4 $\mu g L^{-1}$ with the range of the temporal variability of 2–40% within 3 months.

Benzotriazoles are a class of corrosion inhibitors mainly used in deicing fluids and dishwashing agents. The main representatives 1H-benzotriazole and tolyltriazole are frequently found in wastewater of Swiss WWTP (10 and 1.6 $\mu g L^{-1}$ on average) [127] and in untreated municipal wastewater in the Berlin region with mean dissolved concentrations of 12 $\mu g L^{-1}$ for 1H-benzotriazole and 2.1 $\mu g L^{-1}$ and 1.3 $\mu g L^{-1}$ for 4- and 5-tolyltriazole, respectively [128].

Phthalate acid esters (PAEs) are a class of chemical compounds widely used in different industrial applications, mainly as plasticizers for polyvinyl chloride (PVC) resins, adhesives and cellulose film coatings and with minor applications in cosmetics, medical products, and insecticide carriers. They comprise a large group of compounds, several of them considered as

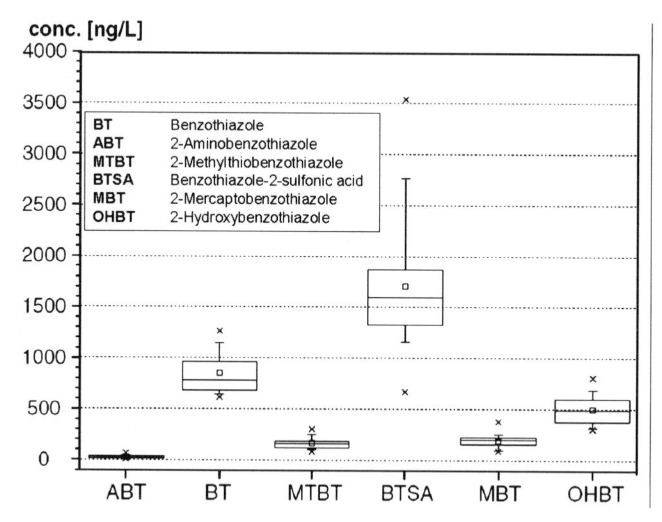


**Fig. 5** Concentrations (ng/L) of the benzothiazoles in the municipal wastewater (influent to Berlin-Puhleben WWTP), summary of 20 composite samples (24 h) collected over 3 months. Adapted from [125]

priority pollutants: dimethyl (DMP), diethyl (DEP), dibutyl (DBP), butylbenzyl (BBP), di(2-ethylhexyl) (DEHP) and di-n-octyl phthalate(DnOP). The worldwide production of PAEs approximates 2.7 million metric tons a year [129] and considerable direct (production of plastic materials) and indirect emission via leaching and volatilization from plastic products after their usage, disposal and incineration, explains their ubiquity in the environment.

In all reported studies, DEHP was found to be a predominant PAE due to its high production (nearly 90% of European plasticizer use) and its physico-chemical properties (low solubility and relatively high $K_{OW}$). Marttineen et al. [130] reported DEHP concentrations of 98–122 $\mu g\,L^{-1}$ in WWTP inlet samples in Finland. Somewhat lower levels were reported by Fauser et al. [131] for inlets to WWTP in Denmark. In five Norwegian WWTP, phthalates (DEHP, BBP, DEP, DMP, and DnOP) were found in raw influent water in concentrations up to 23 $\mu g\,L^{-1}$ with an average of $8.0 \pm 6.4\,\mu g\,L^{-1}$ [96]. However, contrary to other studies, DEHP was the dominant compound in only four out of 10 influent samples, while DEP was the dominating congener in the other six influent samples. The most systematic study on the occurrence of PAEs in the aquatic environment was conducted by Fromme et al. [132]. The levels of DEHP and dibutyl phthalate (DBP) were reported for 116 surface-water samples, 35 sediments from rivers, lakes and channels, 39 sewage effluents and 38 sewage sludges collected in Germany. The phthalate burden was mainly from DEHP, whilst DBP was found in minor concentrations and BBP at concentrations near the detection limit. The concentrations found ranged from 0.3–98 $\mu g\,L^{-1}$ (surface water), 1.7–182 $\mu g\,L^{-1}$ (sewage effluent), 28–154 mg/kg dw (sewage sludge) and 0.2–8.4 mg/kg (sediment). The highest concentrations found were closely related to the input of industrial wastewaters from plastic production and were limited to a few kilometers downstream of the source of contamination.

Bisphenol A (BPA) is used extensively in the production of polycarbonate, epoxy resins, flame-retardants, and many other products. Its global production is more than 1 million tons per year and a significant portion is released into surface waters [133]. In the same study, a high concentration of BPA was confirmed in waste dump water and compost water samples as well as in the liquid manure samples (61–1112 $\mu g\,L^{-1}$). In sewage effluents, concentrations ranged from 18 to 702 $ng\,L^{-1}$ and in surface waters concentrations from 0.5 to 410 $ng\,L^{-1}$.

## 8 Conclusions

The issue of emerging contaminants is closely tied to analytical capabilities. Increased sensitivity in mass spectrometry, as a result of more efficient ionization techniques and better detectors, has allowed detection of virtually any new and potentially harmful contaminant at a very low level. Consequently, a number of new or previously ignored and/or unrecognized contaminants have bean brought under scrutiny and have been detected in different environmental compartments.

Numerous papers reported on the occurrence of a wide range of emerging contaminants in the aquatic environment, being wastewater and treated wastewater (WWTP effluents) the principle source and route of their entry into the environment. However, additional monitoring studies are needed not only to confirm the presence of emerging substances in the aquatic environment but also to allow the refinement of risk assessments in combination with relevant ecotoxicological test data. In relation to the emergence of new pollutants in the environment, the integration of physical/chemical techniques, effect monitoring techniques (e.g., bioassays, functional monitoring, etc.) and ecological monitoring/assessment (community surveys) techniques play a crucial role. The main drawback of the conventional approach is target-compound monitoring, which is often insufficient to assess the environmental relevance of emerging contaminants. An integrated approach combining analytical chemistry and toxicity identification evaluation (TIE) seems to be a more appropriate way to tackle the complex problems of environmental contamination.

## References


1. 6th EU Framework Programme project NORMAN (Network of reference laboratories and related organizations for monitoring and bio-monitoring of emerging environmental pollutants), Contract number 018486
2. Lienert J, Bürki T, Escher BI (2007) Water Sci Technol 56:87
3. Drillia P (2005) J Hazard Mat 122:259
4. Kuspis DA, Krenzelok EP (1996) Vet Hum Toxicol 38:48
5. Halling-Sørensen B, Nors Nielsen S, Lanzky PF, Ingerslev F, Holten Lutzhoft HC, Jørgensen SE (1998) Chemosphere 36:357
6. Daughton CG, Ternes TA (1999) Environ Health Perspect 107:907
7. Mimeault C, Woodhouse AJ, Miao XS, Metcalfe CD, Moon TW, Trudeau VL (2005) Aquat Toxicol 73:44
8. Cleuvers M (2005) Chemosphere 59:199
9. Brooks BW, Chambliss CK, Stanley JK, Ramirez A, Banks KE, Johnson RD, Lewis RJ (2005) Environ Toxicol Chem 24:464
10. Kolpin DW, Furlong ET, Meyer MT, Thurman EM, Zaugg SD, Barber LB, Buxton HT (2002) Environ Sci Technol 36:1202
11. Tolls J (2001) Environ Sci Technol 35:3397
12. Boxall ABA, Blackwell P, Cavallo R, Kay P, Tolls J (2002) Toxicol Lett 131:19
13. Jones OAH, Voulvoulis N, Lester JN (2006) Arch Environ Contam Toxicol 50:297
14. Bendz D, Paxeús NA, Ginn TR, Loge FJ (2005) J Hazard Mat 122:195
15. Alder AC, Bruchet A, Carballa M, Clara M, Joss A, Loffler D, McArdell CS, Miksch K, Omil F, Tuhkanen T, Ternes TA (2006) In: Ternes TA, Joss A (eds) Human Pharma-

ceuticals, Hormones and Fragrances: The Challenge of Micropollutants in Urban Wastewater Management. IWA Publishing, London, p 15
16. Calamari D, Zuccato E, Castiglioni S, Bagnati R, Fanelli R (2003) Environ Sci Technol 37:1241
17. Sedlak DL, Pinkston KE (2001) Water Res Update 120:56
18. Montforts MHMM (2001) In: Kümmerer K (ed) Pharmaceuticals in the environment. Springer, Berlin Heidelberg New York, p 159
19. Castiglioni S, Bagnati R, Fanelli R, Pomati F, Calamari D, Zuccato E (2006) Environ Sci Technol 40:357
20. Miao XS, Bishay F, Chen M, Metcalfe CD (2004) Environ Sci Technol 38:3533
21. Gros M, Petroviæ M, Barceló D (2007) Environ Toxicol Chem 26:1553
22. Terzic S, Senta I, Ahel M, Gros M, Petrovic M, Barcelo D, Müller J, Knepper T, Martí I, Ventura F, Jovancic P, Jabucar D, in press, Sci Total Environ
23. Santos JL, Aparicio I, Alonso E (2007) Environ Int 33:596
24. Radjenovic J, Petroviæ M, Barceló D (2007) Anal Bioanal Chem 387:1365
25. Nakada N, Tanishima T, Shinohara H, Kiri K, Takada H (2006) Water Res 40:3297
26. Heberer T (2002) Toxicol Lett 131:5
27. Ternes TA, Stumpf M, Schuppert B, Haberer K (1998) Vom Wasser 90:295
28. Gros M, Petroviæ M, Barceló D (2006) Talanta 70:678
29. Vieno NM, Tuhkanen T, Kronberg L (2006) J Chromatogr A 1134:101
30. Nikolai LN, McClure EL, MacLeod SL, Wong CS (2006) J Chromatogr A 1131:103
31. Escher BI, Bramaz N, Richter M, Lienert J (2006) Environ Sci Technol 40:7402
32. Watkinson AJ, Murby EJ, Costanzo SD (2007) Water Res 41:4164
33. Clara M, Strenn B, Gans O, Martinez E, Kreuzinger N, Kroiss H (2005) Water Res 39:4797
34. Jones OAH, Voulvoulis N, Lester JN (2001) Environ Technol 22:1383
35. Jjemba PK (2002) Agriculture, Ecosyst Environ 93:267
36. Clara M, Strenn N, Kreuzinger N (2004) Water Res 38:947
37. Gómez MJ, Martínez Bueno MJ, Lacorte S, Fernández-Alba AR, Agúera A (2007) Chemosphere 66:993
38. Lishman L, Smyth SA, Sarafin K, Kleywegt S, Toito J, Peart T, Lee B, Servos Beland M, Seto P (2006) Sci Total Environ 367:544
39. Vieno NM, (2005) Environ Sci Technol 39:8220
40. Tauxe-Wuersch A, De Alencastro LF, Grandjean D, Tarradellas J (2005) Water Res 39:1761
41. Oleksy-Frenzel J, Wischnack S, Jekel M (2000) Fresenius J Anal Chem 366:89
42. Andreozzi R, Raffaele M, Nicklas P (2003) Chemosphere 50:1319
43. Richardson ML, Bowron JM (1985) J Pharm Pharmacol 37:1
44. Khan SJ, Ongerth JE (2004) Chemosphere 54:355
45. Henschel KP, Wenzel A, Diedrich M, Fliedner A (1997) Regulat Toxicol Pharmacol 25:220
46. Ternes TA (1998) Water Res 32:3245
47. EMEA, In: Doc. Ref. EMEA/CHMP/SWP/4447/00; 2006; accessible at http:// www.emea.eu.int/pdfs/human/swp/444700en.pdf ieei, last visit to website: 10.08.2006
48. Kuster M, Lopez de Alda MJ, Rodriguez-Mozaz S, Barceló D (2007) In: Petrovic M, Barceló D (eds) Comprehensive Analytical Chemistry (Analysis, Fate and Removal of Pharmaceuticals in the Water Cycle), vol 50. Elsevier, Amsterdam, p 219
49. Schubert W, Cullberg G, Edgar B, Hedner T (1994) Maturitas 20:155
50. Kuhl H (1990) Maturitas 12:171
51. Turan A (1996) Umweltbundesamt, Berlin, p TEXTE 3/96

52. Ying G-G, Kookana RS, Ru Y-J (2002) Environ Int 28:545
53. Martindale (1982) The extra Pharmacopoeia. The Pharmaceutical Press, London
54. Petrovic M, Eljarrat E, de Alda MJL, Barcelo D (2004) Anal Bioanal Chem 378:549
55. Hanselman TA, Graetz DA, Wilkie AC, Szabo NJ, Diaz CS (2006) J Environ Qual 35:695
56. Lopez de Alda MJ, Barcelo D (2001) Fresenius J Anal Chem 371:437
57. Johnson AC, Williams RJ (2004) Environ Sci Technol 38:3649
58. Baronti C, Curini R, D'Ascenzo G, Di Corcia A, Gentili A, Samperi R (2000) Environ Sci Technol 34:5059
59. Ternes TA, Kreckel P, Mueller J (1999) Sci Total Environ 225:91
60. Desbrow C, Routledge EJ, Brighty GC, Sumpter JP, Waldock M (1998) Environ Sci Technol 32:1549
61. Belfroid AC, Van der Horst A, Vethaak AD, Schafer AJ, Rijs GBJ, Wegener J, Cofino WP (1999) Sci Total Environ 225:101
62. Beck M, Radke M (2006) Chemosphere 64:1134
63. Cui C, Ji S, Ren H (2006) Environ Monitor Assess 121:407
64. D'Ascenzo G, Di Corcia A, Gentili A, Mancini R, Mastropasqua R, Nazzari M, Samperi R (2003) Sci Total Environ 302:199
65. Gomes RL, Birkett JW, Scrimshaw MD, Lester JN (2005) Int J Environ Anal Chem 85:1
66. Koh YKK, Chiu TY, Boobis A, Cartmell E, Lester JN, Scrimshaw MD (2007) J Chromatogr A 1173:81
67. Komori K, Tanaka H, Okayasu Y, Yasojima M, Sato C (2004) Water Sci Technol 50:93
68. Laganà A, Bacaloni A, De Leva I, Faberi A, Fago G, Marino A (2004) Anal Chim Acta 501:79
69. Lee H-B, Peart TE, Svoboda ML (2005) J Chromatogr A 1094:122
70. Li Z, Wang S, Alice Lee N, Allan RD, Kennedy IR (2004) Anal Chim Acta 503:171
71. Lishman L, Smyth SA, Sarafin K, Kleywegt S, Toito J, Peart T, Lee B, Servos M, Beland M, Seto P (2006) Sci Total Environ 367:544
72. Nasu M, Goto M, Kato H, Oshima Y, Tanaka H (2001) Water Sci Technol 43:101
73. Quintana JB, Carpinteiro J, Rodriguez I, Lorenzo RA, Carro AM, Cela R (2004) J Chromatogr A 1024:177
74. Reddy S, Iden CR, Brownawell BJ (2005) Anal Chem 77:7032
75. Roda A, Mirasoli M, Michelini E, Magliulo M, Simoni P, Guardigli M, Curini R, Sergi M, Marino A (2006) Anal Bioanal Chem 385:742
76. Servos MR, Bennie DT, Burnison BK, Jurkovic A, McInnis R, Neheli T, Schnell A, Seto P, Smyth SA, Ternes TA (2005) Sci Total Environ 336:155
77. Zuehlke S, Duennbier U, Heberer T (2005) J Sep Sci 28:52
78. Johnson AC, Belfroid A, Di Corcia A (2000) Sci Total Environ 256:163
79. Nakada N, Tanishima T, Shinohara H, Kiri K, Takada H (2006) Water Res 40:3297
80. Clara M, Kreuzinger N, Strenn B, Gans O, Kroiss H (2005) Water Res 39:97
81. Castiglioni S, Bagnati R, Fanelli R, Pomati F, Calamari D, Zuccato E (2006) Environ Sci Technol 40:357
82. Ternes TA, Stumpf M, Mueller J, Haberer K, Wilken R-D, Servos M (1999) Sci Total Environ 225:81
83. Huang CH, Sedlak DL (2001) Environ Toxicol Chem 20:133
84. EMCDDA (2007) European Monitoring Centre for Drugs and Drug Addiction, Lisbon, http://www.emcdda.europa.eu/html.cfm/index875EN.html#42164
85. UNODC (2007) United Nations Office on Drugs and Crime, Vienna, http://www.unodc.org/pdf/research/wdr07/WDR_2007_executive_summary.pdf

86. Huerta-Fontela M, Galcerán MT, Ventura F (2007) Anal Chem 79:3821
87. Castiglioni S, Zuccato E, Crisci E, Chiabrando C, Fanelli R, Bagnati R (2006) Anal Chem 78:8421
88. Bones J, Thomas KV, Paull B (2007) J Environ Monit 9:701
89. Zucatto E, Chiabrando C, Castiglioni S, Calamari D, Bagnati R, Schiarea S, Fanelli R (2005) Environ Health: Global Access Sci Source 4:1
90. Hummel D, Löffler D, Fink G, Ternes TA (2006) Environ Sci Technol 40:7321
91. Jones-Lepp TL, Alvarez DA, Petty JD, Huckins JN (2004) Arch Environ Contam Toxicol 47:427
92. Postigo C, López de Alda MJ, Barceló D (2007) Anal Chem, in press
93. Pizzolato TM, Lopez de Alda MJ, Barceló D (2007) Trends Anal Chem 26:609
94. Daughton CG (2001) In: Pharmaceuticals and personal care products in the environment: Scientific and regulatory issues. Daughton CG, Jones-Lepp TL (eds) ACS Symposium Series 791. The American Chemical Society, Washington DC, p 116
95. Knepper TP, Eichhorn P, Bonnington LS (2003) Aerobic degradation of surfactants. In: Knepper TP, Barceló D, de Voogt P (eds) Analysis and Fate of Surfactants in the Aquatic Environment. Elsevier, Amsterdam, The Netherlands, p 525
96. Vogelsang C, Grung M, Jantsch TG, Tollefsen KE, Liltved H (2006) Water Res 40:3559
97. Giger W, Alder AC, Ahel M, Schaffner C, Reiser R, Albrecht A, Lotter AF, Sturm M (2002) Proc 1st SedNet Workshop on Chemical Analysis and Risk Assessment of Emerging Contaminants in Sediments and Dredged Material, Barcelona, Spain, p 183
98. Gonzalez S, Petrovic M, barcelo D (2004) J Chromatogr A 1052:111
99. Kvestak R, Terzic S, Ahel M (1994) Mar Chem 46:89
100. Jobling S, Sheahan D, Osborne JA, Matthiessen P, Sumpter JP (1996) Environ Toxicol Chem 15:194
101. Croatian Ordinance on Maximum Permissible Concentrations of Hazardous Contaminants in Waters and Coastal Sea. Narodne novine, No 2, 1984
102. Ahel M, Giger W, Koch M (1994) Water Res 28:1131
103. Vieira VM (2005) Perfluorinated compounds (PFCs). In: Health effects review. International Joint Commission: Department of Environmental Health, Boston University School of Public Health, Boston, MA (http://www.ijc.org/ rel/pdf/health_effects_spring 2005.pdf, last accessed 28th April 2006)
104. Ueno D, Darling C, Alaee M, Campbell L, Pacepavicius G, Teixeira C, Muir D (2007) Environ Toxicol Chem 41:841
105. Armitage J, Cousins IT, Buck RC, Prevedouros K, Russell MH, MacLeod M, Korzeniowski SH (2006) Environ Sci Technol 40:6969
106. Tomy GT, Budakowski W, Halldorson T, Helm PA, Stern GA, Friesen K, Pepper K, Tittlemier SA, Fisk A (2004) Environ Sci Technol 38:6475
107. Lau C, Thibodeaux JR, Hanson RG, Narotsky MG, Rogers JM, Lindstrom AB, Strynar MJ (2006) Toxicol Sci 90:510
108. Kennedy GL, Butenhoff JL, Olsen GW, O'Connor JC, Seacat AM, Perkins RG, Biegel LB, Murphy SR, Farrar DG (2004) Crit Rev Toxicol 34:351
109. Richardson SD (2007) Anal Chem 79:4295
110. Council of the European Communities (1996) Off J Eur Commun L 257:0026–0040
111. Prevedouros K, Cousins IT, Buck RC, Korzeniowski SH (2006) Environ Toxicol Chem 40:32
112. Wania F (2007) Environ Toxicol Chem 41:4529
113. Loganathan BG, Sajwan KS, Sinclair E, Kumar KS, Kannan K (2007) Water Res 41:4611

114. Sinclair E, Kannan K (2006) Environ Sci Technol 40:1408
115. Schultz MM, Higgins CP, Huset CA, Luthy RG, Barofsky DF, Field JA (2006) Environ Sci Technol 40:7350
116. Hohenblum P, Scharf S, Sitka A (2003) Vom Wasser 101:155
117. Alzaga R, Bayona JM (2004) J Chromatogr A 1042:155
118. Schröder HF (2003) J Chromatogr A 1020:131
119. Lange FT, Schmidt CK, Metzinger M, Wenz M, Brauch HJ (2004) Poster at the SETAC-Meeting in Prague (CZ), April 18th–22th
120. Skutlarek D, Exner M, Färber H (2006) Environ Sci Pollut Res 13:299
121. Loos R, Wollgast J, Huber T, Hanke G (2007) Anal Bioanal Chem 387:1469
122. Scott BF, Moody CA, Spencer C, Small JM, Muir DCG, Mabury SA (2006) Environ Sci Technol 40:6405
123. Scott BF, Spencer C, Mabury SA, Muir DCG (2006) Environ Sci Technol 40:7167
124. Yamashita N, Kannan K, Taniyasu S, Horii Y, Petrick G, Gamo T (2005) Mar Pollut Bull 51:658
125. Kloepfer A, Gnirss R, Jekel M, Reemtsma T (2004) Water Sci Technol 50:203
126. Kloepfer A, Jekel M, Reemtsma T (2005) Enviorn Sci Technol 39:3792
127. Voutsa D, Hartmann P, Schaffner C, Giger W (2006) Environ Sci Pollut Res 13:333
128. Weiss S, Jakobs J, Reemtsma T (2006) Environ Sci Technol 40:7193
129. Baurer MJ, Herman R (1997) Sci Total Environ 208:49
130. Marttinen SK, Kettunen RH, Sormunen KM, Rintala JA (2003) Water Res 37:1385
131. Fauser P, Vikelsoe J, Sorensen PB, Carlsen L (2003) Water Res 37:1288
132. Fromme H, Küchler T, Otto T, Pilz K, Müller J, Wenzel A (2002) Water Res 36:1429
133. Staples CA, Dorn PB, Klecka GM, O'Block ST, Harris LR (1998) Chemosphere 36:2149
134. Carballa M, Omil F, Lema JM, Llompart M, García-Jares C, Rodríguez I, Gómez M, Ternes T (2004) Water Res 38:2918
135. MacLeod SL, Sudhir P, Wong CS (2007) J Chromatogr A 1170:23
136. Ternes TA, Hirsch R (2000) Environ Sci Technol 34:2741
137. Li HQ, Jiku F, Schröder HF (2000) J Chromatogr A 889:155
138. Clara M, Scharf S, Scheffknecht, Gans O (2007) Water Res 41:4339
139. Di Corcia A, Cavallo R, Crescenzi C, Nazzari M (2000) Environ Sci Technol 34:3914
140. Crescenzi C, Di Corcia A, Samperi R (1995) Anal Chem 67:1797
141. Eichhorn P, Petroviæ M, Barceló D, Knepper TP (2000) Vom Wasser 95:245
142. Planas C, Guadayol JM, Doguet M, Escalas A, Rivera J, Caixach J (2002) Water Res 36:982
143. Castillo M, Martinez E, Ginebreda A, Tirapu L, Barceló D (2000) Analyst 125:1733
144. Ahel M, Molnar E, Ibric S, Giger W (2000) Water Sci Technol 42:15
145. Ahel M, Giger W (1998) Am Chem Soc Nat Meeting Extended Abst 38:276
146. De Voogt P, Kwast O, Hendriks R, Jonkers CCA (2000) Analysis 28:776
147. Castillo M, Alonso MC, Riu J, Barceló D (1999) Environ Sci Technol 33:1300
148. Tanghe T, Devriese G, Verstraete W (1999) J Environ Qual 28:702
149. Di Corcia A, Samperi R (1994) Environ Sci Technol 28:850
150. González-Barreiro C, Martínez-Carballo E, Sitka A, Scharf S, Gans O (2006) Anal Bioanal Chem 386:2123
151. McLachlan M, Holstrom K, Andursberger M (2007) Environ Sci Technol 41:7260
152. Senthikumar K, Ohi E, Sajwan K, Takasuga T, Kannan K (2007) Bull Environ Contam Toxicol 79:427
153. Rostkowski O, Yamashita N, Man Ka So I, Taniyasu S, Kwan Sing Lam P, Falandysz J, Tae Lee K, Kyu Kim S, Seong Khim J, Hyeon Im S, Newsted JL, Jones P, Kannan K, Giesy JP (2006) Environ Toxicol Chem 25:2374

154. Furdui VI, Stock NL, Ellis DA, Butt CM, Whittle DM, Crozier PW, Reiner EJ, Mabury SA (2007) Environ Sci Technol 41:1554
155. Tanabe S, Madhusree B, Ozturk AA, Tatsukawa R, Miyazaki N, Ozdamar E, Aral O, Samsun O, Ozturk B (1997) Mar Pollut Bull 34:338
156. Houde M, Bujas TAD, Small J, Wells RS, Fair PA, Bossart GD, Solomon KR, Muir DCG (2006) Environ Sci Technol 40:4138
157. Inneke K, de Vijver V, Holsbeek L, Das K, Blust R, Joiris C, De Coen W (2007) Environ Sci Technol 41:315

Hdb Env Chem Vol. 5, Part S/1 (2008): 37–104
DOI 10.1007/698_5_102

Published online: 8 March 2008

# Analysis of Emerging Contaminants of Municipal and Industrial Origin

Meritxell Gros[1] (✉) · Mira Petrovic[1,2] · Damià Barceló[1]

[1]Department of Environmental Chemistry, IIQAB-CSIC, c/Jordi Girona 18–26, 08034 Barcelona, Spain
*megqam@cid.csic.es*

[2]Institució Catalana de Recerca i Estudis Avançats (ICREA), Passeig Lluís Companys 23, 80010 Barcelona, Spain

**Abstract** Besides recognized pollutants, numerous other chemicals are continuously released into the environment as a result of their use in industry, agriculture, consumer goods or household activities. The presence of these substances, known as emerging contaminants, has become an issue of great concern within the scientific community during the last few years. For this reason, the availability of sensitive, accurate and reliable analytical techniques is essential in order to assess their occurrence, removal and fate in the environment.

In this chapter, the state of the art of the analytical techniques used to determine a wide range of emerging contaminants in several environmental matrices will be overviewed.

**Keywords** Emerging contaminants · Instrumental analysis · Sample preparation techniques

**Abbreviations**

| | |
|---|---|
| ADBI | 4-Acetyl-1,1-dimethyl-6-*tert*-butylindane |
| AED | Atomic emission detector |
| AHMI | 6-Acetyl-1,1,2,3,3,5-hexamethylindane |
| AHTN | 7-Acetyl-1,1,3,4,4,6-hexamethyl-1,2,3,4-tetrahydronaphthalene |
| AP | Alkylphenol |
| APCI | Atmospheric pressure chemical ionization |
| APEC | Alkylphenoxy carboxylate |
| APEO | Alkylphenol ethoxylate |
| APPI | Atmospheric pressure photoionization |
| ATII | 5-Acetyl-1,1,2,6-tetramethyl-3-isopropylindane |
| BSA | *N*,*O*-Bis(trimethylsilyl)-acetamide |
| BSTFA | *N*,*O*-Bis(trimethylsilyl)-trifluoroacetamide |
| BTEX | Benzene, toluene, ethylbenzene and xylenes |
| CAPEC | Dicarboxylated alkylphenoxy ethoxylate |
| CAR | Carboxen |
| CDEA | Coconut diethanolamide |
| CID | Collision-induced dissociation |
| CLLE | Continuous liquid–liquid extraction |
| CSIA | Compound-specific stable isotope analysis |
| CW | Carbowax |
| DAI | Direct aqueous injection |
| DEET | *N*,*N*-Diethyl-*m*-toluamide |
| DI-SPME | Direct solid-phase microextraction |
| DMIP | Dummy molecularly imprinted polymer |
| DPMI | 6,7-Dihydro-1,1,2,3,3-pentamethyl-4-(5*H*)-indanone |
| DVB | Divinylbenzene |
| ECD | Electron capture detector |

| | |
|---|---|
| EI | Electron impact |
| ELISA | Enzyme-linked immunosorbent assay |
| EPA | Environmental Protection Agency |
| ESI | Electrospray ionization |
| EU | European Union |
| FAS | Fluorinated alkyl substance |
| FID | Flame ionization detector |
| F NMR | Fluorine nuclear magnetic resonance |
| FTOH | Fluorotelomer alcohol |
| GC | Gas chromatography |
| GCB | Graphitized carbon black |
| GC×GC | Comprehensive two-dimensional gas chromatography |
| GC-MS | Gas chromatography–mass spectrometry |
| GPC | Gel permeation chromatography |
| HHCB | 1,2,4,6,7,8-Hexahydro-4,6,6,7,8,8-hexamethylcyclopenta-γ-2-benzopyrane |
| HLB | Hydrophilic–lipophilic balanced |
| HPLC | High-performance liquid chromatography |
| HS | Headspace |
| HSGC | Headspace gas chromatography |
| HS-SPME | Headspace solid-phase microextraction |
| IA | Immunoaffinity |
| IDA | Information-dependent acquisition |
| IPPC | Integrated Prevention and Control of the Contamination Directive |
| KOH | Potassium hydroxide |
| LAS | Linear alkyl sulphonate |
| LC | Liquid chromatography |
| LC/ESI-MS | Liquid chromatography–electrospray mass spectrometry |
| LLE | Liquid–liquid extraction |
| MAE | Microwave-assisted extraction |
| MCF | Methyl chloroformate |
| MCX | Mixed-mode cation exchange |
| MIMS | Membrane-introduction mass spectrometry |
| MIP | Molecularly imprinted polymer |
| MMLLE | Microporous membrane liquid–liquid extraction |
| MRM | Multiple reaction monitoring |
| MSPD | Matrix solid-phase dispersion |
| MSTFA | *N*-Methyl-*N*-trimethylsilyltrifluoroacetamide |
| MTBE | Methyl *tert*-butyl ether |
| MTBSTFA | *N*-(*tert*-Butyldimethylsilyl)-*N*-methyltrifluoroacetamide |
| NCI | Negative chemical ionization |
| NI | Negative ionization |
| NP | Normal phase |
| NPEC | Nonylphenoxy carboxylate |
| OECD | Organization for Economic Co-operation and Development |
| PA | Polyacrylate |
| PAH | Polycyclic aromatic hydrocarbon |
| PAM-MS | Purge-and-membrane inlet mass spectrometry |
| PBDE | Polybrominated diphenyl ether |
| PCB | Polychlorinated biphenyl |
| PCI | Positive chemical ionization |

| | |
|---|---|
| PCP | Personal care product |
| PDMS | Polydimethylsiloxane |
| PEEK | Polyetheretherketone |
| PFA | Pentafluoropropionic acid anhydride |
| PFDA | Perfluorodecanoic acid |
| PFO | Perfluorooctane sulphonate |
| PFOA | Perfluorooctanoate |
| PI | Positive ionization |
| PID | Photoionization detector |
| PLE | Pressurized-liquid extraction |
| PPY | Polypyrrole |
| PTFE | Polytetrafluoroethylene |
| PTV | Programmable temperature vaporization |
| P&T | Purge and trap |
| Q-LIT | Quadrupole–linear ion trap |
| QqQ | Triple quadrupole |
| Q-TOF | Quadrupole–time of flight |
| RAM | Restricted access material |
| RIA | Radioimmunoassay |
| RP | Reversed phase |
| SAX | Strong anion exchange |
| SEC | Size-exclusion chromatography |
| SFE | Supercritical-fluid extraction |
| SIM | Selected ion monitoring |
| SNUR | Significant new use rule |
| SPE | Solid-phase extraction |
| SPME | Solid-phase microextraction |
| SRM | Selected reaction monitoring |
| TBA | *tert*-Butyl alcohol |
| TBBPA | Tetrabromobisphenol A |
| TBF | *tert*-Butyl formate |
| TBS | *tert*-Butyldimethylsilyl |
| TFC | Turbulent flow chromatography |
| TMS | Trimethylsilyl |
| TMS-DEA | *N*,*N*-Diethyltrimethylamine |
| TrBA | Tri-*n*-butylamine |
| UPLC | Ultra-performance liquid chromatography |
| UV | Ultraviolet |
| VOC | Volatile organic compound |
| WAX | Mixed mode weak anion exchange |
| WWTP | Wastewater treatment plant |

## 1
## Introduction

During the last three decades, the impact of chemical pollution has focused almost exclusively on the conventional “priority” pollutants, which have long been recognized as posing risks to human health, due to their toxicity, carcinogenic and mutagenic effects, and their persistence in the environment. Legislation and long-established standards and certified analytical methods, set by the Environmental Protection Agency (EPA) and the International Organization for Standardization (ISO), are already available for the determination of these priority pollutants. Besides recognized contaminants, numerous other chemicals are continuously released into the environment as a result of their use in industry, agriculture, consumer goods or household activities. The identification, analysis and characterization of the risks posed by these substances, classified as the so-called emerging contaminants, has focused attention and awakened concern among the scientific community during the last few years. This group of compounds, including pharmaceuticals and personal care products, surfactants, gasoline additives, fire retardants and fluorinated organic compounds, among others, is still unregulated. These contaminants may be candidates for future regulation, depending on research on their potential health effects and monitoring data regarding their occurrence.

Several studies have demonstrated that wastewater treatment plants (WWTPs) are major contributors to the presence of emerging contaminants in the environment. As these substances are used in everyday life, they are continuously introduced into the aquatic media via sewage waters mainly through industrial discharges (surfactants, fire retardants), excretion (pharmaceuticals, hormones and contraceptives, personal care products) or disposal of unused or expired substances [1]. Methyl *tert*-butyl ether (MTBE) and other gasoline additives also enter the aquatic environment due to anthropogenic activities, mainly via accidental spills and leakage of corroded tanks at gasoline stations or refineries.

Due to their continuous introduction into the environment, emerging contaminants can be considered as "pseudo-persistent" pollutants, which may be able to cause the same exposure potential as regulated persistent pollutants, since their high transformation and removal rates can be compensated by their continuous input into the environment [2]. Consequently, there is a growing need to develop reliable analytical methods, which enable their rapid, sensitive and selective determination in different environmental compartments at trace levels.

This chapter aims to overview the state of the art of the most recent analytical methodologies developed in the last few years for the analysis of emerging contaminants in environmental samples, using advanced chromatographic techniques and detection systems. Since it is impossible to cover all analytes, we have just focused our attention on selected classes of contaminants, which are currently the most widely studied and ubiquitous in the environment. Trends in sample preparation and instrumental analysis for each group of compounds will be described.

## 2
## Sampling and Sample Preparation

Sample preparation is one of the most important steps within an analytical methodology. Selectivity of stationary phases used for the isolation and preconcentration of target compounds is a key parameter to take into account when analysing emerging contaminants at trace levels from complex environmental samples, since the reduction of co-extracted compounds results in a better sensitivity, achieving lower limits of detection. In the following section, a summary of the trends in stationary phases and materials used for the analysis of emerging contaminants in both aqueous and solid samples will be described.

### 2.1
### Sampling Strategies

Generally, to determine surface waters (river, lake, sea) grab samples are used, whereas for wastewaters composite samples are often collected over sampling periods of 6 h to several days. Some studies reported that the addition of 1% of formaldehyde to water samples prevents degradation of target compounds until analysis. Before sample enrichment, water samples are filtered through glass fibre or cellulose filters. Depending on the nature of the water sample (wastewater, surface water or seawater) and its organic matter content, different pore size filters are used.

In the case of sediments or soil samples, depending on the objective of the study (determination of vertical distribution profiles or concentrations in a surface layer), either core or grab samples are taken. Usually, water is removed and then the solid matrix is stored in the dry state. Removal of water from the sediments before extraction was found to be crucial in obtaining good recoveries [3]. Freeze-drying is an accepted and commonly used procedure for drying solid matrices, but it is not known how this affects the levels of target compounds measured, especially those that are relatively volatile [4].

When small fish, mussels or other bivalves are analysed, several individual species are homogenized to form a pool of tissues, from which sub-samples are taken for extraction. Removal of water is also generally performed by freeze-drying [5].

However, for aqueous matrices, grab samples may not be representative and moreover, a relatively large number of samples must be taken from a given location over the entire duration of sampling [6]. Therefore, a good alternative to overcome this problem could be the use of passive samplers. These devices are based on the free flow of analyte molecules from the sampled medium to a collecting one, as a result of a difference in chemical potentials of the analyte between the two media. Although they have only been applied for the determination of some organic pollutants and pesticides, their application in aqueous and gaseous phases is constantly increasing [6–10].

In passive samplers, the concentration of the analyte is integrated over the whole exposure time, making it immune to accidental or extreme variations of pollutant concentrations [6]. Other advantages against grab sampling are that decomposition of the sample during transport and storage is minimized and that passive sampling and/or extraction methods are simple to perform as, after the isolation and/or enrichment step, no further sample preparation is usually required [6]. Devices used today are based on diffusion through a well-defined diffusion barrier or permeation through a membrane, the former being the most popular ones.

## 2.2 Analysis of Emerging Contaminants in Water Samples

Extraction of target compounds from water matrices is generally achieved by solid-phase extraction (SPE) and solid-phase microextraction (SPME). For SPE, several stationary phases can be used, ranging from mixtures of different polymers (such as divinylbenzene–vinylpyrrolidone) to octadecylsilica ($C_{18}$) or more selective tailor-made materials, such as immunosorbents, molecularly imprinted polymers (MIPs) and restricted access materials (RAMs).

The use of tailor-made materials is very useful when performing single group analysis, as they enhance the selectivity for the compounds of interest in the sample preparation process, reducing the amount of co-extracted material and, as a result, increasing the sensitivity. However, when the aim of the analytical methodology is to analyse a wide spectrum of compounds with different physico-chemical properties, polymeric or $C_{18}$ sorbents are the most recommended ones.

The use of automated on-line systems, which integrate extraction, purification and detection, has increased over the past several years. One option is on-line coupling of SPE and LC, utilizing special sample preparation units, such as PROSPEKT (Spark Holland) and OSP-2 (Merck). This technique has been successfully applied to the analysis of pesticides, estrogens and progestogens in water samples [11–17]. Similarly, on-line coupling of SPE and SPME to GC is a promising approach with good prospects [18, 19].

### 2.2.1 Immunosorbents

The immunosorbents, such as polyclonal antibodies, are immobilized on silica-based supports, activated Sephadex gels, synthetic polymers, sol/gel materials, cyclodextrins, or RAMs and packed into cartridges or pre-columns [20, 21]. Immunoaffinity extraction coupled with LC/ESI-MS has been used for the analysis of pesticides [12, 22–24] and β-estradiol and estrone in wastewater [25]. Immunosorbents have also the potential to be applied to the determination of drugs in aqueous samples. In fact, most on-line immunosorbent applications correspond to pharmaceutical and biomedical trace analysis [26]. Therefore, a high number of pharmaceuticals [27, 28] and hormones [29, 30] have been determined in biological samples using immunoaffinity SPE coupled to on-line LC-MS. With these materials, humic and fulvic acids are not co-extracted and thus no further clean-up is necessary. Moreover, cross-reactivity of the antibody can be advantageous, because it not only extracts a determined substance, but also all compounds within a given class, being then separated and quantified individually by coupling with chromatographic techniques [31].

### 2.2.2 Molecularly Imprinted Polymers (MIPs)

During the last few years, MIPs have appeared as new selective sorbents for SPE of organic compounds in complex materials [32, 33]. Both on-line and off-line MIP-SPE protocols have been developed to determine organic pollutants in environmental waters, mainly pesticides and hormones [34–39].

Molecular imprinting is a rapidly developing technique for the preparation of polymers having specific molecular recognition properties [40–43]. First, the template and the monomer form a stable template–monomer complex prior to polymerization. Then the complex is polymerized in the presence of a cross-linking agent. The resulting MIPs are matrices possessing microcavities with a three-dimensional structure complementary in both shape and chemical functionality to that of the template [44, 45]. After polymerization, the template, which consists of one of the target analytes or related analogues, is removed, generating specific binding sites. Then, the polymer can be used to selectively rebind the template molecule, the analyte or structurally related analogues. The specific binding sites in MIPs are formed by covalent or, more commonly, non-covalent interactions between the imprinting template and the monomer [32].

Apart from their high selectivity for target compounds, MIPs possess other advantages, such as low cost, high stability, ability to be reused without loss of activity, high mechanical strength, durability to heat and pressure and applicability in harsh chemical media [46, 47].

MIPs can be prepared in a variety of physical forms, but the conventional approach is to synthesize the MIP in bulk, grind the resulting polymer and sieve the particles into the desired size ranges [48, 49]. However, this method is tedious and time-consuming, often produces particles that are irregular in size and shape and some interaction sites are destroyed during grinding. In order to overcome these problems, alternative methods have been developed, such as using multi-step swelling procedures, suspension and precipitation polymerization, respectively, to obtain uniform spherical particles [50–55].

In MIP-SPE processes, the sample medium, during the loading step, has an important influence on the recognition properties of the MIP. If the analyte of interest is presented in an aqueous medium, the analyte and other interfering compounds are retained non-specifically on the polymer. Therefore, to achieve the selectivity desired, a clean-up step using organic solvents is required prior to elution [32].

One of the main disadvantages of MIP-SPE is the difficulty in removing the entire template molecule, even after extensive washing, and therefore a leakage of template molecule can occur, which is an obstacle in the determination of target compounds. To overcome this problem, a structural analogue of the target molecule can be imprinted to make a "dummy molecularly imprinted polymer" (DMIP), distinguishing then any leakage of target compound [56].

### 2.2.3 Restricted Access Materials (RAMs)

RAMs are a class of SPE materials that possess a biocompatible surface and a pore size that restricts big molecules from entering the interior extraction phase based on size [26]. Simultaneously, an extraction phase located on the inner pore surface is responsible for isolation of the low molecular weight compounds [26]. Koeber et al. [57] applied this approach in combination with MIP and used an on-line mode to analyse pesticides from environmental samples. There are various references reporting the use of RAMs for direct injection of biological samples [58–60], but few applications have been reported for environmental matrices.

### 2.2.4 Solid-Phase Microextraction (SPME)

Several reviews have been devoted to the application of SPME in environmental analysis [6, 61–66]. SPME is a simple and effective adsorption/absorption and desorption technique which eliminates the need for solvents and combines sampling, isolation and enrichment in one step [66]. Depending on the analyte and matrix, SPME of water samples can be performed in different modes: direct-immersion extraction (for less volatile compounds and relatively clean samples), headspace extraction (for more volatile compounds and dirtier samples), membrane-protected SPME (for the extraction of analytes in heavily polluted samples), in-tube SPME [5, 67] and thin-film microextraction (use of a thin sheet of PDMS membrane) [68].

In-tube SPME has been applied for the determination of a variety of environmental pollutants [69–75] and is based on the use of a fused-silica capillary column as the extraction device. Target analytes in aqueous matrices are directly extracted and concentrated by the coating in the capillary column by repeated withdrawal and expulsion of the sample solution, and can be directly transferred to LC or GC columns for analysis.

The major part of SPME applications has been developed for GC, as the coupling to HPLC is more complex and requires specifically designed interfaces to desorb analytes from the fibres and also because not all fibres can be used for LC, due to solubility and swelling of the fibre coatings in organic solvents [5].

Several fibre coatings are commercially available for the analysis of non-polar organic compounds, such as BTEX, PAHs and pesticides, and polar compounds like phenols, alcohols, etc. [66], including polydimethylsiloxane (PDMS), polyacrylate (PA), divinylbenzene (DVB), Carboxen (CAR) and Carbowax (CW). On the other hand, a polypyrrole (PPY) coating is used to extract polar or ionic analytes [67], which is mainly addressed to the coupling of SPME to LC.

Another way to determine polar compounds by SPME is presented by SPME derivatization, which includes three different approaches: in-coating, direct or on-fibre derivatization. The difference between these techniques is that while in direct derivatization, the derivatizing agent is first added to the sample vial and the derivatives are then extracted by the SPME fibre coating, for on-fibre derivatization, the derivatizing agent is loaded on the fibre, which is subsequently exposed to the sample and extracted [66]. This approach is now widely used for the analysis of organic pollutants in the environment, such as acidic herbicides [76, 77], and has been recently reviewed by Stashenko [78] and Dietz [79].

## 2.3 Analysis of Emerging Contaminants in Solid Samples and Biota

### 2.3.1 Extraction Techniques

Organic contaminants present in solid environmental samples, such as sediments, soils, sludge and biota, are determined by exhaustive extraction with appropriate solvents. Liquid–liquid extraction (LLE), Soxhlet, sonication, pressurized-liquid extraction (PLE), microwave-assisted extraction (MAE) and supercritical-fluid extraction (SFE) are the techniques most commonly used [5]. Also methods based on HS-SPME have been developed to determine volatile and semi-volatile compounds.

Soxhlet has been widely used, as it is considered as the reference method, is inexpensive and is easy to handle. However, new trends are focused on the use of "low-solvent, low-time and low-cost" techniques, amenable to automation, such as PLE, MAE and SFE. These techniques use elevated temperature and pressure, which results in improved mass transfer of the analytes and, consequently, increased extraction efficiency. SFE and MAE are not suitable for highly polar organic compounds or matrices with high water content. Therefore, nowadays PLE, also termed accelerated solvent extraction, is the preferred technique, because it is automated, it consumes low amounts of solvent and because older extraction procedures can be easily adapted. However, it offers some disadvantages, such as its cost, as commercial PLE equipment may be expensive and, moreover, some thermolabile compounds may suffer degradation. A good alternative to PLE would be MAE, as it is more affordable, fast and consumes little solvent, but extracts need to be filtered and microwave heating is uneven and restricted to matrices that adsorb this radiation. SFE with solid-phase trapping has been used for different groups of organic pollutants. Although good results and unique improved selectivity were obtained for selected applications, the method did not find acceptance. This is because the extraction conditions depend on the sample, requiring complicated optimization procedures [5, 80].

### 2.3.2
### Extract Clean-up and Purification

Due to the complexity of samples and the exhaustive extraction techniques used, a substantial number of interfering substances present in the matrix are found in the extracts. Therefore, a clean-up and purification step after extraction is indispensable to remove these compounds and enhance selectivity, in order to reduce ion-suppression effects when working with ESI-MS detection and to improve the separation of analytes from impurities.

#### 2.3.2.1
#### Solid Samples

The conventional approach used is based on solid/liquid adsorption, using either long open columns or disposable cartridges packed with different sorbents, depending on the physico-chemical properties of the analytes of interest. Purification can be also performed by off-line SPE cartridges packed with polymeric materials, $C_{18}$, $NH_2$-, CN-modified silica or anionic exchange materials, by reversed-phase (RP) or normal-phase (NP) liquid chromatography, generally using alumina, silica or Florisil as the packing material, or size-exclusion chromatography (SEC) [5]. When high selectivity for one compound or related analogues is desired, MIPs and RAMs are also appropriate materials to use for the clean-up of crude extracts.

Purification based on two tandem SPE procedures is a widespread approach, which generally consists of the use of anionic exchange cartridges and other polymeric materials. Moreover, when extracts contain high amount of lipids and organic matter, such as sewage sludge and biota, non-destructive and destructive methods are generally used prior to instrumental analysis. The former include gel permeation and column adsorption chromatography, generally using polystyrene–divinylbenzene copolymeric columns. Other neutral adsorbents commonly used are silica gel, alumina and Florisil® [81]. Destructive lipid removal methods consist of sulphuric acid treatment, either directly to the extract or via impregnated silica columns, and saponification of extracts by heating with ethanolic KOH [82].

#### 2.3.2.2 Biota

The analysis of biota, such as fish or mussels, could be an indicator of the water quality, as lipophilic organic contaminants tend to accumulate in the tissues with high lipid content. Isolation of organic compounds from biological tissues is a complicated and laborious task because of the nature of the matrix. Disruption of a cellular structure of biological samples results in an abundance of lipids and proteins. Extraction methods often yield high concentrations of lipids and, therefore, an exhaustive purification is required to achieve the selectivity and sensitivity desired. For this reason, treatment with sulphuric acid and saponification are frequently used for the removal of lipids prior to the purification using the same techniques as for solid samples (RP or NP, LC, SPE, SEC, MIP or RAM). However, in some cases, this step has to be avoided as some target compounds may be destroyed.

A simultaneous extraction and clean-up step was proposed by Eljarrat et al. [83] for the determination of PBDEs in fish. This methodology is based on the inclusion of alumina in the PLE cells, so that both purification and isolation of target analytes is achieved in a single step, speeding up sample preparation considerably.

Another approach to conduct simultaneous disruption and extraction of solid and semi-solid samples involves matrix solid-phase dispersion (MSPD), a technique that combines in one step extraction, concentration and clean-up by blending a small amount of sample with the selected sorbent. It has been successfully applied to the analysis of penicillins, sulphonamides, tetracycline antibiotics [5] and ionic [5, 84, 85] and non-ionic surfactants in fish and mussels.

## 3 Instrumental Analysis and Quantitation

### 3.1 Chromatographic Separation

Both gas chromatography (GC) and liquid chromatography (LC) are techniques par excellence in environmental analysis. Even though the former is more addressed to the analysis of non-polar and volatile compounds (PBDEs and MTBE), non-volatile compounds, such as pharmaceuticals, surfactants, personal care products, estrogens and others, can also be determined after a derivatization step.

### 3.1.1
### Gas Chromatography

GC was one of the first chromatographic separation techniques to be developed, and today is still widely used and has not lost its eminence in the environmental field. The popularity of GC is based on a favourable combination of very high selectivity and resolution, good accuracy and precision, wide dynamic range and high sensitivity. Columns mainly used in GC consist of narrow-bore capillary columns [86–88].

In GC, the three most frequently used injection systems are splitless, on-column and programmable temperature vaporization (PTV). In splitless injection, the transfer of the analytes into the analytical column is controlled by the volume of the liner and by the injected volume. In on-column injection, extracts are directly injected into the column or in a glass insert fitted into a septum-equipped programmable injector kept at low temperature. Finally, PTV is a split/splitless injector which allows the sample to be introduced at a relatively low temperature, thus affording accurate and reproducible sampling. After injection, the PTV is rapidly heated to transfer the vaporized components into the capillary column.

Nowadays, headspace GC (HSGC) and comprehensive two-dimensional GC (GC×GC) have gained popularity in the environmental field. The main advantages presented by the former, against GC, is the ability to increase efficiency and drastically reduce analysis time [89]. On the other hand, GC×GC has a great capability to separate and identify organic compounds in complex environmental samples. This technique has been mainly employed for the determination of MTBE and other oxygenated and aromatic compounds in gasoline-contaminated ground waters [90] and for the determination of PBDEs [91]. In this technique, two GC separations based on distinctly different separation mechanisms are used, with the interface, called modulator, between them. Then, the effluent from the first column is separated into a large number of small fractions, and each of these is subsequently separated on the second column, which is much faster than the first separation. In principle, all kinds of stationary phases can be used in the first dimension of a GC×GC system, but generally, non-polar phases are the preferred ones. Concerning the second dimension, a variety of phases can be selected depending on the desired analyte–stationary phase interactions. However, most applications showed that the combination between a non-polar and (medium) polar phase is by far the most popular option. Concerning column size, samples are generally first separated on a 15–30 m × 0.25–0.32 mm ID × 0.1–1 μm film ($d_f$) column. After modulation, each individual fraction is injected onto a much shorter, narrower column, with dimensions typically 0.5–2 m × 0.1 mm ID × 0.1 μm $d_f$.

### 3.1.2 Liquid Chromatography

Besides the advantages offered by GC, nowadays reversed-phase HPLC is the technique of choice for the separation of polar organic pollutants, silica-bonded columns being preferred [92]. The size parameters of the columns are typically as follows: (1) length in the range 10–25 cm, (2) internal diameter 2.1–4.6 mm and (3) particle sizes 3–5 μm. Gradient elution represents the most common strategy in separation. The mobile phases generally used are acetonitrile, methanol or mixtures of both solvents, obtaining in the latter case shorter retention times and better resolution of the analytes. In order to obtain an efficient retention of the analytes in the column and to improve the sensitivity of MS detection, mobile phase modifiers, buffers and acids are recommended and widely used. The selection of such modifiers strongly depends on the physico-chemical properties of target compounds and their $pK_a$ values. The most common ones include ammonium acetate, ammonium formiate, tri-*n*-butylamine (TrBA), formic acid and acetic acid. Typical concentrations of the salts range from 2 to 20 mM, since it has been observed that higher concentrations could lead to a reduction of the signal intensities [92].

Shortening the analysis times is important for attaining the high sample throughput often required in monitoring studies. This objective can be achieved by shortening the columns and increasing the flow velocity, decreasing the particle size of the stationary phase and finally increasing the temperature, which enhances diffusivity thus allowing working at higher flow rates. These principles are both applied in the Acquity UPLC (ultra-performance liquid chromatography) system, produced by Waters Corporation (Manchester, UK) and in the 1200 Series RRLC (rapid resolution LC) from Agilent Technologies. Both systems use rather short columns (50–100 mm, 4.6 mm ID) packed with sub-2-μm porous particles, allowing very short chromatographic runs. However, the negative effect of using a small particle size is high back-pressure generation (reducing the particle size by a factor of 3 results in an increase in the backpressure by a factor of 27) [92]. Even though the application of UPLC is promising, its application to environmental analysis is still rare. Petrovic et al. [93] developed a UPLC-QqTOF-MS method for screening and confirmation of 29 pharmaceutical compounds belonging to different therapeutic classes in wastewaters, including analgesics and anti-inflammatories, lipid-regulating agents, cholesterol-lowering statin agents, psychiatric drugs, anti-ulcer agents, histamine $H_2$ receptor antagonists, antibiotics and beta-blockers. UPLC, using columns packed with 1.7-μm particles, enabled elution of target analytes in much narrower, more concentrated bands, resulting in better chromatographic resolution and increased peak height. The typical peak width was 5–10 s at the base, permitting very good separation of all compounds in

10 min, which represented an approximate threefold reduction in the analysis time in comparison to conventional HPLC as shown in Fig. 1.

One of the main problems encountered in quantitative LC analysis and a main source of pitfalls is the existence of matrix effects in general, and the ion suppression phenomenon in particular. The ionization suppression or enhancement may severely influence the sensitivity, linearity, accuracy and precision of quantitative LC analysis. Therefore, any study dealing with analysis of complex samples should include a matrix effect study, and if relevant ion suppression (or signal enhancement) occurs, additional procedures should be applied for correction and/or minimization of inaccurate quantification.

There are several strategies to reduce matrix effects, i.e. selective extraction, effective sample clean-up after the extraction, or improvement of the chromatographic separation. Sometimes, these approaches are not the appropriate solutions because they could lead to analyte losses as well as long analysis times [94]. Recently, several strategies have been adopted as standard practices [95–98]. The most often applied approach consists of the use of suitable calibration, such as external calibration using matrix-matched samples, standard addition or internal standard calibration using structurally similar unlabelled pharmaceuticals or isotopically labelled standards. Other approaches include a decrease of the flow that is delivered to the ESI interface, as well as the dilution of sample extracts. However, the most recommended and versatile approach is isotope dilution, which consists of the use of an isotopically labelled standard for each target compound [99]. But such an approach is expensive and in many cases suffers from a lack of isotopically labelled compounds for all target analytes.

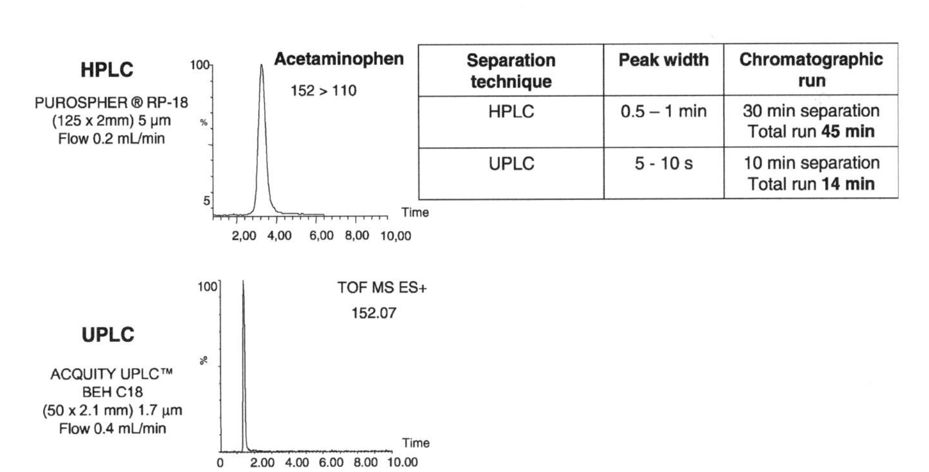


| Separation technique | Peak width | Chromatographic run |
|---|---|---|
| HPLC | 0.5 – 1 min | 30 min separation<br>Total run **45 min** |
| UPLC | 5 - 10 s | 10 min separation<br>Total run **14 min** |

**Fig. 1** UPLC versus HPLC chromatograms for the determination of the analgesic acetaminophen (paracetamol) in the PI mode, showing the reduced peak width and increased peak height achieved with UPLC, which results in an improved sensitivity, reduced spectral overlap in complex mixtures and improved MS spectral data

## 3.2
## Detection Systems

The rapid developments in the field of tandem MS/MS have transformed it into a key technique for environmental analysis, replacing other detectors widely used in the past, such as fluorescence and UV detectors for LC and flame ionization (FID), electron capture (ECD) and photoionization (PID) detectors for GC. While tandem MS/MS is mainly coupled to LC, replacing LC-MS due to its higher sensitivity and selectivity, single mass spectrometry is generally attached to GC, mainly using quadrupole, ion trap (IT) and time of flight (TOF) analysers. The latter is mainly applied when working with GC×GC devices.

With regard to LC-MS/MS, triple quadrupole (QqQ) mass analysers have become the most widely used analytical tool in the determination of emerging contaminants in environmental samples. Triple quadrupole instruments gather a variety of scan functions and modes, such as product ion scan, precursor ion scan, neutral loss and multiple reaction monitoring (MRM) mode. LC-MS/MS (QqQ) has been mostly applied to the determination of target analytes, using the selected reaction monitoring (SRM) mode and reaching typically ng $L^{-1}$ detection limits [92].

Although the sensitivity, selectivity and efficiency of the MRM approach are excellent, qualitative information, needed to support the structural elucidation of compounds other than target analytes, is lost [92]. This drawback can be overcome by using the hybrid MS systems, such as QqTOF or QqLIT. The acceptance of QqTOF-MS for environmental analysis in the last few years has been significantly improved and the number of methods reported in the literature is steadily increasing [92].

QqTOF is mainly used as an unequivocal tool for confirmation of contaminants detected. Its unique characteristic of generating full scan and product ion scan spectra with exact masses is excellent for the elimination of false positives and avoiding interpretation ambiguities. The main field of application is the identification of unknowns and elucidation of structures proposed for transformation products, where the amount of information obtained allows secure identification of compounds [92]. Regarding its quantitative performance, QqTOF has a lower linear dynamic range (over two orders of magnitude) with respect to QqQ instruments (typically > four orders of magnitude) [92]. However, when the application requires a high degree of certainty or is aimed at multiple tasks, such as target analysis combined with qualitative investigation of unknowns, its use could be a viable choice.

Regarding QqLIT, its unique feature is that the same mass analyser Q3 can be run in two different modes, retaining the classical triple quadrupole scan functions such as MRM, product ion, neutral loss and precursor ion while providing access to sensitive ion trap experiments [100] (see Fig. 2). This allows very powerful scan combinations when performing information-

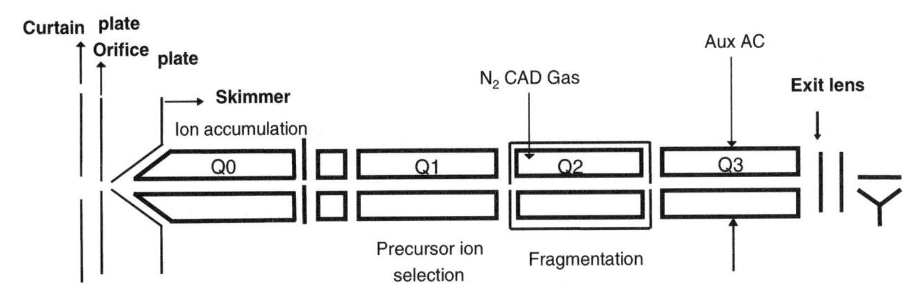


| Scan Type | Q1 | Q2 | Q3 |
|---|---|---|---|
| Q1 Scan | Resolving Scan | RF-only | RF-only |
| Q3 Scan | RF-only | RF-only | Resolving (Scan) |
| Product Ion Scan (PIS) | Resolving (Fixed) | Fragment | Resolving (Scan) |
| Precursor Ion Scan (PI) | Resolving (Scan) | Fragment | Resolving (Fixed) |
| Neutral Loss Scan (NL) | Resolving (Scan) | Fragment | Resolving (Scan Offset) |
| Selected Reaction Monitoring (SRM) | Resolving (Fixed) | Fragment | Resolving (Fixed) |
| Enhanced Product Ion Scan (EPI) | Resolving (Fixed) | Fragment | Trap/Scan |
| $MS^3$ | Resolving (Fixed) | Fragment | Isolation/frag trap/scan |
| Time delayed frag capture Product Ion (TDF) | Resolving (Fixed) | Trap/No frag | Frag/trap/scan |
| Enhanced Q3 single MS (EMS) | RF-only | No frag | Trap/Scan |
| Enhanced Resolution Q3 Single MS (ERMS) | RF-only | No frag | Trap/Scan |
| Enhanced Multiply Charged | RF-only | No frag | Trap/empty/scan |

**Fig. 2** Scheme of the QqLIT instrument (QTRAP, Applied Biosystems/Sciex) and description of the various triple quadrupole and trap operation modes

dependent data acquisition. In the case of small molecules, qualitative and quantitative work can be performed concomitantly on the same instrument. The very fast duty cycle of QqLIT provides a superior sensitivity over that of traditional QqQ and ion trap and allows one to record product ion scan spectra for confirmation purposes without compromising signal-to-noise (S/N) ratio. Also the resolution and accuracy are higher and these peculiarities improve the ion selection capability for complex mixtures, i.e. improve the instrumental selectivity. Although environmental applications are still scarce, a few recent papers reported on the application of a hybrid QqLIT for trace level determination of emerging contaminants, such as perfluorinated chemicals, herbicides and pharmaceuticals [92].

## 3.3 Ionization Sources

For GC-MS instruments, the most common ionization sources employed are electron impact (EI) or chemical ionization, either in negative (NCI) or positive mode (PCI). GC-NCI-MS is mainly used for compounds containing bromine or chlorine ions, such as PBDEs.

As concerns the LC-MS and LC-MS/MS techniques, API interfaces, such as electrospray ionization (ESI) and atmospheric pressure chemical ionization (APCI), are the ones most commonly used. In ESI, a liquid containing target analytes, dissolved in a large amount of solvent, is pushed through a very small, charged and usually metal capillary. The analyte exists as an ion in solution and as charges repel, the liquid pushes itself out of the capillary and forms an aerosol, a mist of small droplets about 10 μm across. An uncharged carrier gas such as nitrogen is sometimes used to help nebulize the liquid and evaporate the neutral solvent in the droplets. As the solvent evaporates, the analyte molecules repel each other and break up the droplets. This process repeats until the analyte is free of solvent and is a lone analyte ion. This process is known as Coulombic fission because it is driven by Coulombic forces between charged molecules. On the other hand, in APCI analytes are already vaporized when introduced into the detector. In this technique, the mobile phase containing eluting analytes is heated to a relatively high temperature (above 400 °C) and sprayed with high flow rates of nitrogen, generating an aerosol cloud which is subjected to a corona discharge to generate analyte ions. These techniques are especially suitable for the determination of low volatility and thermolabile compounds as well as polar substances. ESI is very useful for the analysis of macromolecules because it overcomes the propensity of such molecules to fragment when ionized.

Recently, a new API interface has been developed, the so-called atmospheric pressure photoionization (APPI) interface [101, 102]. APPI is a modification of the APCI source in which the corona is replaced by a gas discharge lamp, emitting radiation in the UV region that is able to selectively ionize the analytes in the presence of the LC mobile phase. Improved performance of APPI can be achieved by adding a dopant, which is a mobile phase additive, like acetone or toluene, which is first ionized itself and then aids ionization of the analytes in further reactions [103]. Compounds like naphthalene, acridine, diphenyl sulphide and 5-fluorouracil could be ionized by an APPI source. Despite being a very new approach, APPI-MS is expected to become an important complementary technique to APCI for low and non-polar analytes in the future [103].

# 4 Emerging Contaminants

## 4.1 Fluorinated Alkyl Substances (FASs)

FASs are a group of compounds of anthropogenic origin used in many industrial and consumer products, such as polymers and surfactants. They have been widely used to synthesize products that resist heat, oil, stains, grease and water, due to their unique properties [104].

FASs include the perfluoroalkyl sulphonates (perfluorooctane sulphonate (PFO) and related chemicals, such as *N*-methyl and *N*-ethyl perfluorooctane-sulphonamidoethanol, and also short- and long-chain perfluoro sulphonate acids), the perfluoroalkyl carboxylates (perfluorooctanoate (PFOA) and fluorotelomer alcohols (FTOHs)) and the short- and long-chain perfluoroalkyl acids (e.g. perfluorodecanoic acid (PFDA) [105]). Other substances, such as PFHS and PFBS, considered as "related substances" to PFOs because they have the same moiety ($C_8F_{17}SO_2$ group), are included in the group of PFAs as, once present in the environment, they may decompose to generate PFOs. Many of the degradation products of FASs have been found in the environment throughout the world, but PFOs and PFOA are the two most widely detected groups. Because of the strong carbon–fluorine (C–F) bond associated with their chemical structure, they are environmentally persistent substances and have been detected in human blood, water, soils, sediments, air and biota [105].

Due to their high production worldwide, in October 2000 the US EPA proposed a significant new use rule (SNUR) for 88 PFO-related substances [105]. On the other hand, PFOs and related substances have also been on the agenda of the Organization for Economic Co-operation and Development (OECD) sincc thc ycar 2000 [105]. In the EU, there is currently no legislation on their use associated with their potential environmental and/or human health effects. However, some legislation which generally applies to the release of substances to the environment may be relevant to the release of PFOs. Therefore, the IPPC Directive 96/61/EC includes fluorine and its compounds in the "indicative list of the main polluting substances to be taken into account if they are relevant for fixing emission limit values". There are several reviews devoted to their analysis in environmental samples [105, 106]. However, these compounds present several difficulties during their analysis, as indicated in the section below.

### 4.1.1 Background Contamination Problems

The analysis of PFAs is rather difficult due to several background contamination problems not only coming from the materials used for sample collection and preparation, but also from the instrumental techniques [104, 107–109]. Therefore, one source of experimental contamination is the use of materials made of, or containing, fluoropolymers, such as polytetrafluoroethylene (PTFE) or perfluoroalkoxy compounds, which should be avoided. Taniyasu et al. [107] performed several experiments to assess possible sources of contamination, from sample collection materials to solvents used. They found that polypropylene sample bottles used for sample collection and storage contained PFOA. In the evaluation of two widely employed SPE cartridges, the Oasis hydrophilic–lipophilic balanced (HLB) and Sep-Pak $C_{18}$, considerable amounts of PFOA, PFOs, PFHS and PFBS were detected, the latter being the one showing higher concentrations. Even purified water was found to be another possible source of contamination. In the light of these concerns, water samples are collected in polyethylene or polypropylene bottles rinsed with methanol and deionized water prior to use. Glass is avoided because analytes tend to bind it and some authors centrifuge water samples, as an alternative to filtration, to avoid possible adsorption of PFOs onto the filter and subsequent loss of analyte [110].

Moreover, during instrumental analysis, especially when working with LC-MS or tandem MS/MS detection, significant instrumental contamination problems can occur. Yamashita et al. [109] determined that the HPLC tubing, internal fluoropolymer parts and autosampler vial septum were potential sources of PFA contamination during LC analysis. Therefore, it is recommended to replace the PTFE HPLC tubing with stainless steel and polyetheretherketone (PEEK). Moreover, the same authors isolated the degasser and solvent selection valves, which contain fluoropolymer coatings and seals from the HPLC system, and the solvent inlet filters were replaced by stainless steel ones. Finally, autosampler vial caps made of Viton fluoropolymers or polyethylene were used, as they reduced considerably the instrumental blank concentrations.

### 4.1.2 Sample Preparation

Fluorinated alkyl substances have been mainly analysed in biological samples and environmental waters [105]. Concerning their determination in aqueous matrices, liquid–liquid extraction (LLE) and solid-phase extraction (SPE) are the traditional methods used for enrichment and isolation of target analytes, mainly using Oasis HLB, octadecyl $C_{18}$ bonded silica and Oasis WAX adsorbents (see Table 1) [105]. On-line direct analysis using diverse preconcentration columns has been proposed by several authors [18, 106, 111–113], to speed up sample preparation.

Only Higgins et al. [114] have determined the presence of fluorinated compounds in sediments. Extraction was performed using a heating sonication bath and afterwards a clean-up procedure with $C_{18}$ SPE cartridges. These compounds have also been determined in sludges by Higgins et al. [114] and Schröder et al. [115]. The former applied the same treatment as for the sediments. The latter compared the efficiency of three extraction techniques (Soxhlet, hot vapour and PLE), PLE being the one yielding better performances. After extraction, crude extracts are purified, generally using SPE with $C_{18}$ cartridges (see Table 2).

**Table 1** Representative methods, indicating the extraction and detection techniques, for the determination of the selected groups of emerging contaminants in environmental waters

| Compounds | Matrix | Extraction method | Purification or derivatization for GC | Detection | GC/LC column | LC mobile phase | LOD (ng/L) | Refs. |
|---|---|---|---|---|---|---|---|---|
| MTBE, degradation products and other gasoline additives | Influent/ effluent wastewaters | P&T | – | GC-EI-MS | | | | [362] |
| | Influent/ effluent wastewaters | HS-SPME | – | GC-EI-MS | | | | [351] |
| | Ground water | P&T with Tenax® silica gel–charcoal at room temperature. Desorption with He at 225 °C | – | GC-EI-MS | Capillary fused silica DB-624 (75 m × 0.53 mm) | | 1–110 | [347] |
| PFOs | Surface water | SPE (Presep-C cartridges) | – | LC-ESI-MS | Zorbax XDB $C_{18}$ (2.1 × 150 mm) | AcN-$H_2O$ (10 mM $NH_4Ac$) | 0.04–0.1 | [111, 112] |
| PFOs, *N*-EtFOSAA | Wastewater | SPE (Waters, Oasis HLB 1 g) | – | LC-ESI-MS/MS | Zorbax SB $C_8$ (3.0 × 150 mm) | A: MeOH/AcN (50%) 0.15% HOAc B: Water 0.15% HOAc | 0.06–0.1 | [363] |

**Table 1** (continued)

| Compounds | Matrix | Extraction method | Purification or derivatization for GC | Detection | GC/LC column | LC mobile phase | LOD (ng/L) | Refs. |
|---|---|---|---|---|---|---|---|---|
| PFNA<br>PFOSA<br>FTOHS | Seawater | SPE (Oasis WAX) | – | LC-ESI-MS/MS | Guard column: XDB-$C_8$ (2.1×12.5 mm) Column: Betasil-$C_{18}$ (2.1×150 mm) | A: $H_2O$ (2 mM $NH_4Ac$) B: MeOH | 1.8 pg/L<br>1pg/L<br>0.01–1 | [107] |
| E1, E2, 17α-E2, EE | Surface water Drinking water STP effluent | SPE (Lichrolut EN) | Derivatization with 10% PFBCl in toluene | GC-NCI-MS | DB5MS (60m×0.32 mm, 0.25 μm) | – | 0.05–0.15 | [185] |
| E1, E2, E3, EE | Ground water | SPE (Oasis HLB) | Derivatization with PFBBR + TMSI (LLE with water and hexane) | GC-NCI-MS/MS | DB5-XLB (60m×0.25 mm, 0.25 μm) | – | 0.2–0.6 | [134] |
| E1, E2, EE | Drinking, ground, surface and wastewater | SPE (Bakerbond $C_{18}$) | For WWTP influent SPE (silica gel) | LC-ESI (NI) MS/MS | RP-$C_8$ Hypersil MO5 (100×2.1 mm, 5 μm) | A: ACN/MeOH B: $H_2O$ | 0.1–2 | [167, 168] |
| E1, E2, E3, EE, DES, E2-17G, E1-3S, E2-17 Acet. | Ground, river and treated waters | Fully automated on-line SPE (PLRP-s) | – | LC-ESI (NI) MS/MS | Purospher STAR-RP18e (125×2 mm, 5 μm Merck) | A: ACN B: $H_2O$ | 0.01–0.38 | [138] |

**Table 1** (continued)

| Compounds | Matrix | Extraction method | Purification or derivatization for GC | Detection | GC/LC column | LC mobile phase | LOD (ng/L) | Refs. |
|---|---|---|---|---|---|---|---|---|
| E1, E2, E3 + PROG + six androgens | Ground and river water | SPE (Carbograph) | – | LC-APCI (PI) MS/MS | Alltima $C_{18}$ (250×4.6 mm, 5 μm Alltech) | A: ACN B: $H_2O$ 5 mM $NH_4Ac$ | 0.5–1 | [364] |
| Antibiotics, β-blockers, psychiatric drugs, anti-inflammatories | Hospital effluent wastewaters | pH adjustment (pH 7) SPE (Oasis HLB) | – | LC-ESI (NI) and (PI) MS/MS | Purospher STAR-RP18e (125×2 mm, 5 μm Merck) | **ESI(+)** A: ACN B: Aq-Formic acid **ESI(–)** A: ACN B: $H_2O$ | 4–47 | [200] |
| Anti-inflammatories, lipid regulators, anti-epileptic, β-blockers, antibiotics and other contaminants | River and wastewaters | Natural water pH SPE Oasis HLB | – | LC-ESI (NI) and (PI) MS/MS | Purospher STAR-RP18e (125×2 mm, 5 μm Merck) | **ESI(+)** A: ACN/MeOH (2:1) B: $NH_4Ac$ 5 m/ HAc **ESI(–)** A: MeOH B: $H_2O$ | 0.5–47 RW 1–60 WW | [2] |
| Analgesics/ anti-inflammatories, lipid regulators, β-blockers, antibiotics, anti-epileptics | Surface water | Sample acidified at pH = 3 SPE Oasis MCX | – | LC-ESI (NI) and (PI) MS/MS | | **ESI(+)** and **ESI(–)** A: MeOH B: 2 mM $NH_4Ac$ | 5–25 | [365] |

**Table 1** (continued)

| Compounds | Matrix | Extraction method | Purification or derivatization for GC | Detection | GC/LC column | LC mobile phase | LOD (ng/L) | Refs. |
|---|---|---|---|---|---|---|---|---|
| Tetracycline and sulphona-mide anti-biotics | Wastewaters | Addition of $Na_2EDTA$ and citric acid (pH<3) SPE Oasis HLB | – | LC-ESI (PI) MS/MS | | **ESI(+)** A: AcN B: 0.1% formic acid | 30–70 | [366] |
| All musk (no metabolites) | Wastewaters | LLE with hexane SEC (Bio Beads SX-3) | Silica purification | GC/EI-MS | VR-5MS (30 m×0.25 mm, 0.25 µm) | | NR | [258] |
| HHCB, AHTN, ATII, ADBI, AHMI, DPMI, MX, MK | WWTP effluent and surface water | SLLE with pentane, DCM, DCM (at pH 2) Dried with sodium sulphate | – | GC/EI-MS | BPX-5 (30 m×0.25 mm, 0.25 µm) | | NR | [234, 235] |
| HHCB, AHTN | Ground water | SPE ($C_{18}$) Eluent: acetone/ hexane (3:17 VR) | Silica purification | GC/EI-MS | XTI-5 (30 m×0.25 mm, 0.25 µm) | | NR | [197] |
| BDE-15, BDE-28, BDE-47, BDE-100, BDE-99, BD-154, BDE-153, BDE-183 | Tap and river water | HF-MMLLE using *n*-undecane as solvent. Extraction time: 60 min; stirring rate: 1200 rpm | – | GC/EI-MS | HP-5 ms (30 m×0.25 mm, 0.25 µm) | | 0.2–0.9 | [320] |
| BDE-47, BDE-100, BDE-99, BDE-85, BDE-154, BDE-153 | River, sea and wastewater | SPME using poly-dimethylsiloxane (PDMS) rods | – | GC-ECD-MS | HP-5 (30 m×0.32 mm, 0.25 µm) | | 0.3–5 | [367] |

**Table 1** (continued)

| Compounds | Matrix | Extraction method | Purification or derivatization for GC | Detection | GC/LC column | LC mobile phase | LOD (ng/L) | Refs. |
|---|---|---|---|---|---|---|---|---|
| α, β, γ-HBCD | Landfill leachate | LLE using DCM<br>SPE<br>Abselut Nexus | – | LC-ESI-MS/MS | Develosil C30-UG-5 (150 mm×2 mm) | **ESI(–)** A: ACN<br>B: $H_2O$ | NR | [368] |
| APEO, APEC, AP, halogenated derivatives | Surface drinking, and wastewaters | SPE<br>$C_{18}$ | – | LC-ESI (NI)/APCI-MS | Lichrospher RP-18 100 (250×4 mm, 5 μm) | **ESI(–)**<br>A: MeOH<br>B: $H_2O$<br>**APCI**<br>A: MeOH/ACN (1:1) B: $H_2O$ | 5–20 μg for river sediment<br>5–25 μm for sewage sludge | [277] |
| AEO, NPEO, CDEA, LAS, NPEGNP, OP | Coastal waters | SPE<br>Lichrolut $C_{18}$ | – | LC-ESI (NI)/APCI-MS | Lichrospher RP-18 100 (250×4 mm, 5 μm) | **AEO, NPEO, CDEA APCI**<br>A: MeOH/ACN (1:1) B: $H_2O$<br>**LAS, NPEC, NP, OP**<br>**ESI(–)**<br>A: MeOH; B: $H_2O$ | 10–150 | [279] |

**Table 2** Representative methods for the determination of the selected groups of emerging contaminants in solid samples, indicating the extraction, purification procedures and detection systems

| Compounds | Matrix | Extraction method | Purification or derivatization for GC | Detection | GC/LC column | LC mobile phase | LOD | Refs. |
|---|---|---|---|---|---|---|---|---|
| MTBE, degradation products and other gasoline additives | Soil | P&T with Tenax® silica gel–charcoal at room temperature. Desorption with He at 225 °C | – | GC-EI-MS | Capillary fused silica DB-624 (75 m×0.53 mm) | | 0.01–1.44 μ/kg | [350] |
| PFOs | Sediments | 3 extractions with 90:10 (v/v) MeOH and 1% HOAc | SPE $C_{18}$ | LC-ESI-MS/MS | Targa Sprite $C_{18}$ (40×2.1 mm) | MeOH-$H_2O$ 2 mM $NH_4Ac$ | 0.04–0.07 ng/L 0.109 ng/g | [114] |
| PFOA, PFHS, *N*-MeFO, SAA, *N*-EtFOSAA, anionic, non-ionic | Sewage sludge | PLE [EtOAc/DMF (8:2), MeOH/$H_3PO_4$ (95:5), MeOH/$H_3PO_4$ (99:1), MeOH/$H_3PO_4$ (99:1)] 150 °C, 10714 kPa | – | LC-ESI-MS | PF-$C_8$ column (150×4.6 mm) filled with spherical perfluorinated RP-$C_8$ material (5 μm) | A: MeOH B: MeOH/$H_2O$ (80:20) (2 mM diethyl ammonium) | 0.6 ng/g | [115] |
| E1, E2, α-E2, E3, MES (+BPA, NP) | River sediment | Ultrasonication (acetone/DCM, 1:1) | LLE with DCM + silica gel fractionation. Derivatization: PFPA | GC-EI-MS | HP-5MS (30 m×0.25 mm, 0.25 μm) | | 0.6–2.5 ng/g | [151] |

**Table 2** (continued)

| Compounds | Matrix | Extraction method | Purification or derivatization for GC | Detection | GC/LC column | LC mobile phase | LOD | Refs. |
|---|---|---|---|---|---|---|---|---|
| E1, E2, EE, MES | Sludge | Ultrasonication (MeOH + acetone) | GPC Biobeads SX-3 SPE (silica gel) Derivatization: MSTFA/TMSI/ DTE (1000:2:2, v/v/w) | GC-(IT)-MS/MS | XTI-5 (30 m×0.25 mm, 0.25 μm) | | 2–4 ng/g | [149] |
| 17G, E2–3, 17diS E1, E2 | Estuary sediment | Sonication (MeOH) | SPE (Lichrolut EN + BondElut $C_{18}$) + NP-LC fractionation | LC-ESI (NI)-TOF-MS | Betasil C18 (150×2.1 mm, 3 μm, Keystone Scientific) | A: AcN B: $H_2O$ | 0.03–0.04 ng/g | [152] |
| E1, E2, E3, EE, DES (+ progestins) | River sediment | Sonication (acetone: methanol, 1:1) | SPE ($C_{18}$) | LC-ESI (NI)-MS | Lichrospher 100 RP-18 (250×4 mm, 3 μm, Merck) | A: AcN B: $H_2O$ | 1–2 ng/g | [153] |
| Tetracycline, macrolide and sulphonamide antibiotics | Agricultural soils | PLE MeOH/citric acid (1:1, v/v) adjusted to pH = 4.7 with NaOH | Dilute PLE extracts to MeOH content < 10%. Purification with SAX-Oasis HLB in tandem | LC-ESI (PI)-MS/MS | X-terra MS-$C_{18}$ (100×2.1 mm, 3.5 μm, Merck) | A: MeOH B: Aq. formic acid | 8–22 μg/L | [194] |

**Table 2** (continued)

| Compounds | Matrix | Extraction method | Purification or derivatization for GC | Detection | GC/LC column | LC mobile phase | LOD | Refs. |
|---|---|---|---|---|---|---|---|---|
| Tetracycline, sulphonamides, fluoro-quinolone antibiotics and trimethoprim | Arable soils fertilized with manure | **TCs, SAs and TMP**<br>MeOH/EDTA-McIlvaine buffer pH = 6 (90:10, v/v)<br>**FQs**<br>AcN acidified with 2% HCOOH | **TCs, SAs and TMP**<br>SPE $C_{18}$<br>**FQs**<br>LLE with hexane | **TCs, SAs and TMP**<br>LC-ESI (PI) MS/MS<br>**FQs**<br>LC-ESI (PI) MS | **TCs, SAs and TMP**<br>Luna (Pheno-menex) $C_8$ (150×2 mm, 5 μm)<br>**FQs**<br>Luna (Pheno-menex) $C_8$ (150×3 mm, 5 μm) | **TCs, SAs and TMP**<br>A: ACN<br>B: $H_2O$<br>C: 0.5% HCOOH 10 mM $NH_4OAc$<br>**FQs**<br>A: ACN 0.01% HCOOH<br>B: $H_2O$ 0.01% HCOOH | 1.6–18 (ng/mL) | [369] |
| Analgesics and anti-inflammatories, lipid regulators, antibiotics and ivermectin | River sediment | Ultrasound<br>**Acidic compounds**<br>Acetone/HAc (20:1, v/v) + ethyl acetate<br>**Antibiotics**<br>MeOH/acetone + ethyl acetate | Dilute extracts<br>**Acidic compounds**<br>Acidify at pH = 2<br>SPE Oasis MCX<br>**Antibiotics**<br>Acidify at pH = 3<br>SPE Lichrolut EN + $C_{18}$<br>**Ivermectin** | **Acidic compounds**<br>LC-ESI (NI) MS/MS<br>**Antibiotics**<br>LC-ESI (PI) MS | All compounds<br>Lichrospher RP-18 (125×3 mm, 5 μm, Merck) | **Acidic compounds**<br>A: ACN<br>B: $H_2O$<br>pH = 2.9 (with HAc)<br>**Antibiotics**<br>A: Eluent B + AcN<br>B: 20 mM $NH_3$ at pH = 5.7 with HAc | **Acidic compounds**<br>0.4–20 ng/g<br>**Antibiotics**<br>3–20 ng/g | [195] |

**Table 2** (continued)

| Compounds | Matrix | Extraction method | Purification or derivatization for GC | Detection | GC/LC column | LC mobile phase | LOD | Refs. |
|---|---|---|---|---|---|---|---|---|
| | | | Add $NH_4Ac$ buffer<br>SPE Lich-rolut EN | | | **Ivermectin**<br>A: ACN 10% B<br>B: 15 mM $NH_4$ AC + HAc (pH = 4) | | |
| All musks and metabolites (except DPMI) | Activated sludge | LLE with hexane | Silica purification | GC-MS/MS<br>GC-EI-MS | DB-1 (60 m×0.25 mm, 0.25 μm) | | NR | [265, 370] |
| HHCB, AHTN, ATII, ADBI, AHMI, DPMI, MX, MK, MA, MM, MT | Digested sludge | Dried with sodium sulphate<br>Soxhlet extraction with DCM<br>Sulphur removed with copper in flask during extraction | Silica/alumina purification (layered)<br>SEC (Bio Beads S-X3)<br>Silica/alumina purification | GC-EI-MS | HP-5MS (30 m×0.25 mm) | | NR | [261] |
| All musks | Sludge | SFE with acetone/DCM (1:1) | Silica purification<br>Sulphur removed with copper | GC-NCI/MS<br>GC-EI-MS | HP-5MS (30 m×0.25 mm, 0.25 μm) | | NR | [371] |

**Table 2** (continued)

| Compounds | Matrix | Extraction method | Purification or derivatization for GC | Detection | GC/LC column | LC mobile phase | LOD | Refs. |
|---|---|---|---|---|---|---|---|---|
| Mono-hepta-BDEs (39 compounds) | Marine and river sediment | PLE (Cu + $Al_2O_3$ 1:2) using DCM:C6 (1:1) as solvent | – | GC-NCI-MS | HP-5MS (30 m×0.25 mm, 0.25 $\mu$m) | | 1–46 pg/g | [326] |
| α, β, γ-HBCD | Sediments | Soxhlet (acetone:C6, 3:1) | LLE with $H_2SO_4$ + GP + $SiO_2$ | LC-ESI (NI) MS | Luna $C_{18}$ (150×2 mm, 5 μm, Merck) | A: AcN + 10 mM $NH_4OAc$<br>B: $H_2O$ + 10 mM $NH_4OAc$ | NR | [372] |
| Di-hexa BDEs + deca-BDEs (14 compounds) | Sewage sludge | PLE (DCM:C6, 1:1) | $H_2SO_4$ + $SiO_2$<br>$H_2SO_4$ + $Al_2O_3$ | **Di-hexa BDE:** GC-MS/MS<br>**Deca-BDE:** GC-NCI-MS | NR | | NR | [373] |
| Mono-deca BDEs (40 compounds), total HBCD | Fish tissue | PLE ($Al_2O_3$, DCM:C6, 1:1) | – | GC-NCI-MS | HP-5MS (30 m×0.25 mm, 0.25 μm) | | 2–19 pg/g (wet-weight) | [306] |
| Tri-deca BDEs (27 compounds) | Fish tissue | PLE (DCM) | GPC + $SiO_2$ | GC-NCI-MS | NR | | NR | [374] |
| Non-ionic surfactants, NPEO, AEO, CDEA | Sewage sludge | Sonication (DCM/MeOH, 3:7) | SPE $C_{18}$ | LC-ESI (NI)/APCI-MS | Lichrospher RP-18 100 (250×4 mm, 5 μm) | **ESI** (–)<br>A: MeOH<br>B: $H_2O$<br>**APCI**<br>A: ACN<br>B: $H_2O$ | 5–25 μg/kg | [277] |

**Table 2** (continued)

| Compounds | Matrix | Extraction method | Purification or derivatization for GC | Detection | GC/LC column | LC mobile phase | LOD | Refs. |
|---|---|---|---|---|---|---|---|---|
| APEO, APEC, AP, halogenated derivatives | River sediment, sludge | Sonication (DCM/MeOH, 3:7) | SPE $C_{18}$ | LC-ESI (NI)/APCI-MS | Lichrospher RP-18 100 (250×4 mm, 5 µm) | **ESI**(–) A: MeOH B: $H_2O$ **APCI** A: MeOH/ ACN (1:1) B: $H_2O$ | 20–100 µ/kg | [277] |
| Ionic surfactants LAS, SPC | Marine sediment | Soxhlet (MeOH) | SPE $C_{18}$ | LC-FL | Lichrosorb RP-18 (250×4.6 mm, 10 µm) | A: MeOH/$H_2O$ (80:20) with 1.25 mM tetraethyl-ammonium B: $H_2O$ | 5–10 µ/kg | [375] |

### 4.1.3 Instrumental Analysis

Fluorinated surfactants can be detected by $^{19}$F NMR, gas and liquid chromatography–mass spectrometry and liquid chromatography coupled to tandem mass spectrometry [105], the latter two being the most widely employed.

$^{19}$F NMR spectroscopy is a non-specific method, as it determines the presence of $CF_2$ and $CF_3$ moieties [116, 117]. Moody et al. [117] compared the results achieved by this technique with LC-MS/MS, showing discrepancies between the two methods. With $^{19}$F NMR the total content of perfluorinated compounds was higher than that calculated by LC-MS/MS, attributed to the presence of other surfactants in the samples which yielded a similar $^{19}$F NMR spectrum to perfluoroalkanesulphonates and perfluorocarboxylates [105].

Gas chromatography–mass spectrometry can be used for the direct determination of neutral and volatile FASs, such as sulphonamides or fluorotelomer alcohols, which have high vapour pressures [105]. Perfluorocarboxylates have been quantitatively determined by GC-MS after derivatization of the carboxylates to their methyl esters [116, 117]. However, PFOs was not able to be detected by such a method [117]. Although perfluoroalkane sulphonate esters may be formed during the derivatization step, the esters are unstable because of the excellent leaving group properties of perfluoroalkane sulphonates [105]. Thus, despite the fact that some fluorinated surfactants can be analysed by GC-MS, this technique is not so useful for multi-residue analysis of all groups of PFAs [105]. The drawbacks offered by both $^{19}$F NMR and GC-MS and the multiple advantages presented by LC-MS and LC-MS/MS, in terms of sensitivity and selectivity, have made these techniques the preferred tools for the instrumental analysis of PFAs in environmental samples. Other detectors coupled to LC include fluorescence detection for the determination of perfluorocarboxylic acids [118], ion-exclusion chromatography with conductimetric detection for perfluorocarboxylic acid and perfluorosulphonates [119, 120] and LC with conductimetric detection for perfluorosulphonates [121].

Electrospray ionization (ESI) working in the negative ion (NI) mode is the interface most widely used for the determination of anionic perfluorinated surfactants. APCI is not suitable for the determination of PFOs due to their ionic nature. The ESI interface has also been optimized for the determination of neutral compounds, such as the sulphonamides PFOSA, Et-PFOSA and *t*-Bu-PFOs [122]. Takino et al. [110] developed a method based on an APPI interface, which would alleviate matrix effects found with ESI interfaces.

Chromatographic separation of fluorinated compounds has been mainly carried out using both RP-$C_{18}$ and RP-$C_8$ materials. However, RP-$C_{18}$ presented some interferences, enhancing analyte signals and, therefore, the

**Fig. 3** LC-ESI(NI)-MS chromatograms obtained in the SIM mode for a standard solution containing **a** perfluorocarboxylic acids and **b** sulphonates and neutral FASs. Reprinted with permission from [376] ▶

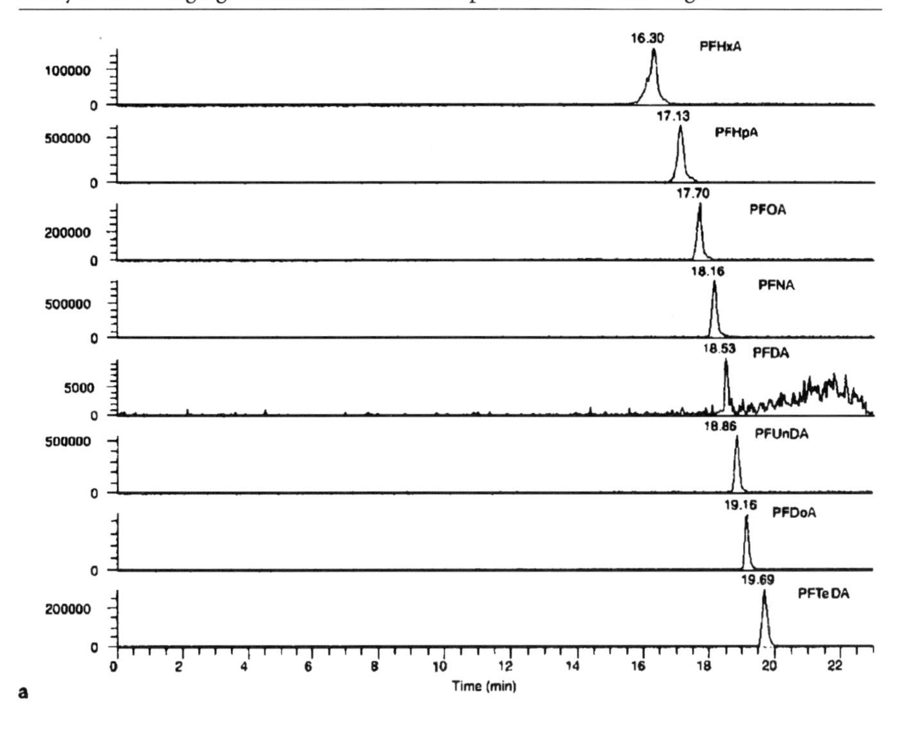
16.30
PFHxA
100000
0
17.13
PFHpA
500000
0
17.70
PFOA
200000
0
18.16
PFNA
500000
0
18.53
PFDA
5000
0
18.86
PFUnDA
500000
0
19.16
PFDoA
0
19.69
PFTeDA
200000
0
0 2 4 6 8 10 12 14 16 18 20 22
Time (min)
a

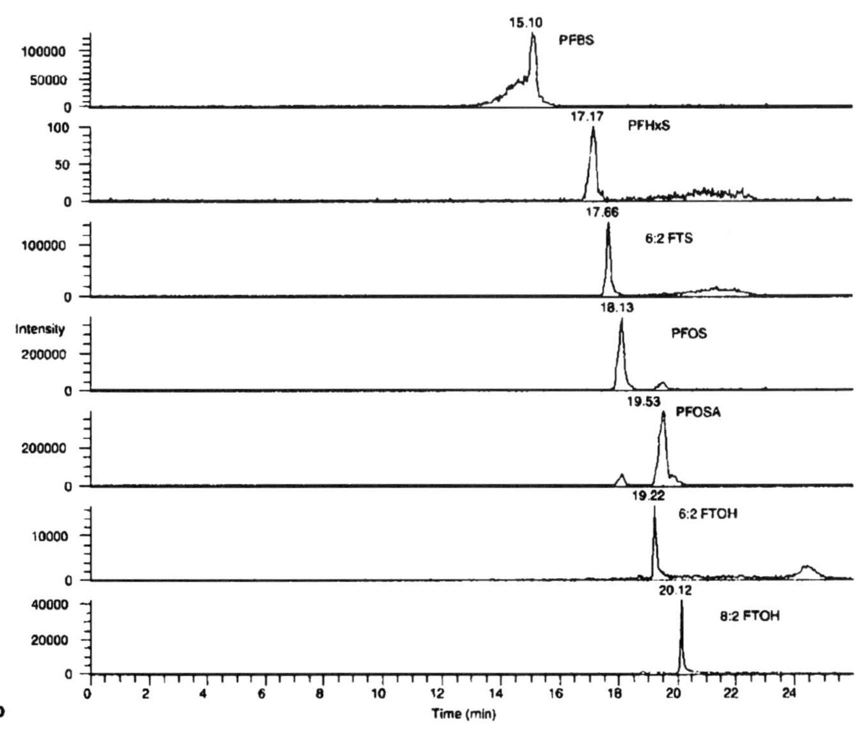
15.10
PFBS
100000
50000
0
17.17
PFHxS
100
50
0
17.66
6:2 FTS
100000
0
18.13
PFOS
Intensity
200000
0
19.53
PFOSA
200000
0
19.22
6:2 FTOH
10000
0
20.12
8:2 FTOH
40000
20000
0
0 2 4 6 8 10 12 14 16 18 20 22 24
Time (min)
b

RP-$C_8$ ones are more recommended. Nevertheless, using RP-$C_{18}$ branched isomers can be distinguished, while RP columns with shorter alkyl chains ($C_8$) are not so efficient. This effect can be minimized by increasing the LC column temperature from 30 to 40 °C [110, 112, 123]. Comparison of the retention times of a $C_8$ column and an end-capped $C_8$ one indicated that the interaction of FASs with the residual silanol groups in the non-end-capped column played an important role in providing a good separation of the analytes [115].

Moreover, in reversed-phase LC columns, the FAS standards display a characteristic chromatographic pattern with two unresolved signals or shoulders adjacent to the major signal (see Fig. 3). This is due to the fact that most commercially available standards are mixtures of linear and branched isomers (approximately 70% linear), which contain impurity isomers with the same alkyl chain lengths. It is assumed that the response factor for branched and linear isomers is equivalent and that the standard mixtures are representative of those identified in the samples [124]. Regarding mobile phases, mixtures of acetonitrile–water and methanol–water, often modified with ammonium acetate (1.0–20 mM) are the ones most commonly employed.

In the fragmentation pattern of FASs, the deprotonated molecules $[M-H]^-$ are the predominant ions. Typical ions and fragmentations monitored for PFOs and related substances correspond to $[SO_3]^-$, $[FSO_3]^-$ and $[M-SO_3]^-$ ions. For PFOSA and PFOA, $[SO_2N]^-$ and $[MCOOH]^-$ ions are the most abundant ones, respectively [105].

## 4.2 Steroid Estrogens, Pharmaceuticals and Personal Care Products

### 4.2.1 Steroid Estrogens (Hormones and Contraceptives)

Estrogens have often been identified as the compounds responsible for the estrogenic effects that have been observed in different wildlife species, such as intersex in carp, high levels of plasma vitellogenin in fish, etc. [125].

Chemical analysis has focused on the investigations of free estrogens, both natural (estradiol, estrone and estriol) and synthetic (basically ethynyl estradiol, mestranol and diethylstilberol). In contrast, conjugated estrogens and halogenated derivatives have been seldom studied, maybe due to their lower estrogenic effect and recent identification.

#### 4.2.1.1 Sample Preparation

There are multiple reviews devoted to the analysis of esteroid estrogens in environmental samples [25, 126–133]. An important precaution that should be taken into account when analysing steroid estrogens in tap water, or water samples that could contain chlorine, is the addition of sodium thiosulphate immediately after collection in order to avoid losses of target analytes [134].

Extraction of estrogens from water samples has usually been carried out by off-line SPE using either disks or, most frequently, cartridges (see Table 1), with octadecyl $C_{18}$-bonded silica, polymeric graphitized carbon black (GCB) and Oasis HLB being the most widely employed cartridges [134–136]. On the other hand, many works are based on the use of on-line SPE [129, 137, 138], using the same extraction materials as indicated for off-line SPE. To elute compounds trapped in the SPE cartridges, methanol is the solvent generally used. However, Isobe et al. [136] determined that adding 5 mM of TEA to 10 mL of methanolic solution, as an ion pair reagent, improved the efficiency of elution, thus achieving higher recoveries for conjugates which were not effectively removed by only using methanol.

Other widely employed materials to isolate steroid estrogens from water samples are molecularly imprinted polymers (MIPs) [25, 38, 139]. Some recent works have also proposed the use of SPME, using fibre and in-tube SPME, in combination with either LC or GC instruments [140, 141, 143].

As concerns the determination of esteroid estrogens in solid samples, the analytical methods are generally adapted from those developed for water samples, incorporating additional purification steps of crude extracts prior to instrumental analysis [144]. Extraction techniques more commonly used are pressurized liquid extraction (PLE) [145, 146], microwave-assisted extraction (MAE) [147] and, more frequently, ultrasonication [148–153], using methanol [148, 152], methanol/acetone [145, 146, 149, 153], acetone/dichloromethane [151], ethyl acetate [154, 155] or dichloromethane/water [150] as extraction solvents. Some of the most representative methods are summarized in Table 2.

Purification of extracts is generally carried out by liquid–liquid extraction (LLE) [156–158], HPLC fractionation [156, 159–162], gel permeation chromatography (GPC) [158], immunoaffinity (IA) extraction [25] or SPE using Florisil [136, 157], $C_{18}$ sorbents [132, 156, 159, 160], silica gel [163–169] and restricted access materials (RAMs).

#### 4.2.1.2
#### Instrumental Analysis

In the past, the techniques most commonly used for the environmental analysis of estrogens have been immunoassays and, to a greater extent, GC-MS. The former are simple and sensitive but they can have false positive results due to the influence of coexisting materials present in the sample matrix. On the other hand, GC-MS and GC-MS/MS are also highly sensitive methods, but derivatization is required prior to analysis [141]. Moreover, these methodologies are mainly based on the determination of unconjugated (i.e. free) estrogens, unless intermediate hydrolysis steps are performed [136, 170]. LC-MS and especially LC-MS/MS are the preferred tools nowadays [171, 172], which allow the determination of both conjugated and free estrogens without derivatization and hydrolysis.

Enzyme-linked immunosorbent assay (ELISA) and radioimmunoassay (RIA) are by far the most common *bioassays* used for the determination of estrogens. Several recent works have reported their application in the analysis of estrogens in environmental matrices, such as water [173–176], sludge and manure, although they have been more extensively used for the analysis of biological samples in clinical studies. Their main advantages are ease of use, relatively simple protocol and fairly good sensitivity. Bioassays are also used to measure the estrogenic (endocrine disrupting) activity of sample extracts or of chemicals. The in vitro and in vivo assays available for this purpose have been recently reviewed [177, 178]. Many bioassays show potential for development as biosensors [179, 180].

On the other hand, GC separation has been performed with a variety of capillary columns (DB5-MS, XTI-5, HP Ultra II, etc.), using helium as carrier gas. Both conventional MS and MS/MS detection have been accomplished in most instances in the electron impact (EI) mode at 70 eV. The use of negative ion chemical ionization (NICI) has been reported on fewer occasions [134, 165, 181–184]. However, it has been observed that the highest sensitivity for the GC-NICI-MS methods is obtained when estrogens have pentafluorobenzyl (PFB) [181, 182], pentafluorobenzoyl [184, 185] and other fluorine-containing derivatives.

Derivatization is generally carried out in the – OH groups of the steroid ring, performed by silylation with reagents such as *N*,*O*-bis(trimethylsilyl)-acetamide (BSA), *N*-methyl-*N*-trimethylsilyltrifluoroacetamide (MSTFA), *N*,*O*-bis(trimethylsilyl)-trifluoroacetamide (BSTFA), or *N*-(*tert*-butyldimethylsilyl)-*N*-methyltrifluoroacetamide (MTBSTFA), which lead to the formation of trimethylsilyl (TMS) and *tert*-butyldimethylsilyl (TBS) derivatives [186]. Some authors reported breakdown of some TMS derivatives with various solvent–reagent combinations, pyridine and dimethylformamide being the most suitable ones [186–188].

LC has been performed by octadecyl silica stationary phases. As mobile phases, mixtures of water/methanol and, more frequently, water/acetonitrile have normally been used, sometimes with added modifiers such as 0.1% acetic acid, 0.2% formic acid or 20 mM ammonium acetate. The interfaces most widely employed are electrospray ionization (ESI) in the negative ion (NI) mode and, to a lesser extent, atmospheric pressure chemical ionization (APCI) in the positive ionization (PI) mode. These API interfaces have been applied in a variety of MS analysers, including quadrupole, ion-trap, orthogonal-acceleration time-of-flight (oaTOF), and combinations of them. Single and triple quadrupole analysers have been the most widely used for the analysis of estrogens, the latter being preferred nowadays. Some works

**Table 3** MRM transitions monitored for the determination of steroid estrogens and pharmaceuticals in environmental samples using LC-ESI-MS/MS (QqQ) instruments

| Group of substances | Compound | MRM 1 | MRM 2 |
|---|---|---|---|
| Steroid estrogens | Estriol | 287>171 Loss of $C_6H_{12}O_2$ | 287>145 Loss of $C_8H_{14}O_2$ |
| | Estradiol | 287>145 Loss of $C_8H_{14}O$ | 281>183 Loss of $C_5H_{12}O$ |
| | Estrone | 269>145 Loss of $C_8H_{12}O$ | 269>143 Loss of $C_8H_{14}O$ |
| | Ethynyl estradiol | 295>145 Loss of $C_9H_{12}O$ | 295>159 Loss of $C_{10}H_{14}O$ |
| *Anti-inflammatory/ analgesic/antiphlogistic* | Ibuprofen | 205>161 Loss of $CO_2$ | – |
| | Ketoprofen | 253>209 $[M\text{-}H\text{-}CO_2]^-$ | 253>197 |
| | Naproxen | 229>185 $[M\text{-}H\text{-}CO_2]^-$ | 229>170 $[M\text{-}H\text{-}C_3H_2O_2]^-$ |
| | Indomethacin | 356>312 $[M\text{-}H\text{-}CO_2]^-$ | 356>297 $[M\text{-}H\text{-}C_3H_2O_2]^-$ |
| | Diclofenac | 294>250 $[M+H\text{-}H_2O]^+$ | 294>214 |
| | Acetaminophen | 152>110 Loss of $CH_2CO$ 150>107 Loss of $COCH_3$ | 152>93 – |
| | Fenoprofen | 241>197 | 241>93 |
| | Mefenamic acid | 240>196 Loss of $CO_2$ | 240>180 $[M\text{-}H\text{-}CO_2\text{-}CH_3]^-$ |
| | Propyphenazone | 231>189 $[M\text{-}C_3H_7+H]^+$ | 231>201 |
| | Phenylbutazone | 309>160 $[M\text{-}(C_6H_5\text{-}N\text{-}(C_4H_9))]^+$ | 309>181 |
| | | 362>276 | 362>316 |
| *Lipid regulating agents* | Bezafibrate | 360>274 Loss of $C_4H_6O_2$ | 360>154 Loss of $C_{12}H_{14}O_3$ |
| | Clofibric acid | 213>127 $[C_6O_4ClO]^-$ | 213>85 |
| | Gemfibrozil | 249>121 $[M\text{-}H\text{-}C_7H_{12}O_2]^-$ | – |
| *Psychiatric drugs* | Carbamazepine | 237>194 Loss of HNCO | 237>192 |
| | Fluoxetine | 310>44 $[M\text{-}F_3C_7H_4OC_8H_8]^+$ | 310>148 $[M\text{-}F_3C_7H_4O]^+$ |
| | Paroxetine | 330>192 $[M\text{-}C_7H_5NO_3]^+$ | 330>123 $[M\text{-}C_{12}H_4NOF]^+$ |
| | Diazepam | 285>257 $[M\text{-}CO+H]^+$ | 285>154 |

**Table 3** (continued)

| Group of substances | Compound | MRM 1 | MRM 2 |
|---|---|---|---|
| *Macrolide antibiotics* | Erythromycin-$H_2O$ | 716>522 $[M\text{-}DS\text{-}2H_2O+H]^+$ | 716>558 $[M\text{-}DS\text{-}H_2O+H]^+$ |
| | Clarythromycin | 750>116 $[CL\text{-}OCH_3+H]^+$ | 750>592 $[M\text{-}DS+H]^+$ |
| | Roxythromycin | 838>158 $[DS+H]^+$ | 838>680 $[M\text{-}DS+H]^+$ |
| | Oleandomycin | 689>545 $[M\text{-}oleandrose+H]^+$ | 689>158 $[DS+H]^+$ |
| | Tylosin | 916>723 $[M\text{-}MY+H]^+$ | 916>174 $[DS\text{-}O\text{-}MY+H]^+$ |
| *Tetracycline antibiotics* | Chlortetracycline | 479>444 | 479>462 |
| | Doxycycline | 445>428 | 445>410 |
| | Oxytetracycline | 461>426 | 461>443 |
| | Tetracylcline | 445>410 $[M\text{-}H_2O\text{-}NH_3+H]^+$ | 445>427 $[M\text{-}H_2O+H]^+$ |
| *Quinolone antibiotics* | Ciprofloxacin | 332>314 $[M\text{-}H_2O+H]^+$ | 332>288 $[M\text{-}H_2O\text{-}CO_2+H]^+$ |
| | Ofloxacin | 362>344 $[M\text{-}H_2O+H]^+$ | – |
| | Norfloxacin | 320>302 $[M\text{-}H_2O+H]^+$ | 320>302 $[M\text{-}CO_2+H]^+$ |
| | Enrofloxacin | 360>342 $[M\text{-}H_2O+H]^+$ | 360>316 $[M\text{-}CO_2+H]^+$ |
| *Sulphonamide antibiotics* | Sulphamethoxazole | 254>156 $[H_2NPhSO_2]^+$ | 254>92 $[H_2NPhO]^+$ |
| | Sulphamethazine | 279>186 $[M\text{-}H_2NPh]^+$ | 279>124 [aminodimethyl-pyridine+H]$^+$ |
| | Sulphadiazine | 251>156 $[H_2NPhSO_2]^+$ | 251>108 $[H_2NPhO]^+$ |
| *Penicillins* | Dicloxacillin | 487>160 $[F1+H]^+$ | 487>311 $[F2+H]^+$ |
| | Nafcillin | 432>171 [ethoxynaphthyl]$^+$ | 432>199 [ethoxynaphtyl-carbonyl]$^+$ |
| | Amoxycillin | 366>208 $[M\text{-}NH_3+H]^+$ | 366>113 $[F1+H]^+$ |
| | Oxacillin | 419>144 [phenylisoxazolyl+H]$^+$ | 419>243 [aminodimethyl-pyridine+H]$^+$ |
| | Penicillin G | 352>160 $[F1+H]^+$ | 352>176 $[F2+H]^+$ |
| | Penicillin V | 368>114 $[F1\text{-}CO_2+H]^+$ | 368>160 $[F1+H]^+$ |

**Table 3** (continued)

| Group of substances | Compound | MRM 1 | MRM 2 |
|---|---|---|---|
| *Other antibiotics* | Chloramphenicol | 323>152 [nitrobenzyl alcohol carbanion]$^-$ | 323>176 [194-$H_2O$]$^-$ |
| | Trimethoprim | 291>230 [M-2$CH_3O$]$^+$ | 291>213 [M-trimethoxy-phenyl]$^+$ |
| β-blockers | Atenolol | 267>190 [M-$H_2O$-$NH_3$-isopropyl+2H]$^+$ | 267>145 [190-CO-$NH_3$]$^+$ |
| | Sotalol | 273>255 [M-$H_2O$+H]$^+$ | 273>213 [M-$C_3H_9N$+H]$^+$ |
| | Metoprolol | 268>133 [$C_6H_{15}NO_2$]$^+$ | 268>159 [$C_8H_{17}NO_2$]$^+$ |
| | Propranolol | 260>116 [(*N*-isopropyl-*N*-2-hydroxypropyl-amine)+H]$^+$ | 260>183 |
| *Other drugs* | Salbutamol | 240>166 [M+H-$(CH_3)$2C-$CH_2$-$H_2O$]$^+$ | 240>148 [166-$H_2O$]$^+$ |
| | Ranitidine | 315>176 [M-$C_8H_{12}NO$]$^+$ | 315>130 [M-$C_8H_{12}NO$-$NO_2$]$^+$ |
| | Omeprazole | 346>136 [M-$H_3CO$-($C_7H_4N_2$)-SO-$CH_2$]$^+$ | 346>198 [M-$H_3CO$-$C_7H_4N_2$]$^+$ |

are available using Q-TOF analysers [152], but this technique has not been routinely employed yet.

In most cases, the base peak selected for quantitation of estrogens in SIM and MRM modes, when operating with an ESI (NI) and APCI (PI) interface, corresponds to the deprotonated molecule $[M-H]^-$ and to the $[M+H-H_2O]^+$ ion ($[M+H]^+$ for estrone). In Table 3, the most common fragmentations monitored in LC-MS/MS analysis, using triple quadrupole instruments, are summarized for the most studied steroid estrogens.

### 4.2.2 Pharmaceuticals

A large number of reports and reviews are devoted to the occurrence, fate and risk assessment of pharmaceuticals in the environment [92, 93, 127, 189–

193]. While their occurrence in the aquatic environment has been extensively studied, data regarding their presence in solid samples are still scarce, veterinary antibiotics being the ones most commonly investigated in such matrices [194–199].

Most of the analytical methods available in the literature are focused on the analysis of particular therapeutic groups. However, the general trend in recent years is the development and application of generic methods that permit simultaneous analysis of multiple-class compounds [2, 99, 200–209]. Multi-residue methods provide wider knowledge about their occurrence, necessary for further understanding of their removal, partition and ultimate fate in the environment. Nevertheless, simultaneous analysis of compounds from diverse groups with different physico-chemical properties requires a compromise in the selection of experimental conditions for all analytes studied.

#### 4.2.2.1 Sample Preparation

In such multi-residue methods, simultaneous extraction of all target analytes in one single SPE step from water samples is the approach most widely employed [190]. Another option consists of the combination of two SPE materials operating either in series or classifying target compounds into two or more groups, according to their physico-chemical properties [190]. In both situations Oasis HLB or $C_{18}$ cartridges are the most widely employed materials for pre-concentration and extraction of target compounds. For the former, neutral sample pH is advisable to achieve good recoveries for all compounds, whereas for $C_{18}$, sample pH adjustment prior to extraction is required depending on the acidic, neutral or basic nature of the analytes. The less common cartridges employed are Lichrolut ENV+, Oasis MCX and StrataX. While these materials generally need sample pH adjustment and sometimes special elution conditions (mixtures of methanol/ammonia, acidified or basified methanol), Oasis HLB provides good performances at neutral sample pH and elution with pure organic solvents, generally methanol (see Table 2).

When these methods include the determination of antibiotics, some precautions have to be taken into account during the analytical procedure. As tetracycline, sulphonamides and polypeptide antibiotics form complexes with metal ions, the addition of some chelating agent before SPE, such as $Na_2EDTA$, is recommended to avoid important losses during analysis. When analysing tetracycline, it should be highly recommended to use PTFE instead of glass materials, since they tend to bind to the glass, resulting in significant losses [93, 189, 190]. Additional problems are the formation of keto–enol tautomers in alkaline aqueous solutions [210] and the formation of 4-epimer isomers in acidic ones [211]. For this reason, it is advisable to work at neutral sample pH.

MIPs and immunosorbents could be a useful tool to provide high selectivity for target analytes when performing single group analysis. Although these materials have been widely employed to selectively isolate clenbuterol, aniline β-agonists, tetracycline and sulphonamide antibiotics, β-agonists and β-antagonists from biological samples, few applications have been reported for environmental matrices [212–215].

With regard to their analysis in solid samples, most of the methods available in the literature are based on sonication and PLE as the extraction technique followed by a clean-up procedure. The extraction solvents used generally consist of pure organic solvents, such as methanol and acetonitrile, or mixtures of polar solvents with water, acidified water (acetic acid, orthophosphoric acid), or buffers (citric acid) in different proportions. An important issue to consider is that when extracting tetracycline and macrolide antibiotics by PLE, temperature control is required, since temperatures higher than room temperature can cause their transformation into epi- or anhydrous forms for TCs. Moreover, values higher than 100 °C promote the degradation of macrolides [127].

For the extraction of tetracycline antibiotics, special precautions have to be taken into account. As they tend to form complexes with metal ions, extraction solvents consist of mixtures with organic solvent, generally methanol, with citric acid and McIlvaine buffer (mixture of citric acid with $Na_2HPO_2$), also containing $Na_2EDTA$ [194].

After extraction, a purification step is required and is generally performed by SPE, using the same cartridges and conditions as the analysis of pharmaceuticals in water samples. Sample extracts are therefore diluted with an appropriate volume of MilliQ water, until the organic solvent content is below 10%, in order to avoid losses of target compounds during SPE [194]. Cartridges mainly used consist of Oasis HLB (see Table 2). However, some authors use either SAX or MCX [189] cartridges in tandem with the polymeric Oasis HLB [194], in order to remove negatively charged humic material (in the SAX material) and organic matter (in the MCX cartridge), and therefore selectively retain target compounds in the Oasis HLB material. When SAX cartridges are employed, samples are acidified at pH values ranging from 2 to 3 to ensure an efficient removal of the humic material (see Table 2).

Elution of target compounds from SPE cartridges is achieved with a large variety of organic solvents, according to the physico-chemical properties of the compounds analysed, methanol and acetonitrile being the most common ones (see Tables 1 and 2).

#### 4.2.2.2
#### Instrumental Analysis

LC-MS/MS is the instrumental method of choice due to its versatility, specificity and selectivity, replacing GC-MS and LC-MS [190]. GC-MS can only be successfully applied for a limited number of non-polar and volatile pharmaceutical compounds, requiring a time-consuming derivatization step for the determination of polar pharmaceuticals [216–219]. Among LC-MS/MS techniques, triple quadrupole (QqQ) and ion trap (IT) instruments are in common use [92], the former being the most widely used, working in selected reaction monitoring (SRM) mode and typically reaching ng/L detection limits. More recent approaches in LC-MS/MS are linear ion traps (LITs), new generation triple quadrupoles, and hybrid instruments, such as quadrupole–time of flight (QqTOF) and quadrupole–linear ion trap (QqLIT) [92, 220].

The main applications of QqTOF instruments are focused on the elucidation of structures proposed for transformation products or are used as a complementary tool to confirm positive findings obtained by a QqQ screening method. Recently, Eichhorn et al. [221] reported on the structural elucidation of the metabolites of the antimicrobial trimethoprim. Stolker et al. [203], Marchese et al. [222], Petrovic et al. [93] and Gómez et al. [223] used QqTOF to identify the presence of various pharmaceuticals in environmental waters. Recently, Pozo et al. [224] evaluated the potential of a QqTOF instrument to confirm positive findings in the analysis of penicillin and quinolone antibiotics in surface and ground water samples. An example of the analysis of selected pharmaceuticals in an urban wastewater by UPLC-QqTOF-MS is shown in Fig. 4.

As concerns QqLIT, Seitz et al. [225] developed a method for the determination of diclofenac, carbamazepine and iodinated X-ray contrast media using direct analysis (among other contaminants), reaching LODs of 10 ng/L. Nikolai et al. [226] used QqLIT operating in QqQ mode for stereoisomer quantification of β-blockers in wastewater. On the other hand, Gros et al. [212] developed an analytical methodology for trace analysis of eight β-blockers in wastewaters using MIPs for pre-concentration of target compounds combining different functions of QqQ. Quantitative analysis was performed using a 4000QTRAP tandem mass spectrometer in SRM mode. Using the information-dependent acquisition (IDA) function in the software, a large amount of data for unequivocal identification and confirmation of the target compounds were generated at high sensitivity. An example of an IDA experiment for the determination of atenolol in an influent wastewater sample is shown in Fig. 5.

Regarding LC, reversed-phase LC is mainly used, $C_{18}$ columns being the preferred ones. Only one method, targeted to acidic drugs, was based on ion-pair reversed-phase LC with a Phenyl–Hexyl column [227]. As mobile phases, acetonitrile, methanol, or mixtures of both solvents are normally used. In order to improve the sensitivity of MS detection and give an efficient retention, mobile phase modifiers, buffers and acids are widely employed, with ammonium acetate, tri-*n*-butylamine (TrBA), formic acid and acetic acid being the more common ones. Typical concentrations of salts range from 2 to

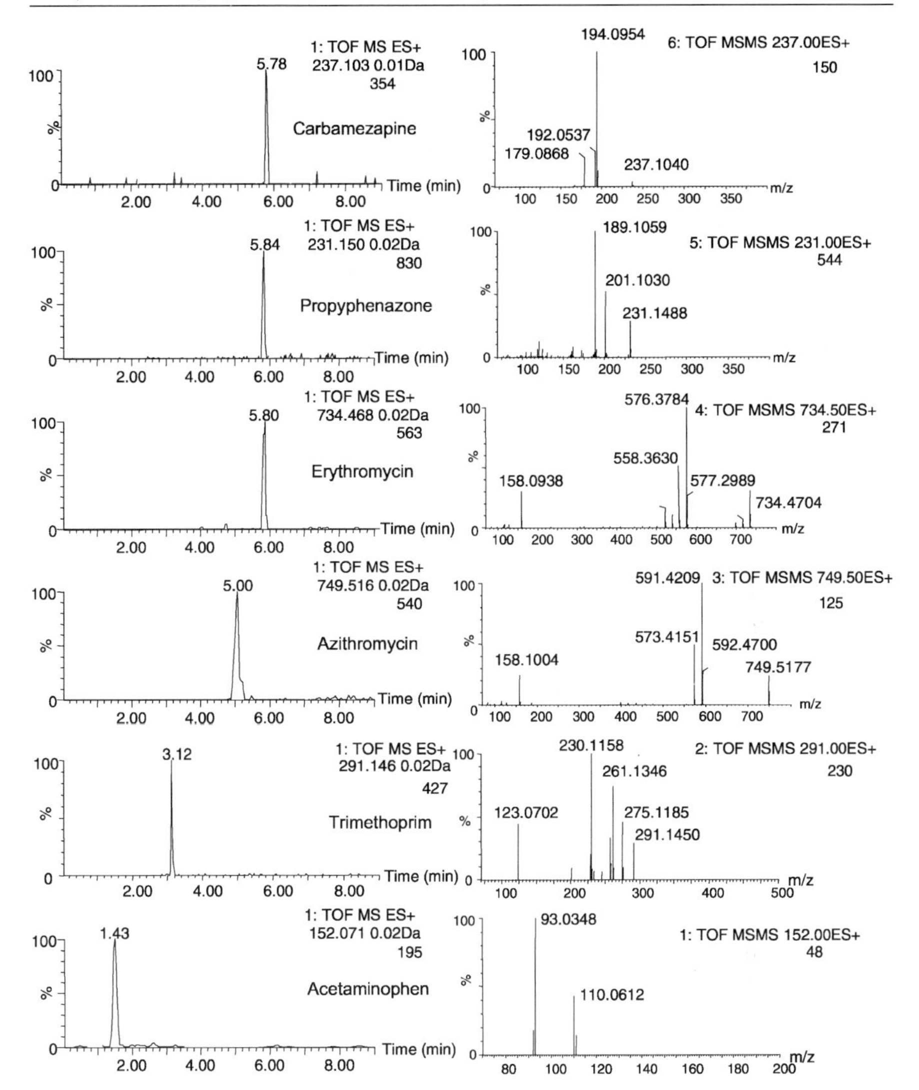


**Fig. 4** Confirmation of several pharmaceuticals in an urban wastewater. *Left panel*: narrow window extracted ion chromatograms (nwXICs) of $[M+H]^+$ obtained in the TOF mode for $m/z$ 152.071 (acetaminophen), $m/z$ 291.146 (trimethoprim), $m/z$ 749.516 (azithromycin), $m/z$ 734.468 (erythromycin), $m/z$ 231.150 (propyphenazone) and $m/z$ 237.103 (carbamazepine). *Right panel*: product ion spectra obtained in the Q-TOF mode

20 mM, since it has been observed that higher concentrations could lead to a reduction of signal intensities [190].

Shortening the analysis time is important for attaining the high sample throughput often required in monitoring studies. This can be achieved by

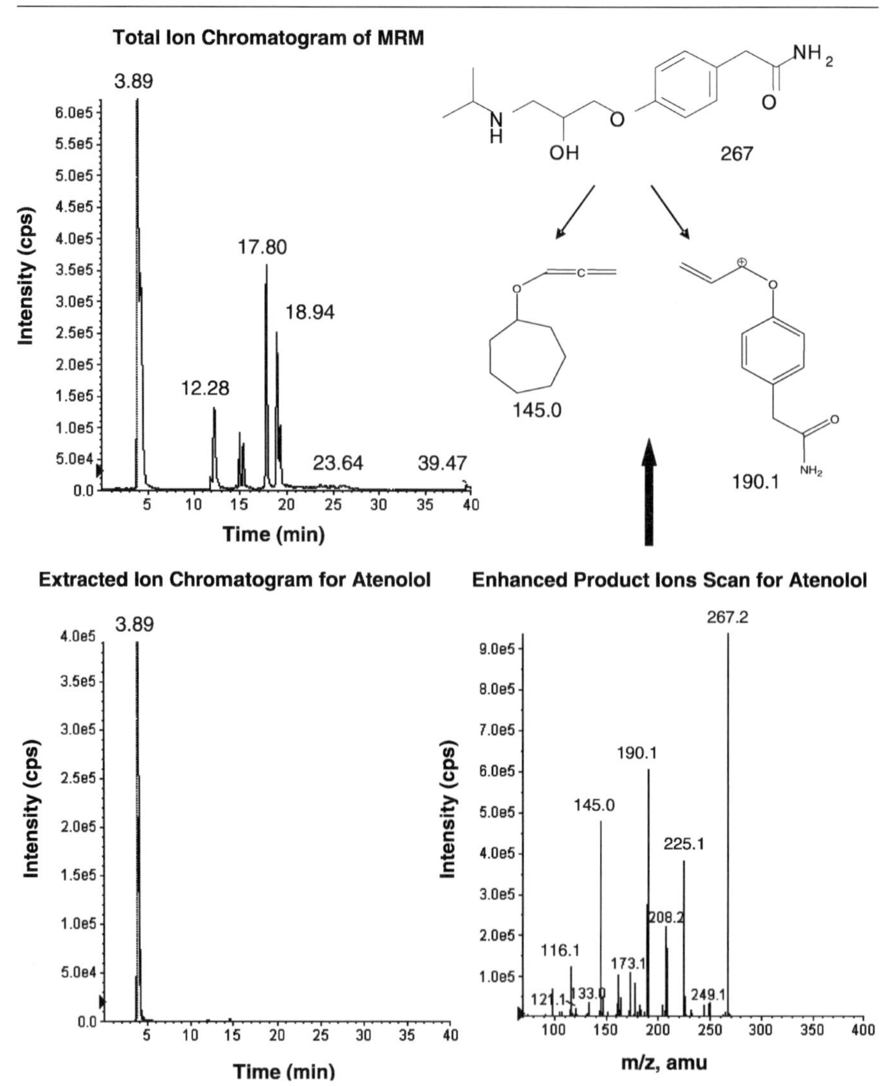


**Fig. 5** Information-dependent acquisition (IDA) experiment for the determination of atenolol in an influent wastewater sample

using short columns and increased flow velocity, decreasing the particle size of stationary phases or increasing temperature. These approaches are applied in two newly developed instruments, UPLC (ultra-performance LC) and by RRLC (rapid resolution LC). For the moment, only one publication presented by Petrovic et al. [93] describes the use of UPLC coupled to a QqTOF system for the multi-residue analysis of 29 pharmaceuticals in environmental waters. Compounds more frequently detected in multi-residue methods and their MRM transitions are summarized in Table 3.

### 4.2.3 Personal Care Products (PCPs)

This group of compounds includes synthetic musk fragrances (nitro and polycyclic musk fragrances), antimicrobials (triclosan and its metabolites and triclocarban), sunscreen agents (ultraviolet filters), insect repellents (*N*,*N*-diethyl-*m*-toluamide, known as DEET) and parabens (*p*-hydroxybenzoic esters), which are basically substances used in soaps, shampoos, deodorants, lotions, toothpaste and other PCPs. The nitro musk fragrances were the first to be produced and include musk xylene, ketone, ambrette, moskene and tibetene. In the environment, the nitro substituents can be reduced to form amino metabolites of these compounds. The polycyclic musk fragrances, which are used in higher quantities than nitro musks, include 1,2,4,6,7,8-hexahydro-4,6,6,7,8,8-hexamethylcyclopenta-γ-2-benzopyrane (HHCB), 7-acetyl-1,1,3,4,4,6-hexamethyl-1,2,3,4-tetrahydronaphthalene (AHTN), 4-acetyl-1,1-dimethyl-6-*tert*-butylindane (ADBI), 6-acetyl-1,1,2,3,3,5-hexamethylindane (AHMI), 5-acetyl-1,1,2,6-tetramethyl-3-isopropylindane (ATII) and 6,7-dihydro-1,1,2,3,3-pentamethyl-4-(5*H*)-indanone (DPMI). Parabens are the most common preservatives used in personal care products and in pharmaceuticals and food products. This group of substances includes methylparaben, propylparaben, ethylparaben, butylparaben and benzylparaben.

These substances have been analysed in various environmental matrices, such as water, sediments, sewage sludge and aquatic biota. The hydrophobicity of many of these compounds indicates their potential for bioaccumulation [228].

#### 4.2.3.1 Sample Preparation

Methods used for the extraction of PCPs from water samples are based on liquid–liquid extraction (LLE) [1, 52–67], continuous liquid–liquid extraction (CLLE), SPE [219, 229–231] and SPME [232, 233]. When LLE and CLLE are applied, various organic solvents are used for the extraction of target compounds, dichloromethane, pentane [234, 235], hexane [236–238], toluene [239, 240], cyclohexane [233] and petroleum ether [241], and mixtures of them in appropriate proportions, being the most widely employed (see Table 2). Extraction of target compounds using these techniques is performed either at ambient pH or by acidifying the sample, generally to values ranging from pH 2 to 3 [219, 228]. For the extraction of UV filters, LLE with cyclohexane at pH 3 is the most common procedure [228].

For SPE, a wide range of sorbents are used, including $C_{18}$ [219, 230, 231, 242–248] at ambient and acidic (pH<3) sample pH, Abselut Nexus [249, 250] (Varian, Palo Alto, CA, USA), Isolute ENV+ [231], Oasis MAX [241], Bio Beads

SM-2 [251–253] (Bio-Rad Laboratories, Hercules, CA, USA), XAD-2 [254] (Supelco, St. Louis, MO, USA), SDB-XC [255, 256] and XAD-4/XAD-8 [254, 257]. Elution of target compounds from these materials is achieved with a large variety of organic solvents, according to the physico-chemical properties of the compounds analysed, with acetone, methanol, toluene, hexane, mixtures of dichloromethane/acetone and methanol, hexane/acetone or hexane/ethyl acetate and acetone/ethyl acetate being the most widely used [228]. When analysing antimicrobials with Oasis MAX, the sample is acidified (pH 3) prior to extraction, washed with methanol/sodium acetate solution and eluted with pure methanol. For parabens, few methods are reported relevant to environmental matrices, but their analysis is mainly based on SPE extraction using Oasis HLB.

Sometimes, when using these techniques, sample purification prior to instrumental analysis is necessary, generally using SPE with silica and alumina [228]. The most common techniques used for their extraction from sewage sludge include PFE [197, 231, 241, 244, 245, 252, 258, 259], SFE [230, 241] (using $CO_2$), sonication, Soxhlet [240, 260–263], LLE [264, 265] and MAE [266]. Sometimes, before extraction of target compounds, copper is added to remove sulphur content in the samples. Generally, after extraction, a purification step with silica columns or size-exclusion chromatography (SEC) followed by Bio Beads SX-3 or silica columns is required. Hexane, ethyl acetate, acetone, cyclohexane and mixtures of them are the solvents mainly used for the elution of target compounds [228].

On the other hand, SPME has also been a widespread technique for the extraction of PCPs in environmental waters and solid samples, using either direct (DI-SPME) or headspace (HS-SPME) methods [228, 248, 267, 268]. The materials most commonly used are polydimethylsiloxane (100 μm) (PDMS) for DI-SPME, and PDMS-DVB (65 μm), Carboxen-PDMS (75 μm), Carbowax-DVB (65 μm) and Carbowax-PDMS (65 μm) for both types of SPME, PDMS-DVB being the one yielding higher recoveries [228].

The extraction techniques used for the analysis of biota samples are the same as those used for solid samples but after extraction, removal of the lipid content is essential, generally performed by SEC in tandem with Bio Beads SX-3 cartridges. For the determination of nitro musks, lipids cannot be removed destructively with $H_2SO_4$ since important losses of target compounds could occur.

#### 4.2.3.2 Instrumental Analysis

Synthetic musk fragrance standards and deuterated musk xylene and AHTN standards are commercially available for use as recovery or injection standards. There have been reports of problems with the use of the deuterated AHTN (AHTN-$d_3$) due to the occurrence of proton exchange during sample processing [228]. A variety of other recovery and injection standards have been used for the analysis of synthetic musk fragrances, including pentachloronitrobenzene, deuterated polycyclic aromatic hydrocarbons (PAHs), and various labelled and unlabelled polychlorinated biphenyls (PCBs).

PCPs are most commonly analysed by GC-EI-MS, but GC-NCI-MS is more sensitive for nitro musk fragrances. These compounds have also been analysed by GC-FID, GC-ECD, and high-resolution and ion-trap tandem mass spectrometry (MS/MS). Common GC phases are 5% phenylmethylpolysiloxane and dimethylpolysiloxane [228].

Triclosan and its chlorinated metabolites are also determined by GC-EI-MS with and without derivatization, LC-MS and LC-MS/MS. When derivatizing, *N*,*N*-diethyltrimethylamine (TMS-DEA), *N*,*O*-bis(trimethysilyl)trifluoroacetamide (BSTFA), pentafluorinated triclosan and *tert*-butyldimethylsilyl triclosan are the ether derivatives generated after reaction with methyl chloroformate (MCF), pentafluoropropionic acid anhydride (PFA) and *N*-*tert*-butyldimethylsilyl-*N*-methyltrifluoroacetamide (MTBSTFA), respectively [228].

GC-based techniques dominate the analysis of UV filters and insect repellents, using DB-5 and 5% polyphenylmethylsilicone columns, respectively. Almost all UV filters are amenable to GC except octyl triazone, avobenzone, 4-isopropyldibenzoylmethane and 2-phenylbenzimidazole-5-sulphonic acid, some of them being determined by HPLC-UV. Although there are few methods published dealing with the analysis of parabens in environmental samples, the methods reported are based on LC-MS/MS under NI conditions using a $C_{18}$ column.

## 4.3 Surfactants

A number of books and reviews are already available on the determination of surfactants in wastewaters, sludges, sediments and biological samples, using GC-MS, LC-MS or LC-MS/MS techniques [4, 269–271]. Among the various surfactant classes, both non-ionic and ionic substances are the most widely employed in both industry (e.g. alcohol ethoxylates (AEOs), alkylphenol ethoxylates (APEOs) and different fatty amine or acid ethoxylates [269]) and household applications (linear alkylbenzene sulphonates (LASs)).

From the environmental point of view, APEOs and LASs are the ones deserving especial attention due to their ubiquity and ecotoxicological relevance. Sixty percent of APEOs that enter mechanical or biological sewage or sewage sludge treatment plants are subsequently released into the environment, 85% being in the form of the potentially estrogenic metabolic products, alkylphenols (APs), alkylphenol carboxylates (APECs) and alkylphenol dicarboxylates (CAPECs) [272–275]. Moreover, numerous studies have confirmed that alkylphenolic compounds can mimic endogenous hormones. APEOs and their biodegradation products are transformed into halogenated by-products during chlorination disinfection in wastewater or drinking water treatment plants, in the presence of bromide ion [276, 277].

### 4.3.1 Sample Preparation

Both ionic and non-ionic surfactants are generally isolated from water samples by SPE, at natural sample pH, Lichrolut $C_{18}$ cartridges (Merck, Darmstadt, Germany) being the most widely employed. For halogenated derivatives, SPE using Lichrolut $C_{18}$ is also widely employed [278]. Elution is usually performed using pure solvents, with methanol the most common one [5].

Analysis of surfactants and their halogenated derivatives from solid samples is challenging due to their strong adsorption on the soil/sludge particles by hydrophobic and electrostatic interactions. Most of the methods available in the literature are based on sonication and PLE as the extraction technique followed by a clean-up procedure, generally using SPE $C_{18}$, ENV+, strong anion exchange (SAX) or polymeric cartridges [5, 279–281]. The former has been widely employed for the analysis of LASs, NPEOs and their degradation products nonylphenol carboxylates (NPECs) and NPs, AEOs, and coconut diethanolamides (CDEAs) [282]. On the other hand, PLE methods have been optimized for LASs, NPEOs and their neutral and acidic metabolites, AEOs and alkylamine ethoxylates (ANEOs) [282]. Pure solvents, such as methanol and dichloromethane, and mixtures of organic solvents (hexane/acetone or methanol/dichloromethane) are mainly used for the extraction of surfactants from solid matrices (see Table 2). Other methods based on extraction with pressurized (supercritical) hot water as well as SFE with solid-phase trapping, using $CO_2$ and methanol or water as modifier, have been described in the literature for the simultaneous extraction of several surfactant classes [282].

### 4.3.2 Instrumental Analysis

Commercial mixtures of surfactants comprise several tens to hundreds of homologues, oligomers and isomers. For LASs, mixtures of secondary isomers with alkyl chain lengths of 10–13 carbons are available.

GC and LC coupled to MS detection systems are now the commonly used methods to identify and quantitate surfactants in both aqueous and solid matrices. Although GC-MS is adopted in many analytical methodologies, it cannot be applied for the direct determination of several classes of surfactants since derivatization of low volatility compounds is required. This is why, in surfactants analysis, GC-MS methods are partially substituted with LC-MS or LC-MS/MS [269, 283]. However, most of the methods available focus on one or two classes of surfactants which are similar in nature, generally including their main degradation products. Only recently, several efforts have been made to develop generic methods that allow simultaneous determination of a broad range of surfactant types.

Gas chromatography–mass spectrometry has been widely used for the analysis of alkylphenolic compounds and anionic surfactants (LASs). Alkylphenolic substances, which mainly include the most volatile compounds AP, APEO, AEO and ANEO with fewer than four ethoxy groups, and the rest of the non-ionic surfactants can be determined without derivatization, while for anionic surfactants derivatization prior to analysis is required [284]. Derivatization is usually performed by transforming parent compounds to the corresponding trimethylsilyl ethers, methyl ethers, acetyl esters and pentafluorobenzoyl or heptafluorobutyl esters [5, 285, 286]. After derivatization, NPEO derivatives can be analysed by GC-MS in the EI or NCI modes [130]. GC-CI-MS, using ammonia as reagent gas for the detection of $NPE_nC$, gave intense ammonia–molecular ion adducts of the methyl esters, at *m/z* 246, 310, 354 and 398 for $NPE_1C$, $NPE_2C$, $NPE_3C$ and $NPE_4C$, respectively, with little or no secondary fragmentation [5]. Moreover, GC-CI-MS spectra of the NPECs with isobutene as reagent gas showed characteristic hydride-ion-abstracted fragment ions shifted 1 Da from those in the corresponding EI mass spectra. On-line direct GC injection-port derivatization with ion-pair reagents (tetraalkylammonium salts) has also been reported [287].

As concerns liquid chromatography, even though LC-MS/MS is more specific and sensitive than LC-MS, the majority of studies dealing with the analysis of surfactants in environmental samples are based on LC-MS [128, 270]. However, several papers describing the application of tandem MS to the unambiguous identification and structural elucidation of alkylphenolic compounds have been published [275, 288–291].

The analysis of LASs by LC-MS operating in the ESI and NI modes is particularly attractive due to their anionic character. MS analysis of commercial LAS mixtures shows four ions at *m/z* 297, 311, 325 and 339, corresponding to deprotonated molecules of $C_{10}$–$C_{13}$ LAS homologues [282]. With increasing cone voltage using in-source collision-induced dissociation (CID), the spectra show additional fragment ions at *m/z* 183 and 80, which were assigned to styrene-4-sulphonate and $[SO_3]^-$. The analysis of APEOs by LC-MS in the PI mode yields a characteristic pattern of equally spaced signals with mass differences of 44 Da (one ethoxy unit), which is a diagnostic fingerprint for this group of compounds. Using an ESI interface and aprotic solvent, APEOs predominantly give evenly spaced sodium adducts $[M + Na]^+$ [270], which are relatively stable and generally no further structurally significant fragmentation is provided in the mass spectrum. Some authors used ammonium acetate as mobile phase in order to enhance the formation of ammonium adducts over sodium or proton adducts, which give fragments in CID processes, enabling a more specific detection of APEOs [275].

On the other hand, alkylphenoxy carboxylates ($APE_nC$) are generally determined by ESI operating in the NI mode, and less frequently by the PI mode [282]. For the analysis by NI, two types of ions, one corresponding to the deprotonated molecule and the other corresponding to deprotonated alkylphenols, are obtained. For the determination of AEOs, some authors used LC-MS operating in APCI mode [282] to analyse AEOs with alkyl chains from $C_{10}$ to $C_{14}$ and from $C_{10}$ to $C_{18}$.

Like their non-halogenated analogues, halogenated APEOs show a great affinity for alkali metal ions when analysed by LC-MS in ESI mode, and they give exclusively evenly spaced (44 Da) sodium adduct peaks $[M + Na]^+$ with no further structurally significant fragmentation [277]. Fully de-ethoxylated degradation products, octylphenol (OP) and nonylphenol (NP), were detected under NI conditions with both APCI and ESI interfaces. However, sensitivity was higher when using an ESI source than an APCI one [5].

Diagnostic ions used for the analysis of XAPEOs under NI conditions using LC-MS corresponded to the cleavage of the alkyl moiety ($CH_2$ group), leading to a sequential loss of $m/z$ 14, the most abundant fragments being at $m/z$ 167 for $^{35}Cl$ and $m/z$ 169 for $^{37}Cl$.

In LC-tandem MS, compounds analysed under NI conditions (AP, APEC and their halogenated derivatives) were analysed by ESI-MS/MS, while for APEO, detected in the PI mode, no fragmentation was obtained using an ESI source. These compounds were determined by APCI-MS/MS. Using ESI-MS/MS, the CID spectrum of NP shows fragments at $m/z$ 147, 133, 110 and 93, attributed to the progressive fragmentation of the alkyl chain [5]. For APEC, an intense signal at $m/z$ 219 is observed for NPEC, produced after the loss of the carboxylated (ethoxy) chain, and other peaks at $m/z$ 133 and 147, due to the sequential fragmentation of the alkyl chain [128, 275, 288]. LC-tandem MS was also used to determine halogenated surfactants, obtaining the same product ions as for LC-MS, with $m/z$ 167 for $^{35}Cl$ and $m/z$ 169 for $^{37}Cl$, with a relative ratio of intensities of 3.03, being the most abundant fragment ions.

LC-ESI-IT-MS and LC-(PI)-APCI-IT-MS have been used to determine LASs and SPCs, and APEOs, AEOs and cationic surfactants, respectively, in several environmental matrices [292–296]. These instruments permit $MS^n$, which makes them suitable for identification and quantitation purposes. On the other hand, MALDI-TOF and MALDI-Q-IT have been used to determine APEOs [297, 298]. Ayorinde et al. [292] used α-cyano-4-hydroxycinnamic acid as a matrix to determine NPEO (with 2–120 ethoxy units).

## 4.4 Polybrominated Diphenyl Ethers (PBDEs)

Polybrominated flame retardants are chemicals used in large quantities as they are added to polymers, which are used in plastics, textiles, electronic circuitry and other materials, to prevent fires, due to their fire retarding properties [299]. Several studies have reported that these substances tend to bioaccumulate in biota and humans due to their lipophilicity [300–311]. Moreover, PBDEs are suspected to cause endocrine dysfunction by interfering with thyroid hormone metabolism [312, 313]. In 2003, the European Union banned the use of the PBDE commercial mixtures PentaBDE and OctaBDE. Nowadays, the only remaining unregulated PBDE mixture in production is DecaBDE [314].

### 4.4.1 Sample Preparation

Analytical methods developed for the determination of PBDEs are very similar to those used for PCBs, due to their similarity in physico-chemical properties. As they are non-polar compounds, their occurrence has been widely reported in solid samples, such as sewage sludge, soil and sediments. For this reason, the determination of PBDEs in liquid samples is mainly focused on the analysis of human milk or plasma, while few studies have analysed them in natural and sewage waters [81].

BDE congeners typically measured in human tissues are associated primarily with the PentaBDE mixture, and to some extent with the OctaBDE mixture. One of the greatest challenges to measuring PBDEs in environmental samples has been developing methods to accurately quantify BDE 209. While analytical methods are readily available for quantifying tribrominated through heptabrominated congeners found in the PentaBDE and OctaBDE mixtures, the analysis of brominated compounds has proven to be difficult. Currently, there are several reviews available in the scientific literature devoted to the analysis of PBDEs in different environmental matrices [81, 82, 299].

The techniques used are mainly based on liquid–liquid extraction (LLE) [315–319], with mixtures of non-polar and polar solvents. Recently, headspace solid-phase microextraction (HS-SPME) and microporous membrane liquid–liquid extraction (MMLLE) have been proposed as suitable techniques [320]. Other techniques used consist of saponification with ethanolic KOH, especially for their analysis in human milk [299]. Similar procedures involving protein denaturation with HCl/isopropanol and extraction with hexane/methyl *tert*-butyl ether have been used for the determination of neutral and phenolic brominated compounds from human serum [321].

Extraction of PBDEs from solid and biological samples is generally performed using non-polar solvents, such as hexane, toluene, dichloromethane or hexane/acetone mixtures. Binary solvent mixtures, combining a non-polar and a polar solvent, are most commonly used for their known extraction efficiency, especially for biota and wet sediment samples, as non-polar solvents are not able to penetrate the organic matter and therefore desorb contaminants. Soxhlet [322–324], supercritical-fluid extraction (SFE) [325], accelerated solvent extraction [326, 327] and microwave-assisted extraction (MAE) are the techniques mainly used [328].

Extracts obtained using these techniques need a clean-up step prior their analysis by chromatographic techniques. Therefore, extracts from sediments, sewage sludge or soil samples may contain sulphur that has to be removed as it could disturb the GC analysis. Typical methods used for this purpose are treatment with copper powder, silica modified with $AgNO_3$ in a multilayer silica column, desulphuration with mercury or reaction with tetrabutylammonium sulphite [81, 82, 299]. In the case of Cu powder, it is generally added in the Soxhlet beaker or PLE cell.

On the other hand, in the case of sewage sludge, extracts contain a high amount of lipids and organic matter, which should be removed prior to instrumental analysis, by either non-destructive or destructive methods. The former include gel permeation and column adsorption chromatography, using polystyrene–divinylbenzene copolymeric columns and dichloromethane or mixtures of dichloromethane/hexane and ethyl acetate/cyclohexane as eluents. Other neutral adsorbents commonly used are silica gel, alumina and Florisil® [323, 329]. Destructive lipid removal methods consist of sulphuric acid treatment, either directly to the extract or via impregnated silica columns, and saponification of extracts by heating with ethanolic KOH. Since PBDE concentrations are generally related to the amount of lipids, the lipid content is often measured gravimetrically prior to the clean-up step, or determined separately by a total lipid determination [299, 323].

It is important to remark that when analysing BDE 209 special precautions should be taken, as it is sensitive to UV light and it may also adsorb to small dust particles. Therefore, incoming sunlight into the laboratory should be blocked and all glassware covered with aluminium foil, to prevent dust particles and UV light entering either the solutions or samples. The use of isooctane for the extraction should be avoided due to the insolubility of BDE 209 in this solvent. Moreover, it is recommended not to evaporate extracts until dryness because it may not completely re-dissolve after that step even when using toluene.

### 4.4.2
### Instrumental Analysis

Like perfluorinated alkyl substances, standards available for PBDE determination consist of a mixture of several congeners of different degrees of bromination. As reported by Stapleton [314], about 160 of the 209 possible BDE congeners are currently commercially available. Isotopically labelled standards to be used for internal standard calibration purposes are scarce, and therefore some authors have used $^{13}C$-labelled bromobiphenyls and chlorinated diphenyl ethers as an alternative.

Owing to their vapour pressure and polarity, GC coupled to ECD, NCI-LRMS and EI-LRMS detectors has become a standard analytical separation method for the analysis of PBDEs. The three most common injection techniques for PBDEs are split/splitless, on-column and programmable temperature vaporization (PTV) injection. When working with split/splitless injection, the high inlet temperature can lead to thermal degradation and discrimination of higher molecular weight PBDEs, particularly the fully brominated BDE 209. This problem can be solved by using on-column injection, which consists of the direct injection of the sample, dissolved in a carrier solvent, onto the head of the column [314, 330]. PTV inlets have become a more popular choice for injection over the past 5 years, where higher injection volumes can be used, thus improving detection limits.

Both on-column and PTV injections require the use of a guard column, composed either of untreated silica with active silanol groups or deactivated fused silica. Short DB columns (10–15 m) with thin (0.1 μm) stationary phases are the most commonly used and the ones providing higher sensitivity for measuring the entire range (low to high bromine substitution). However, longer columns are not well suited for higher molecular weight PBDEs, as they can degrade [314]. Again, BDE 209 should receive special attention, due to its susceptibility to degrade at higher temperatures in the GC system.

ECNI-LRMS provides higher sensitivity than EI-LRMS, the LODs for the former being at least one order of magnitude lower than for the latter. However, EI-LRMS provides higher specificity and accuracy in quantification, as isotopically labelled standards can be used for the isotope dilution approach.

GC/ECNI-LRMS mass spectra for all PBDEs rely upon selective ion monitoring (SIM) of $Br^-$ ions [$^{79}Br$ and $^{81}Br$]. By contrast, EI provides more structural information, giving the molecular ions and the sequential losses of bromine atoms (molecular clusters for mono- to tri-BDEs and $[MBr_2]^+$ for tetra- to hepta-BDEs).

The presence of potential interferences in the NCI and EI approaches has been widely studied [314, 331, 332]. In general, EI-MS is affected by chlorinated interferences, especially PCBs, as analytical procedures developed for PBDE analysis are mainly based on the methods already available for PCBs. Thus, purified extracts may contain both PCBs and PBDEs. Alaee et al. [332] found that the isotopic cluster of $[M-Cl_2]^+$ from heptachlorinated biphenyls contains the same mass fragments found in tetrabrominated diphenyl ethers $[M-Br_2]^+$ and resolving powers of 25 000 ($m/\Delta m$) were required to differentiate them.

Such interferences are illustrated in Figs. 6 and 7, where the chromatograms obtained following the injection of a PBDE standard mixture and PCB standard mixtures are depicted. As can be observed, some hepta-CBs (CB-180) and octa-CBs (CB-199) elute with tetra-BDEs. Furthermore, some octa-CBs (CB-194) elute with penta-BDEs [82].

When using NICI-LRMS, such chlorinated interferences do not occur, but due to the presence of different brominated compounds, such as MeO-BDEs,

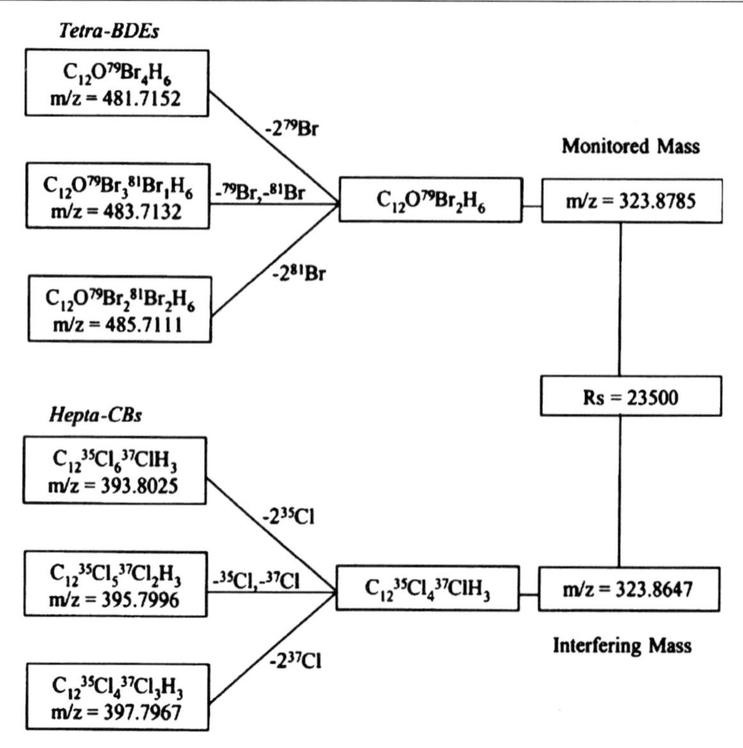


**Fig. 6** Interferences between tetra-BDEs and hepta-CBs. Reprinted with permission from Elsevier [331]

can produce the same fragment ion and confound analysis of PBDEs. Several papers have reported the co-elution of 2,2′4,4′,5′5-hexabromobiphenyl (PBB 153) and TBBPA with BDE 154 and of tetrabromobisphenol A with BDE 153 [81, 323, 333–336] on 15- and 30-cm capillary columns. Moreover, naturally produced brominated compounds, such as halogenated bipyrroles and brominated phenoxyanisoles, can be considered as potential interferences.

High-resolution instruments operating in the EI mode offer the best selectivity for PBDE measurements, with a mass resolution of approximately 10 000, resulting in fewer co-eluting interferences [337]. Moreover, they also allow the use of isotope dilution with $^{13}$C-labelled BDE standards due to the reduction of interferences.

Tandem mass spectrometers using ion traps have also been reported for the analysis of PBDEs [338, 339], offering the advantage of increased sensitivity at low mass resolution because analytes are fragmented twice, minimizing the chance of isobaric interferences and reducing background noise. In this equipment, precursor ions, which are typically $[M]^+$ or $[M - Br_2]^+$, are fragmented yielding $[M - COBr]^-$ ions.

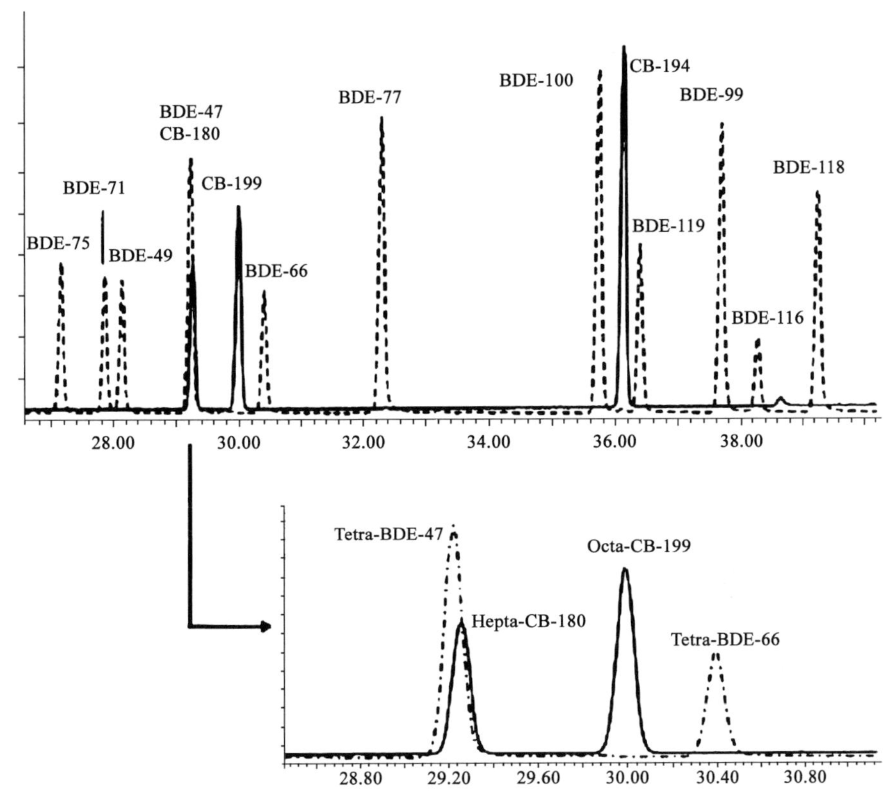


**Fig. 7** TIC obtained following the co-injection of PBDE and PCB standard mixtures. Hepta- and octa-CBs eluted within the chromatographic window are defined for tetra- and penta-BDEs. BDE-47 and CB-180 eluted at the same retention time. Reprinted with permission from Elsevier [331]

HR-TOF mass spectrometers have also been used to determine PBDEs in environmental samples, with detection limits comparable to those of most other MS techniques [340, 341]. Alternative analytical techniques are LC-MS, LC-MS/MS [342, 343] and GC×GC [336, 340]. The former two are promising, but use atmospheric pressure photoionization (APPI), as PBDEs do not ionize well with either ESI or APCI. When working with APPI, both negative and positive ionization modes are suitable for their analysis, depending on the degree of bromine substitution. However, the analysis of metabolites, such as hydroxylated BDEs (OH-BDEs), can be successfully conducted when operating in ESI mode. Finally, GC×GC could be very useful to avoid the co-elution problems found in standard GC-MS methods [344].

## 4.5 Methyl *tert*-Butyl Ether (MTBE) and Other Gasoline Additives

MTBE, and gasoline additives in general, are not usually analysed in wastewaters, but this section was included as they are an important group of compounds to be considered when dealing with emerging contaminants. Fuel oxygenates have been added to gasoline since the 1970s, mainly as octane enhancers that increase the combustion efficiency and reduce toxic air emissions, such as lead compounds or carbon monoxide. Since the ban on tetraalkyl lead compounds, MTBE has become the most commonly used oxygenate and the one with the highest production volume worldwide [345].

Among fuel additives, MTBE is the ether with higher solubility and lower sorption and Henry's law constant, enhancing its higher mobility (nearly as fast as that of ground water) and the difficulty in removing it from water by aeration or degradation processes [346]. For this reason, as well as its intense use, MTBE has become one of the most frequently detected volatile organic compounds (VOCs) in ground water which can be adsorbed on subsurface solids [346].

Besides the health effects, toxicity and carcinogenicity at high concentrations [347], there is much interest in the aesthetic implications of MTBE in drinking water. Taste and odour thresholds for this compound in water have been reported at very low concentrations, approximately 25–60 μg/L for flavour and 40–70 μg/L for odour at 25 °C [347]. For this reason, the US Environmental Protection Agency (EPA) established a drinking water advisory for aesthetic concerns at 20–40 μg/L [347]. To date, there are no regulations for MTBE in water, air or soil in Europe but some countries are establishing their own guidelines.

Analytical methodologies dealing with the analysis of MTBE also include the determination of its main degradation products, *tert*-butyl alcohol (TBA) and *tert*-butyl formate (TBF), as well as other gasoline additives present in fuel, such as the oxygenate dialkyl ethers, for example ethyl *tert*-butyl ether, *tert*-amyl methyl ether and diisopropyl ether, and the aromatic compounds benzene, toluene, ethylbenzene and xylene (BTEX).

### 4.5.1 Analysis in Environmental Samples

There are some reviews devoted to the analysis of MTBE and other gasoline additives in environmental samples [346, 348, 349]. Even though MTBE is more likely to be present in ground and surface waters as well as soil samples, due to its physico-chemical properties (high mobility and solubility), some studies also revealed its presence in wastewaters [350, 351].

The most crucial step in trace analysis of VOCs is definitely enrichment and sampling. For MTBE analysis, samples do not need to be preserved, as biodegradation is very slow [352]. However, special precautions have to be taken in VOC analysis to avoid losses and prevent contamination. Bottles used to collect samples are filled to the top, avoiding air bubbles passing through the sample, to prevent volatilization of target compounds [347].

As to enrichment techniques, some methodologies, including direct aqueous injection (DAI), membrane-introduction mass spectrometry (MIMS), headspace (HS) analysis, purge and trap (P&T), solid-phase microextraction (SPME) by direct immersion or headspace compound-specific stable isotope analysis (CSIA), which is an emerging tool in environmental sciences, have been proposed and discussed by [353, 354] as appropriate methods to be used. These techniques are recommended when VOCs are found at lower concentrations and they mainly operate coupled to an instrumental technique. As VOCs, fuel oxygenates are almost exclusively analysed by GC and MS detection. Other detectors, such as flame ionization (FID), photoionization (PID) and atomic emission (AED), can also be used, but MS is the preferred one due to its higher sensitivity and selectivity [350]. In Tables 1 and 2, some of the most representative methods for the analysis of MTBE and other gasoline additives in water and solid samples, respectively, are described.

The selection of one technique or another depends on the type of matrix analysed, the concentration range and the need for compliance with the regulations [350]. P&T and SPME were the methods that obtained the best accuracy in a MTBE inter-laboratory study with 20 European participating laboratories and, when coupled with mass spectrometry, were the ones offering the best results according to the quality state assurance/quality control requirements [350, 355]. When P&T is used, VOCs are purged from water with helium, and generally they are subsequently adsorbed onto a Tenax® silica gel–charcoal trap. After sample loading, trapped components are desorbed at high temperatures and transferred directly to the GC-MS system [347].

For the analysis of MTBE and gasoline additives in solid samples, the same techniques as for water samples (P&T, SPME, etc.) are used [350]. Pressurized-liquid extraction (PLE) has also been used for the determination of higher concentrations (mg/kg) of BTEX (Application note 324) in soils using hexane/acetone (1:1). A semi-automatic purge-and-membrane inlet mass spectrometric (PAM-MS) instrument [377] provided good sensitivity and accuracy for some BTEX compounds and MTBE. Among the ifferent types of P&T instruments assembled for the analysis of VOCs in solid matrices [356–361], closed-system P&T is directed to determine low concentrations ($<200$ μg/kg), as indicated in the EPA Method 5035 [350].

Quantitative analysis of MTBE, its degradation products and other gasoline additives is performed by operating the mass spectrometer in EI mode, generally at 70 eV. In order to increase sensitivity and selectivity, samples are injected in time scheduled SIM mode. Due to the rather high energy transfer in the EI ionization mode, fuel oxygenates do not yield molecular ions. Typical fragments obtained correspond to the α-cleavage $[M-CH_3]^+$ or $[M-CH_5]^+$, taken as base peaks in the mass spectra [347]. Typical columns used in the GC separation are fused-silica capillary DB-624 columns (75 m × 0.53 mm ID) with a 3-μm film thickness.

## 5 Conclusions

Among modern analytical techniques, GC and LC, coupled to both MS and tandem MS, are the key techniques for the determination of emerging contaminants in complex environmental samples. These techniques, combined with appropriate sample preparation procedures, allow the detection of target compounds at the low environmental levels. Furthermore, the introduction of new chromatographic techniques, such as fast LC, fast GC, and GC×GC, has improved the analysis of complex mixtures. However, current analytical methods only focus their attention on parent target compounds and rarely include metabolites and transformation products. The question is whether chemical analysis of only target compounds is sufficient to assess contaminants present in the environment. Recent developments in the mass spectrometry field, such as the introduction of Q-TOF and Q-LIT instruments, allow the simultaneous determination of both parent and transformation products. Exact mass measurements provided by Q-TOF and the ability to combine several scan functions are a powerful tool to provide a more accurate identification of target compounds in complex samples, as well as to enable structural elucidation of unknown compounds. However, general screening for unknown substances is time-consuming and expensive, and is often shattered by problems, such as lack of standards and mass spectral libraries. Therefore, effect-related analysis, focused on relevant compounds, nowadays seems to be a more appropriate way to assess and study environmental contamination problems.

**Acknowledgements** This work was financially supported by the European Union EMCO project (INCO-CT-2004-509188) and by the Spanish Ministerio de Ciencia y Tecnología (EVITA project CTM2004-06265-C03-01).

## References

1. Petrovic M, Gonzalez S, Barcelo D (2003) TrAC-Trends Anal Chem 22:685
2. Gros M, Petrovic M, Barcelo D (2006) Talanta 70:678
3. Shang DY, Ikonomou MG, McDonald RW (1999) J Chromatogr A 849:467
4. Petrovic M, Barcelo D (2002) Chromatographia 56:535
5. Barcelo D, Petrovic M, Eljarrat E, Lopez De Alda MJ, Kampioti A (2004) Chromatography 69B(6):987
6. Namiesnik J, Zabiegaa B, Kot-Wasik A, Partyka M, Wasik A (2005) Anal Bioanal Chem 381:279

7. Kozdron-Zabiegala B, Przyjazny A, Namiesnik J (1996) Indoor Built Environ 5:212
8. Belardi RP, Pawliszyn JB (1989) Water Pollut Res J Canada 24:1
9. Kot A, Zabiegala B, Namiesnik J (2000) TrAC-Trends Anal Chem 19:446
10. Lauridsen FS (2005) Environ Pollut 136:503
11. Lacorte S, Barcelo D (1996) Anal Chem 68:2464
12. Ferrer I, Hennion MC, Barcelo D (1997) Anal Chem 69:4508
13. Ferrer I, Pichon V, Hennion MC, Barcelo D (1997) J Chromatogr A 777:91
14. Ferrer I, Barcelo D (1999) J Chromatogr A 854:197
15. Renner T, Baumgarten D, Unger KK (1997) Chromatographia 45:199
16. Aguilar C, Ferrer I, Borrull F, Marce RM, Barcelo D (1998) J Chromatogr A 794:147
17. Hogenboom AC, Hofman MP, Jolly DA, Niessen WMA, Brinkman UAT (2000) J Chromatogr A 885:377
18. Slobodnik J, Oztezkizan O, Lingeman H, Brinkman UAT (1996) J Chromatogr A 750:227
19. Slobodnik J, Ramalho S, Van Baar BLM, Louter AJH, Brinkman UAT (2000) Chemosphere 41:1469
20. Weller MG (2000) Fresenius J Anal Chem 366:635
21. Delaunay N, Pichon V, Hennion MC (2000) J Chromatogr B Biomed Sci Appl 745:15
22. Bean KA, Henion JD (1997) J Chromatogr A 791:119
23. Martin-Esteban A, Fernandez P, Stevenson D, Camara C (1997) Analyst 122:1113
24. Pichon V, Chen L, Hennion MC, Daniel R, Martel A, Le Goffic F, Abian J, Barcelo D (1995) Anal Chem 67:2451
25. Ferguson PL, Iden CR, McElroy AE, Brownawell BJ (2001) Anal Chem 73:3890
26. Rodriguez-Mozaz S, Lopez de Alda MJ, Barcelo D (2007) J Chromatogr A 1152:97
27. Deinl I, Angermaier L, Franzelius C, MacHbert G (1997) J Chromatogr B Biomed Appl 704:251
28. Nedved ML, Habibi-Goudarzi S, Ganem B, Henion JD (1996) Anal Chem 68:4228
29. Creaser CS, Feely SJ, Houghton E, Seymour M (1998) J Chromatogr A 794:37
30. Rhemrev-Boom MM, Yates M, Rudolph M, Raedts M (2001) J Pharm Biomed Anal 24:825
31. Delaunay-Bertoncini N, Hennion MC (2004) J Pharm Biomed Anal 34:717
32. Qiao F, Sun H, Yan H, Row KH (2006) Chromatographia 64:625
33. Pichon V (2007) J Chromatogr A 1152:41
34. Dong X, Wang N, Wang S, Zhang X, Fan Z (2004) J Chromatogr A 1057:13
35. Zhu X, Yang J, Su Q, Cai J, Gao Y (2005) J Chromatogr A 1092:161
36. Pap T, Horvath V, Tolokan A, Horvai G, Sellergren B (2002) J Chromatogr A 973:1
37. Turiel E, Martin-Esteban A, Fernandez P, Perez-Cond C, Camara C (2001) Anal Chem 73:5133
38. Watabe Y, Kubo T, Nishikawa T, Fujita T, Kaya K, Hosoya K (2006) J Chromatogr A 1120:252
39. Watabe Y, Kondo T, Morita M, Tanaka N, Haginaka J, Hosoya K (2004) J Chromatogr A 1032:45
40. Whitcombe MJ, Martin L, Vulfson EN (1998) Chromatographia 47:457
41. Dickert FL, Lieberzeit P, Tortschanoff M (2000) Sens Actuators B 65:186
42. Bolisay LD, Culver JN, Kofinas P (2006) Biomaterials 27:4165
43. Wei HS, Tsai YL, Wu JY, Chen H (2006) J Chromatogr B 836:57
44. Shea KJ, Sasaki DY (1989) J Am Chem Soc 111:3442
45. Rimmer S (1998) Chromatographia 47:470
46. Lavignac N, Allender CJ, Brain KR (2004) Anal Chim Acta 510:139
47. Vlatakis G, Andersson LI, Miller R, Mosbach K (1993) Nature 361:645

48. Baggiani C, Anfossi L, Baravalle P, Giovannoli C, Tozzi C (2005) Anal Chim Acta 531:199
49. Sellergren B, Shea KJ (1995) J Chromatogr A 690:29
50. Haginaka J, Kagawa C (2002) J Chromatogr A 948:77
51. Hosoya K, Yoshizako K, Shirasu Y, Kimata K, Araki T, Tanaka N, Haginaka J (1996) J Chromatogr A 728:139
52. Pang X, Cheng G, Li R, Lu S, Zhang Y (2005) Anal Chim Acta 550:13
53. Mayes AG, Mosbach K (1996) Anal Chem 68:3769
54. Downey JS, McIsaac G, Frank RS, Stöver HDH (2001) Macromolecules 34:4534
55. Ho KC, Yeh WM, Tung TS, Liao JY (2005) Anal Chim Acta 542:90
56. Venn RF, Goody RJ (1999) Chromatographia 50:407
57. Koeber R, Fleischer C, Lanza F, Boos KS, Sellergren B, Barceló D (2001) Anal Chem 73:2437
58. Lamprecht G, Kraushofer T, Stoschitzky K, Lindner W (2000) J Chromatogr B Biomed Sci Appl 740:219
59. El Mahjoub A, Staub C (2000) J Chromatogr B Biomed Sci Appl 742:381
60. Yu Z, Westerlund D, Boos KS (1997) J Chromatogr B Biomed Appl 704:53
61. Gorecki T, Namienik J (2002) TrAC-Trends Anal Chem 21:276
62. Vrana B, Allan IJ, Greenwood R, Mills GA, Dominiak E, Svensson K, Knutsson J, Morrison G (2005) TrAC-Trends Anal Chem 24:845
63. Koester CJ, Moulik A (2005) Anal Chem 77:3737
64. Koester CJ, Simonich SL, Esser BK (2003) Anal Chem 75:2813
65. Lord H, Pawliszyn J (2000) J Chromatogr A 885:153
66. Ouyang G, Pawliszyn J (2006) Anal Bioanal Chem 386:1059
67. Wu J, Yu X, Lord H, Pawliszyn J (2000) Analyst 125:391
68. Bruheim I, Liu X, Pawliszyn J (2003) Anal Chem 75:1002
69. Eisert R, Pawliszyn J (1997) Anal Chem 69:3140
70. Globig D, Weickhardt C (2005) Anal Bioanal Chem 381:656
71. Wu J, Tragas C, Lord H, Pawliszyn J (2002) J Chromatogr A 976:357
72. Gou Y, Eisert R, Pawliszyn J (2000) J Chromatogr A 873:137
73. Gou Y, Pawliszyn J (2000) Anal Chem 72:2774
74. Gou Y, Tragas C, Lord H, Pawliszyn J (2000) J Microcolumn Sep 12:125
75. Takino M, Daishima S, Nakahara T (2001) Analyst 126:602
76. Lee MR, Lee RJ, Lin YW, Chen CM, Hwang BH (1998) Anal Chem 70:1963
77. Rodriguez I, Rubi E, Gonzalez R, Quintana JB, Cela R (2005) Anal Chim Acta 537:259
78. Stashenko EE, Martinez JR (2004) TrAC-Trends Anal Chem 23:553
79. Dietz C, Sanz J, Camara C (2006) J Chromatogr A 1103:183
80. Camel V (2002) Anal Bioanal Chem 372:39
81. Covaci A, Voorspoels S, de Boer J (2003) Environ Int 29:735
82. Eljarrat E, Barcelo D (2004) TrAC-Trends Anal Chem 23:727
83. Eljarrat E, De La Cal A, Raldua D, Duran C, Barcelo D (2004) Environ Sci Technol 38:2603
84. De Voogt P, Kwast O, Hendriks R, Jonkers N (2000) Analysis 28:776
85. Zhao M, Van Der Wielen F, De Voogt P (1999) J Chromatogr A 837:129
86. Syage JA, Nies BJ, Evans MD, Hanold KA (2001) J Am Soc Mass Spectrom 12:648
87. Cochran JW (2002) J Chromatogr Sci 40:254
88. Hada M, Takino M, Yamagami T, Daishima S, Yamaguchi K (2000) J Chromatogr A 874:81
89. Santos FJ, Galceran MT (2002) TrAC-Trends Anal Chem 21:672
90. Gaines RB, Ledford EB Jr, Stuart JD (1998) J Microcolumn Sep 10:597

91. Hyotylainen T, Kallio M, Hartonen K, Jussila M, Palonen S, Riekkola ML (2002) Anal Chem 74:4441
92. Petrovic M, Gros M, Barcelo D (2007) In: Petrovic M, Barcelo D (eds) Comprehensive analytical chemistry. Elsevier, Amsterdam, p 157
93. Petrovic M, Gros M, Barcelo D (2006) J Chromatogr A 1124:68
94. Stuber M, Reemtsma T (2004) Anal Bioanal Chem 378:910
95. Alder L, Luderitz S, Lindtner K, Stan HJ (2004) J Chromatogr A 1058:67
96. Benijts T, Lambert W, De Leenheer A (2004) Anal Chem 76:704
97. Kloepfer A, Quintana JB, Reemtsma T (2005) J Chromatogr A 1067:153
98. Van De Steene JC, Mortier KA, Lambert WE (2006) J Chromatogr A 1123:71
99. Vanderford BJ, Snyder SA (2006) Environ Sci Technol 40:7312
100. Hopfgartner G, Husser C, Zell M (2003) J Mass Spectrom 38:138
101. Raffaelli A, Saba A (2003) Mass Spectrom Rev 22:318
102. Hanold KA, Fischer SM, Cormia PH, Miller CE, Syage JA (2004) Anal Chem 76:2842
103. Zwiener C, Frimmel FH (2004) Anal Bioanal Chem 378:851
104. Taniyasu S, Kannan K, Horii Y, Hanari N, Yamashita N (2003) Environ Sci Technol 37:2634
105. Villagrasa M, López de Alda MJ, Barceló D (2006) Anal Bioanal Chem 386:953
106. Schultz MM, Barofsky DF, Field JA (2006) Environ Sci Technol 40:289
107. Taniyasu S, Kannan K, Man KS, Gulkowska A, Sinclair E, Okazawa T, Yamashita N (2005) J Chromatogr A 1093:89
108. Karrman A, Van Bavel B, Järnberg U, Hardell L, Lindstrøm G (2005) Anal Chem 77:864
109. Yamashita N, Kannan K, Taniyasu S, Horii Y, Okazawa T, Petrick G, Gamo T (2004) Environ Sci Technol 38:5522
110. Takino M, Daishima S, Nakahara T (2003) Rapid Commun Mass Spectrom 17:383
111. Saito N, Sasaki K, Nakatome K, Harada K, Yoshinaga T, Koizumi A (2003) Arch Environ Contam Toxicol 45:149
112. Saito N, Harada K, Inoue K, Sasaki K, Yoshinaga T, Koizumi A (2004) J Occup Health 46:49
113. Pocurull E, Aguilar C, Alonso MC, Barcelo D, Borrull F, Marce RM (1999) J Chromatogr A 854:187
114. Higgins CP, Field JA, Criddle CS, Luthy RG (2005) Environ Sci Technol 39:3946
115. Schroder HF (2003) J Chromatogr A 1020:131
116. Moody CA, Field JA (1999) Environ Sci Technol 33:2800
117. Moody CA, Field JA (2000) Environ Sci Technol 34:3864
118. Ohya T, Kudo N, Suzuki E, Kawashima Y (1998) J Chromatogr B Biomed Appl 720:1
119. Abe T, Baba H, Itoh E, Tanaka K (2001) J Chromatogr A 920:173
120. Abe T, Baba H, Soloshonok I, Tanaka K (2000) J Chromatogr A 884:93
121. Hori H, Hayakawa E, Yamashita N, Taniyasu S, Nakata F, Kobayashi Y (2004) Chemosphere 57:273
122. Kuehl DW, Rozynov B (2003) Rapid Commun Mass Spectrom 17:2364
123. Kuklenyik Z, Reich JA, Tully JS, Needham LL, Calafat AM (2004) Environ Sci Technol 38:3698
124. Hansen KJ, Johnson HO, Eldridge JS, Butenhoff JL, Dick LA (2002) Environ Sci Technol 36:1681
125. Sumpter JP, Johnson AC (2005) Environ Sci Technol 39:4321
126. Kuster M, Lopez de Alda MJ, Barcelo D (2004) TrAC-Trends Anal Chem 23:790
127. Diaz-Cruz MS, Lopez de Alda MJ, Barcelo D (2003) TrAC-Trends Anal Chem 22:340
128. Petrovic M, Eljarrat E, Lopez de Alda MJ, Barcelo D (2002) J Chromatogr A 974:23

129. Lopez de Alda MJ, Barcelo D (2001) J Chromatogr A 938:145
130. Petrovic M, Eljarrat E, Lopez de Alda MJ, Barcelo D (2001) TrAC-Trends Anal Chem 20:637
131. Ying GG, Kookana RS, Ru YJ (2002) Environ Int 28:545
132. Hanselman TA, Graetz DA, Wilkie AC (2003) Environ Sci Technol 37:5471
133. Kuster M, Lopez de Alda M, Rodriguez-Mozaz S, Barcelo D (2007) In: Petrovic M, Barcelo D (eds) Comprehensive analytical chemistry. Elsevier, Amsterdam, p 219
134. Fine DD, Breidenbach GP, Price TL, Hutchins SR (2003) J Chromatogr A 1017:167
135. Liu R, Zhou JL, Wilding A (2004) J Chromatogr A 1022:179
136. Isobe T, Shiraishi H, Yasuda M, Shinoda A, Suzuki H, Morita M (2003) J Chromatogr A 984:195
137. Lopez de Alda MJ, Barcelo D (2001) J Chromatogr A 911:203
138. Rodriguez-Mozaz S, Lopez de Alda MJ, Barcelo D (2004) Anal Chem 76:6998
139. Tozzi C, Anfossi L, Giraudi G, Giovannoli C, Baggiani C, Vanni A (2002) J Chromatogr A 966:71
140. Penalver A, Pocurull E, Borrull F, Marce RM (2002) J Chromatogr A 964:153
141. Mitani K, Fujioka M, Kataoka H (2005) J Chromatogr A 1081:218
142. Braun P, Moeder M, Schrader S, Popp P, Kuschk P, Engewald W (2003) J Chromatogr A 988:41
143. Zang X, Luo R, Song N, Chen TK, Bozigian H (2005) Rapid Commun Mass Spectrom 19:3259
144. Kuster M, Lopez de Alda MJ, Barceló D (2005) Handbook of environmental chemistry, vol 2. Springer, Heidelberg
145. Petrovic M, Tavazzi S, Barcelo D (2002) J Chromatogr A 971:37
146. Cespedes R, Petrovic M, Raldua D, Saura U, Pina B, Lacorte S, Viana P, Barcelo D (2004) Anal Bioanal Chem 378:697
147. Liu R, Zhou JL, Wilding A (2004) J Chromatogr A 1038:19
148. Peck M, Gibson RW, Kortenkamp A, Hill EM (2004) Environ Toxicol Chem 23:945
149. Ternes TA, Andersen H, Gilberg D, Bonerz M (2002) Anal Chem 74:3498
150. Williams RJ, Johnson AC, Smith JJL, Kanda R (2003) Environ Sci Technol 37:1744
151. Peng X, Wang Z, Yang C, Chen F, Mai B (2006) J Chromatogr A 1116:51
152. Reddy S, Brownawell BJ (2005) Environ Toxicol Chem 24:1041
153. Lopez de Alda MJ, Gil A, Paz E, Barcelo D (2002) Analyst 127:1299
154. Ying GG, Kookana RS (2003) Environ Sci Technol 37:1256
155. Ying GG, Kookana RS, Dillon P (2003) Water Res 37:3785
156. Desbrow C, Routledge EJ, Brighty GC, Sumpter JP, Waldock M (1998) Environ Sci Technol 32:1549
157. Ingrand V, Herry G, Beausse J, De Roubin MR (2003) J Chromatogr A 1020:99
158. Larsson DGJ, Adolfsson-Erici M, Parkkonen J, Pettersson M, Berg AH, Olsson PE, Forlin L (1999) Aquat Toxicol 45:91
159. Belfroid AC, Van Der Horst A, Vethaak AD, Schafer AJ, Rijs GBJ, Wegener J, Cofino WP (1999) Sci Total Environ 225:101
160. Johnson AC, Belfroid A, Di Corcia A (2000) Sci Total Environ 256:163
161. Huang CH, Sedlak DL (2001) Environ Toxicol Chem 20:133
162. Rodgers-Gray TP, Jobling S, Morris S, Kelly C, Kirby S, Janbakhsh A, Harries JE, Waldock MJ, Sumpter JP, Tyler CR (2000) Environ Sci Technol 34:1521
163. Ternes TA, Stumpf M, Mueller J, Haberer K, Wilken RD, Servos M (1999) Sci Total Environ 225:81
164. Servos MR, Bennie DT, Burnison BK, Jurkovic A, McInnis R, Neheli T, Schnell A, Seto P, Smyth SA, Ternes TA (2005) Sci Total Environ 336:155

165. Kuch HM, Ballschmiter K (2000) Fresenius J Anal Chem 366:392
166. Beck IC, Bruhn R, Gandrass J, Ruck W (2005) J Chromatogr A 1090:98
167. Zuehlke S, Dunnbier U, Heberer T, Fritz B (2004) Ground Water Monit Rem 24:78
168. Zuehlke S, Duennbier U, Heberer T (2005) J Sep Sci 28:52
169. Quintana JB, Rodil R, Reemtsma T (2004) J Chromatogr A 1061:19
170. Gomes RL, Birkett JW, Scrimshaw MD, Lester JN (2005) Int J Environ Anal Chem 85:1
171. Shimada K, Mitamura K, Higashi T (2001) J Chromatogr A 935:141
172. Lopez de Alda MJ, Diaz-Cruz S, Petrovic M, Barcelo D (2003) J Chromatogr A 1000:503
173. Atkinson S, Atkinson MJ, Tarrant AM (2003) Environ Health Perspect 111:531
174. Schneider C, Scholer HF, Schneider RJ (2005) Anal Chim Acta 551:92
175. Hintemann T, Schneider C, Scholer HF, Schneider RJ (2006) Water Res 40:2287
176. Barel-Cohen K, Shore LS, Shemesh M, Wenzel A, Mueller J, Kronfeld-Schor N (2006) J Environ Manage 78:16
177. Soto AM, Maffini MV, Schaeberle CM, Sonnenschein C (2006) Best Pract Res Clin Endocrinol Metab 20:15
178. Clode SA (2006) Best Pract Res Clin Endocrinol Metab 20:35
179. Rodriguez-Mozaz S, Marco MP, Lopez de Alda MJ, Barcelo D (2004) Anal Bioanal Chem 378:588
180. Rodriguez-Mozaz S, Lopez de Alda MJ, Barcelo D (2006) Talanta 69:377
181. Nakamura S, Hwee Sian T, Daishima S (2001) J Chromatogr A 919:275
182. Cathum S, Sabik H (2001) Chromatographia 53:s-394
183. Xiao XY, McCalley DV, McEvoy J (2001) J Chromatogr A 923:195
184. Lerch O, Zinn P (2003) J Chromatogr A 991:77
185. Kuch HM, Ballschmiter K (2001) Environ Sci Technol 35:3201
186. Shareef A, Angove MJ, Wells JD (2006) J Chromatogr A 1108:121
187. Shareef A, Parnis CJ, Angove MJ, Wells JD, Johnson BB (2004) J Chromatogr A 1026:295
188. Labadie P, Budzinski H (2005) Environ Sci Technol 39:5113
189. Díaz-Cruz MS, Barceló D (2006) Anal Bioanal Chem 386:973
190. Gros M, Petrovic M, Barcelo D (2006) Anal Bioanal Chem 386:941
191. Diaz-Cruz MS, Barcelo D (2005) TrAC-Trends Anal Chem 24:645
192. Fatta D, Achilleos A, Nikolaou A, Meric S (2007) TrAC-Trends Anal Chem 26:515
193. Farre M, Petrovic M, Barcelo D (2007) Anal Bioanal Chem 387:1203
194. Jacobsen AM, Halling-Sørensen B, Ingerslev F, Hansen SH (2004) J Chromatogr A 1038:157
195. Loffler D, Ternes TA (2003) J Chromatogr A 1021:133
196. Schlusener MP, Spiteller M, Bester K (2003) J Chromatogr A 1003:21
197. Ternes TA, Bonerz M, Herrmann N, Loffler D, Keller E, Lacida BB, Alder AC (2005) J Chromatogr A 1067:213
198. Turiel E, Martin-Esteban A, Tadeo JL (2006) Anal Chim Acta 562:30
199. Hernandez F, Sancho JV, Ibanez M, Guerrero C (2007) TrAC-Trends Anal Chem 26:466
200. Gomez MJ, Petrovic M, Fernandez-Alba AR, Barcelo D (2006) J Chromatogr A 1114:224
201. Castiglioni S, Bagnati R, Calamari D, Fanelli R, Zuccato E (2005) J Chromatogr A 1092:206
202. Hao C, Lissemore L, Nguyen B, Kleywegt S, Yang P, Solomon K (2006) Anal Bioanal Chem 384:505

203. Stolker AAM, Niesing W, Hogendoorn EA, Versteegh JFM, Fuchs R, Brinkman UAT (2004) Anal Bioanal Chem 378:955
204. Kasprzyk-Hordern B, Dinsdale RM, Guwy AJ (2007) J Chromatogr A 1161:132
205. Nebot C, Gibb SW, Boyd KG (2007) Anal Chim Acta 598:87
206. Zhang ZL, Zhou JL (2007) J Chromatogr A 1154:205
207. Botitsi E, Frosyni C, Tsipi D (2007) Anal Bioanal Chem 387:1317
208. Trenholm RA, Vanderford BJ, Holady JC, Rexing DJ, Snyder SA (2006) Chemosphere 65:1990
209. Roberts PH, Bersuder P (2006) J Chromatogr A 1134:143
210. Naidong W, Roets E, Busson R, Hoogmartens J (1990) J Pharm Biomed Anal 8:881
211. Bryan PD, Hawkins KR, Stewart JT, Capomacchia AC (1992) Biomed Chromatogr 6:305
212. Gros M, Pizzolato TM, Petrovic M, Lopez de Alda MJ, Barcelo D (2007) J Chromatogr A, in press; doi:10.1016/jchroma.2007.10.052
213. Bravo JC, Garcinuno RM, Fernandez P, Durand JS (2007) Anal Bioanal Chem 388:1039
214. O'Connor S, Aga DS (2007) TrAC-Trends Anal Chem 26:456
215. Chapuis F, Mullot JU, Pichon V, Tuffal G, Hennion MC (2006) J Chromatogr A 1135:127
216. Kolpin DW, Furlong ET, Meyer MT, Thurman EM, Zaugg SD, Barber LB, Buxton HT (2002) Environ Sci Technol 36:1202
217. Metcalfe CD, Koenig BG, Bennie DT, Servos M, Ternes TA, Hirsch R (2003) Environ Toxicol Chem 22:2872
218. Weigel S, Berger U, Jensen E, Kallenborn R, Thoresen H, Huhnerfuss H (2004) Chemosphere 56:583
219. Bendz D, Paxéus NA, Ginn TR, Loge FJ (2005) J Hazard Mater 122:195
220. Perez S, Barcelo D (2007) Trends Anal Chem 26:494
221. Eichhorn P, Ferguson PL, Perez S, Aga DS (2005) Anal Chem 77:4176
222. Marchese S, Gentili A, Perret D, D'Ascenzo G, Pastori F (2003) Rapid Commun Mass Spectrom 17:879
223. Gomez MJ, Malato O, Ferrer I, Aguera A, Fernandez-Alba AR (2007) J Environ Monit 9:719
224. Pozo OJ, Guerrero C, Sancho JV, Ibanez M, Pitarch E, Hogendoorn E, Hernandez F (2006) J Chromatogr A 1103:83
225. Seitz W, Schulz W, Weber WH (2006) Rapid Commun Mass Spectrom 20:2281
226. Nikolai LN, McClure EL, MacLeod SL, Wong CS (2006) J Chromatogr A 1131:103
227. Quintana JB, Reemtsma T (2004) Rapid Commun Mass Spectrom 18:765
228. Peck AM (2006) Anal Bioanal Chem 386:907
229. Bester K, Huhnerfuss H, Lange W, Rimkus GG, Theobald N (1998) Water Res 32:1857
230. McAvoy DC, Schatowitz B, Jacob M, Hauk A, Eckhoff WS (2002) Environ Toxicol Chem 21:1323
231. Aguera A, Fernandez-Alba AR, Piedra L, Mezcua M, Gomez MJ (2003) Anal Chim Acta 480:193
232. Winkler M, Headley JV, Peru KM (2000) J Chromatogr A 903:203
233. Artola-Garicano E, Borkent I, Hermens JLM, Vaes WHJ (2003) Environ Sci Technol 37:3111
234. Ricking M, Schwarzbauer J, Hellou J, Svenson A, Zitko V (2003) Mar Pollut Bull 46:410
235. Dsikowitzky L, Schwarzbauer J, Littke R (2002) Org Geochem 33:1747
236. Winkler M, Kopf G, Hauptvogel C, Neu T (1998) Chemosphere 37:1139

237. Gatermann R, Huhnerfuss H, Rimkus G, Attar A, Kettrup A (1998) Chemosphere 36:2535
238. Gatermann R, Huhnerfuss H, Rimkus G, Wolf M, Franke S (1995) Mar Pollut Bull 30:221
239. Bester K (2005) Arch Environ Contam Toxicol 49:9
240. Bester K (2003) Water Res 37:3891
241. Lee HB, Peart TE, Sarafin K (2003) Water Qual Res J Canada 38:683
242. Paxeus N (2004) Water Sci Technol 50:253
243. Standley LJ, Kaplan LA, Smith D (2000) Environ Sci Technol 34:3124
244. Difrancesco AM, Chiu PC, Standley LJ, Allen HE, Salvito DT (2004) Environ Sci Technol 38:194
245. Simonich SL, Begley WM, Debaere G, Eckhoff WS (2000) Environ Sci Technol 34:959
246. Simonich SL, Federle TW, Eckhoff WS, Rottiers A, Webb S, Sabaliunas D, De Wolf W (2002) Environ Sci Technol 36:2839
247. Sakkas VA, Giokas DL, Lambropoulou DA, Albanis TA (2003) J Chromatogr A 1016:211
248. Lambropoulou DA, Giokas DL, Sakkas VA, Albanis TA, Karayannis MI (2002) J Chromatogr A 967:243
249. Osemwengie LI, Gerstenberger SL (2004) J Environ Monit 6:533
250. Osemwengie LI, Steinberg S (2001) J Chromatogr A 932:107
251. Lindstrom A, Buerge IJ, Poiger T, Bergqvist PA, Muller MD, Buser HR (2002) Environ Sci Technol 36:2322
252. Poiger T, Buser HR, Müller MD, Balmer ME, Buerge IJ (2003) Chimia 57:492
253. Buerge IJ, Buser HR, Müller MD, Poiger T (2003) Environ Sci Technol 37:5636
254. Peck AM, Hornbuckle KC (2004) Environ Sci Technol 38:367
255. Boyd GR, Palmeri JM, Zhang S, Grimm DA (2004) Sci Total Environ 333:137
256. Boyd GR, Reemtsma H, Grimm DA, Mitra S (2003) Sci Total Environ 311:135
257. Van Stee LLP, Leonards PEG, Van Loon WMGM, Hendriks AJ, Maas JL, Struijs J, Brinkman UAT (2002) Water Res 36:4455
258. Yang JJ, Metcalfe CD (2006) Sci Total Environ 363:149
259. Burkhardt MR, ReVello RC, Smith SG, Zaugg SD (2005) Anal Chim Acta 534:89
260. Zeng X, Sheng G, Xiong Y, Fu J (2005) Chemosphere 60:817
261. Stevens JL, Northcott GL, Stern GA, Tomy GT, Jones KC (2003) Environ Sci Technol 37:462
262. Morales-Munoz S, Luque-Garcia JL, Ramos MJ, Fernandez-Alba A, De Castro MDL (2005) Anal Chim Acta 552:50
263. Morales-Munoz S, Luque-Garcia JL, Ramos MJ, Martinez-Bueno MJ, De Castro MDL (2005) Chromatographia 62:69
264. Kupper T, Berset JD, Etter-Holzer R, Furrer R, Tarradellas J (2004) Chemosphere 54:1111
265. Herren D, Berset JD (2000) Chemosphere 40:565
266. Morales S, Canosa P, Rodriguez I, Rubi E, Cela R (2005) J Chromatogr A 1082:128
267. Felix T, Hall BJ, Brodbelt JS (1998) Anal Chim Acta 371:195
268. Llompart M, Garcia-Jares C, Salgado C, Polo M, Cela R (2003) J Chromatogr A 999:185
269. Gonzalez S, Barcelo D, Petrovic M (2007) TrAC-Trends Anal Chem 26:116
270. Lee HB (1999) Water Qual Res J Canada 34:3
271. Petrovic M, Barcelo D (2001) J Mass Spectrom 36:1173
272. Di Corcia A, Cavallo R, Crescenzi C, Nazzari M (2000) Environ Sci Technol 34:3914
273. Ahel M, Giger W, Koch M (1994) Water Res 28:1131

274. Di Corcia A, Costantino A, Crescenzi C, Marinoni E, Samperi R (1998) Environ Sci Technol 32:2401
275. Jonkers N, Knepper TP, De Voogt P (2001) Environ Sci Technol 35:335
276. Ventura F, Figueras A, Caixach J, Espadaler I, Romero J, Guardiola J, Rivera J (1988) Water Res 22:1211
277. Petrovic M, Diaz A, Ventura F, Barcelo D (2001) Anal Chem 73:5886
278. Petrovic M, Barcelo D (2000) Anal Chem 72:4560
279. Petrovic M, Fernandez-Alba AR, Borrull F, Marce RM, Mazo EG, Barcelo D (2002) Environ Toxicol Chem 21:37
280. Petrovic M, Lacorte S, Viana P, Barcelo D (2002) J Chromatogr A 959:15
281. Gonzalez S, Petrovic M, Barcelo D (2004) J Chromatogr A 1052:111
282. Petrovic M, Barcelo D (2004) TrAC-Trends Anal Chem 23:762
283. Gonzalez S, Petrovic M, Barcelo D (2007) Chemosphere 67:335
284. Suter MJF, Reiser R, Giger W (1996) J Mass Spectrom 31:357
285. Bennie DT, Sullivan CA, Lee HB, Peart TE, Maguire RJ (1997) Sci Total Environ 193:263
286. Lee HB, Peart TE (1995) Anal Chem 67:1976
287. Ding WH, Chen CT (1999) J Chromatogr A 862:113
288. Hao C, Croley TR, March RE, Koenig BG, Metcalfe CD (2000) J Mass Spectrom 35:818
289. Schroder HF (2001) J Chromatogr A 926:127
290. Houde F, DeBlois C, Berryman D (2002) J Chromatogr A 961:245
291. Petrovic M, Barcelo D, Diaz A, Ventura F (2003) J Am Soc Mass Spectrom 14:516
292. Ayorinde FO, Elhilo E (1999) Rapid Commun Mass Spectrom 13:2166
293. Ayorinde FO, Eribo BE, Johnson JH Jr, Elhilo E (1999) Rapid Commun Mass Spectrom 13:1124
294. Andreu V, Pico Y (2004) Anal Chem 76:2878
295. Cantero M, Rubio S, Perez-Bendito D (2006) J Chromatogr A 1120:260
296. Cantero M, Rubio S, Perez-Bendito D (2004) J Chromatogr A 1046:147
297. Hanton SD, Parees DM, Zweigenbaum J (2006) J Am Soc Mass Spectrom 17:453
298. Willetts M, Clench MR, Greenwood R, Mills G, Carolan V (1999) Rapid Commun Mass Spectrom 13:251
299. Covaci A, Voorspoels S, Ramos L, Neels H, Blust R (2007) J Chromatogr A 1153:145
300. Luo Q, Cai ZW, Wong MH (2007) Sci Total Environ 383:115
301. Xiang CH, Luo XJ, Chen SJ, Yu M, Mai BX, Zeng EY (2007) Environ Toxicol Chem 26:616
302. Labandeira A, Eljarrat E, Barcelo D (2007) Environ Pollut 146:188
303. Streets SS, Henderson SA, Stoner AD, Carlson DL, Simcik MF, Swackhamer DL (2006) Environ Sci Technol 40:7263
304. Law K, Halldorson T, Danell R, Stern G, Gewurtz S, Alaee M, Marvin C, Whittle M, Tomy G (2006) Environ Toxicol Chem 25:2177
305. Gama AC, Sanatcumar P, Viana P, Barcelo D, Bordado JC (2006) Chemosphere 64:306
306. Eljarrat E, De La Cal A, Raldua D, Duran C, Barcelo D (2005) Environ Pollut 133:501
307. Hites RA (2004) Environ Sci Technol 38:945
308. Weber H, Heseker H (2004) Fresenius Environ Bull 13:356
309. Schecter A, Pavuk M, Papke O, Ryan JJ, Birnbaum L, Rosen R (2003) Environ Health Perspect 111:1723
310. Ikonomou MG, Rayne S, Addison RF (2002) Environ Sci Technol 36:1886
311. Guillamon M, Martinez E, Eljarrat E, Lacorte S (2002) Organohalogenated Compounds 55:199

312. Helleday T, Tuominen KL, Bergman A, Jenssen D (1999) Mutat Res Genet Toxicol Environ Mutagen 439:137
313. Meerts IATM, Letcher RJ, Hoving S, Marsh G, Bergman A, Lemmen JG, Van Der Burg B, Brouwer A (2001) Environ Health Perspect 109:399
314. Stapleton HM, Keller JM, Schantz MM, Kucklick JR, Leigh SD, Wise SA (2007) Anal Bioanal Chem 387:2365
315. Darnerud PO, Atuma S, Aune M, Cnattingus S, Wenroth ML, Wicklund-Glynn A (1998) Organohalogenated Compounds 35:411
316. Ohta S, Ishizuka D, Nishimura H, Nakao T, Aozasa O, Shimidzu Y, Ochiai F, Kida T, Nishi M, Miyata H (2002) Chemosphere 46:689
317. Booij K, Zegers BN, Boon JP (2002) Chemosphere 46:683
318. Hovander L, Malmberg T, Athanasiadou M, Athanassiadis I, Rahm S, Bergman A, Wehler EK (2002) Arch Environ Contam Toxicol 42:105
319. Sjodin A, Hagmar L, Klasson-Wehler E, Kronholm-Dlab K, Jakobsson E, Bergman A (1999) Environ Health Perspect 107:643
320. Fontanals N, Barri T, Bergstrom S, Jonsson JA (2006) J Chromatogr A 1133:41
321. Stapleton HM, Harner T, Shoeib M, Keller JM, Schantz MM, Leigh SD, Wise SA (2006) Anal Bioanal Chem 384:791
322. de Boer J, Allchin CR, Law R, Zegers BN, Boon JP (2001) Trends Anal Chem 20:591
323. De Boer J, Wester PG, van den Horst A, Leonards PEG (2003) Environ Pollut 122:63
324. Nylund K, Asplund L, Jansson B, Jonsson P, Litzen K, Sellstrom U (1992) Chemosphere 24:1721
325. Hartonen K, Bowadt S, Hawthorne SB, Riekkola ML (1997) J Chromatogr A 774:229
326. De La Cal A, Eljarrat E, Barcelo D (2003) J Chromatogr A 1021:165
327. Samara F, Tsai CW, Aga DS (2006) Environ Pollut 139:489
328. Yusa M, Pardo O, Pastro A, de la Guardia M (2006) Anal Chim Acta 557:304
329. Law RJ, Allchin CR, Bennett ME, Morris S, Rogan E (2002) Chemosphere 46:673
330. Bjorklund J, Tollback P, Hiarne C, Dyremark E, Ostman C (2004) J Chromatogr A 1041:201
331. Eljarrat E, De la Cal A, Barcelo D (2003) J Chromatogr A 1008:181
332. Alaee M, Backus S, Cannon C (2001) J Sep Sci 24:465
333. Zhu LY, Hites RA (2002) Environ Sci Technol 38:2779
334. Hale RC, La Guardia MJ, Harvey E, Gaylor MO, Mainor TM (2006) Chemosphere 64:181
335. Wise SA, Poster DL, Schantz MM, Kucklick JR, Sander LC, Lopez De Alda M, Schubert P, Parris RM, Porter BJ (2004) Anal Bioanal Chem 378:1251
336. Korytar P, Covaci A, Leonards PEG, De Boer J, Brinkman UAT (2005) J Chromatogr A 1100:200
337. Alaee M, Sergeant DB, Ikonomou MG, Luross JM (2001) Chemosphere 44:1489
338. Polo M, Gomez-Noya G, Quintana JB, Llompart M, Garcia-Jares C, Cela R (2004) Anal Chem 76:1054
339. Wang D, Cai Z, Jiang G, Wong MH, Wong WK (2005) Rapid Commun Mass Spectrom 19:83
340. Focant JF, Sjodin A, Patterson DG Jr (2003) J Chromatogr A 1019:143
341. Cajka T, Hajslova J, Kazda R, Poustka J (2005) J Sep Sci 28:601
342. Debrauwer L, Riu A, Jouahri M, Rathahao E, Jouanin I, Antignac JP, Cariou R, Le Bizec B, Zalko D (2005) J Chromatogr A 1082:98
343. Hua W, Bennett ER, Letcher RJ (2005) Environ Int 31:621
344. Stapleton HM (2006) Anal Bioanal Chem 386:807

345. Johnson R, Pankow J, Bender D, Price C, Zogorski J (2000) Environ Sci Technol 34:210A
346. Squillace PJ, Pankow JF, Korte NE, Zogorski JS (1997) Environ Toxicol Chem 16:1836
347. Rosell M, Lacorte S, Ginebreda A, Barcelo D (2003) J Chromatogr A 995:171
348. Rosell M, Lacorte S, Barcelo D (2006) TrAC-Trends Anal Chem 25:1016
349. Atienza J, Aragon P, Herrero MA, Puchades R, Maquieira A (2005) Crit Rev Anal Chem 35:317
350. Rosell M, Lacorte S, Barcelo D (2006) J Chromatogr A 1132:28
351. Achten C, Kolb A, Puttmann W, Seel P, Gihr R (2002) Environ Sci Technol 36:3652
352. Schmidt TC, Duong HA, Berg M, Haderlein SB (2001) Analyst 126:405
353. Schmidt TC (2003) TrAC-Trends Anal Chem 22:776
354. Atienza J, Aragon P, Herrero MA, Puchades R, Maquieira A (2005) Crit Rev Anal Chem 35:317
355. Schuhmacher R, Fuhrer M, Kandler W, Stadlmann C, Krska R (2003) Anal Bioanal Chem 377:1140
356. Bellar T (1991) US Environmental Protection Agency, Environmental Monitoring Systems Laboratory, Cincinnati
357. Bianchi A, Varney MS (1989) J High Resolut Chromatogr 12:184
358. Bianchi AP, Varney MS, Phillips J (1991) J Chromatogr 542:413
359. Amaral OC, Olivella L, Grimalt JO, Albaiges J (1994) J Chromatogr A 675:177
360. Zuloaga O, Etxebarria N, Fernandez LA, Madariaga JM (2000) Anal Chim Acta 416:43
361. Campillo N, Vinas P, Lopez-Garcia I, Aguinaga N, Hernandez-Cordoba M (2004) Talanta 64:584
362. Tanabe A, Tsuchida Y, Ibaraki T, Kawata K, Yasuhara A, Shibamoto T (2005) J Chromatogr A 1066:159
363. Boulanger B, Vargo JD, Schnoor JL, Hornbuckle KC (2005) Environ Sci Technol 39:5524
364. Lagana A, Fago G, Marino A, Santarelli D (2001) Anal Lett 34:913
365. Stolker AAM, Niesing W, Fuchs R, Vreeken RJ, Niessen WMA, Brinkman UAT (2004) Anal Bioanal Chem 378:1754
366. Yang S, Cha J, Carlson K (2005) J Chromatogr A 1097:40
367. Montes R, Rodriguez I, Rubi E, Cela R (2007) J Chromatogr A 1143:41
368. Suzuki S, Hasegawa A (2006) Anal Sci 22:469
369. Martinez-Carballo E, Gonzalez-Barreiro C, Scharf S, Gans O (2007) Environ Pollut 148:570
370. Berset JD, Bigler P, Herren D (2000) Anal Chem 72:2124
371. Lee HB, Sarafin K, Peart TE, Svoboda ML (2003) Water Qual Res J Canada 38:667
372. Morris S, Allchin CR, Zegers BN, Haftka JJH, Boon JP, Belpaire C, Leonards PEG, Van Leeuwen SPJ, De Boer J (2004) Environ Sci Technol 38:5497
373. Fabrellas B, Sanz P, Larrazabal D, Abad E (2000) Organohalogenated Compounds 45:160
374. Stapleton HM, Brazil B, Holbrook RD, Mitchelmore CL, Benedict R, Konstantinov A, Potter D (2006) Environ Sci Technol 40:4653
375. Leon VM, Gonzalez-Mazo E, Gomez-Parra A (2000) J Chromatogr A 889:211
376. Voogt P, Saez M (2006) Trends Anal Chem 25:326
377. Ojala M, Mattila I, Tarkiainen V, Sarme T, Ketola RA, Maattanen A, Kosiainen R, Kotiaho T (2001) Anal Chem 73:3624

Hdb Env Chem Vol. 5, Part S/1 (2008): 105–142
DOI 10.1007/698_5_105

Published online: 21 March 2008

# Acute and Chronic Effects of Emerging Contaminants

Tvrtko Smital

Laboratory for Molecular Ecotoxicology,
Division for Marine and Environmental Research,
Rudjer Boskovic Institute, Bijenicka 54, 10000 Zagreb, Croatia
*smital@irb.hr*



**Abstract** Acute or chronic toxicity profiling represents one of the critical elements for scientifically reliable characterization and prioritization of potentially hazardous contaminants. The very same is true for so-called emerging contaminants, regardless of the definition used in defining various aspects of “emerging”, including substances that belong to new chemical classes, new types of use, new effects, mechanism of action, source, or exposure route. From the (eco)toxicological perspective, however, there are two essential drawbacks which prevent efficient characterization of risk posed to humans and the environment by the presence of emerging contaminants. First is related to the fact that the potential of analytical chemistry to measure contaminants currently exceeds our understanding of their potential environmental effects. Secondly, for most emerging contaminants there is currently little information regarding their potential toxicological significance in ecosystems, particularly the effects from long-term low-level environmental exposures. Based on these facts a brief overview of acute and chronic toxic effects on human and wildlife, reported for various classes of emerging contaminants, is presented in this chapter. The most demanding research unknowns, methodological drawbacks, and priorities will be highlighted, and finally, future strategies needed for efficient (eco)toxicological characterization of emerging contaminants will be suggested.

**Keywords** Acute and chronic toxicity · (Eco)toxicological characterization · Emerging contaminants

**Abbreviations**

| | |
|---|---|
| AFOs | Animal feeding operations |
| ALS | Amyotrophic lateral sclerosis |
| ARGs | Antibiotic resistance genes |
| BADGE | Bisphenol A diglycidyl ether |
| BPA | Bisphenol A |
| CHE | The Collaborative on Health and the Environment |
| DES | Diethylstilbestrol |
| ELS | Early life-stages |
| GDS | Genotoxic disease syndrome |
| HAdV | Human adenoviruses |
| HEV | Hepatitis E virus |
| HPV | High Production Volume |
| MATC | Maximum acceptable toxicant concentration |
| MXR | Multixenobiotic resistance |
| NOAA | US National Centers for Coastal Ocean Science |
| OSPAR | Oslo and Paris Convention for the Protection of the Marine Environment of the North-East Atlantic |
| PBDEs | Polybrominated diphenyl ethers |
| PCNs | Polychlorinated naphthalenes |
| PCPs | Personal care products |
| PFCs | Perfluorochemicals |
| POPs | Persistent organic pollutants |
| PVC | Poly-vinyl chloride |
| QDs | Quantum dots |
| REACH | Registration, Evaluation, and Authorization of Chemicals |
| STP | Sewage treatment plant |
| US FDA | US Food and Drug Administration |
| USCDC | US Centers for Disease Control |
| USEPA | US Environmental Protection Agency |
| WWF | World Wide Fund |

# 1 Introduction

Cancer, reproductive disorders, impaired neurological development, allergies – these are the types of health effects that make headlines. That puts corresponding chemicals "culprits" on the top of any list of emerging contaminants: potentially toxic substances whose effects or presence are poorly known, often because these chemicals have only begun to enter the human water or food supply. On the other hand, humans and wildlife are constantly exposed to a variety of contaminants present at low levels. These include both new chemicals, with previously unknown effects and those with well known acute (short-term exposure) human and ecological health effects. The result has been new research on emerging contaminants and an increased emphasis on methods of analyzing health effects of contaminants. The area in which several advances have recently been made is related to long-term health effects of chemical exposure. Other studies are now examining the impacts of organic compounds which may interfere with the endocrine systems of living organisms. Another active area of research is focused on how chemicals interact with each other and the natural environment. Finally, researchers are continuing to find new chemicals that bioaccumulate in the food chain. Such chemicals can be present in water at very low levels, however, they accumulate to higher concentrations in living tissue, substantially magnifying any health effects.

Three components have been usually considered to be critical for a chemical to be classified as highly hazardous contaminant: (1) persistence (structural stability resulting in long environmental half-lives); (2) lypophilicity (resulting in bioconcentration and possible biomagnification in the food chain); and (3) proven acute or chronic toxicity. However, all of these criteria need certain reconsideration – for example, continual release of some contaminants by the sewage treatment plants (STPs) give them a "pseudo-persistance" in aquatic environments; some drugs are actively transported in cells regardless of their lipid-water partition coefficients; finally, chemicals may act as indirect toxicants (such as nanoparticles or antibiotics, for example). Nevertheless, toxicity remains one of the cornerstones for scientifically reliable classification and hazard prioritization. From the (eco)toxicological perspective, however, two serious drawbacks appears to be essential in preventing efficient and reliable characterization of risk posed to humans and the environment by the presence of emerging contaminants.

Firstly, due to recent improvements in analytical chemistry, the types of chemicals that can be detected are increasing, and the limits of concentration at which they can be detected are continuously lowered. Our ability to measure contaminants currently exceeds our understanding of their potential environmental effects. Proving the link between real environmental exposure levels and acute or chronic toxic effects to humans and/or wildlife is an expensive, time-consuming, and complex research endeavor. Evaluating ecological effects of environmental contamination extends beyond observing co-occurrence of contaminants and adverse effects to documenting cause-and-effect relationships. Research to characterize cause-and-effect relationships requires documentation of contaminant uptake, modes of action, and biological endpoints. Numerous substances that act through specific or sensitive mechanisms of action (e.g., mediated by receptors or other mechanisms) may have effects on the environment or sensitive human populations at concentrations well below those previously considered to be safe. Clearly, traditional (eco)toxicological methods are not adequate to address the complexity of emerging environmental contaminants. It is a new challenge for toxicologists to effectively identify and assess the potential impact of these substances on human and ecological receptors, so that appropriate decisions can be made that balance the societal and environmental benefits and risks.

Secondly, for most emerging contaminants, there is currently little information regarding their potential toxicological significance in ecosystems, particularly effects from long-term, low-level environmental exposures. Furthermore, the fact is that we know very little about the vast majority of the chemicals we use. In the EU, more than 100 000 chemicals were reported to be on the market in 1981, which was the first and only time that the chemicals used in the EU were listed[1]. For 99% of chemicals (by volume), information on properties, uses, and risks is sketchy. Chemicals produced in high volumes (above 1000 tons per year) have been examined more closely, and there are still no data for about 21% of them. Another 65% come with insufficient data. Similar figures would be anticipated for the US and Japan (Table 1). Therefore, the raise of emerging contaminants may be only an inevitable consequence of this disproportion.

**Table 1** Estimated numbers or proportions of indexed, commercially available, regulated/inventoried, and/or toxicologically characterized chemicals [172]

| | |
|---|---|
| No. of chemicals indexed in the CAS Registry | >26 000 000 |
| No. of commercially available chemicals | 8 400 000 |
| No. of regulated and/or inventoried chemicals | 240 000 |
| No. of chemicals marketed in the US/EU | 100 000 |
| No. of bioactive compounds in various R&D phases | >150 000 |
| Proportion of chemicals (by volume) with known properties and risks | 1% |
| Proportion of high volume (>1000 t) chemicals sufficiently characterized | 79% |
| Proportion of high volume (>1000 t) chemicals insufficiently characterized | 65% |

[1] Public availability of data on EU high production volume chemicals, European Chemicals Bureau, Joint Research Centre, European Commission (http://ecb.jrc.it/Data-Availability-Documents/datavail.doc).

In an attempt to illustrate these critical drawbacks in this chapter we will try to present a brief overview of acute and chronic effects to human and wildlife, reported for various classes of emerging contaminants present in waste waters and aquatic environments in general. In addition, we will highlight the most demanding research unknowns, methodological drawbacks and priorities, and, finally, address future strategies needed for efficient (eco)toxicological characterization of potentially harmful substances.

## 2 Emerging Contaminants from (Eco)toxicological Perspective

### 2.1 Definition(s) – Emerging Contaminants vs. Emerging Concerns

"Emerging contaminants" can be broadly defined as any synthetic or naturally occurring chemical or any microorganism that is not commonly monitored in the environment, but has the potential to enter the environment and cause known or suspected adverse ecological and/or human health effects. In some cases, release of emerging chemical or microbial contaminants to the environment has likely occurred for a long time, but may not have been recognized until new detection methods were developed. In other cases, synthesis of new chemicals or changes in use and disposal of existing chemicals can create new sources of emerging contaminants. Not all of these substances can accurately be described as emerging contaminants or pollutants. Some of them are found naturally in our surface waters; others are natural substances which are concentrated by anthropogenic activities; and still others are man-made chemicals that do not occur in nature. Those pollutants that are truly new, those that have just gained entry into the environment, are relatively rare in comparison to known chemicals already being released into aquatic environments, and are often confused with those whose presence has just been detected but which have long been present [1]. The term "emerging" is also used to describe not the pollutant itself, but rather a new "emerging concern", i.e. newly demonstrated toxic effect and/or mechanism of action of an old pollutant [2]. This approach is highly legitimate and is often favored among toxicologists in comparison to classifications and definitions based on chemical entities. In reality, however, scientists and regulators will have to deal with both, "emerging contaminants" and "emerging concerns", and this artificial partition is certainly not critical for principal understanding of the problem and its possible solutions.

Furthermore, once a substance is called an emerging contaminant, the longevity of its emerging contaminant status in the view of scientists and the public is largely determined by whether the biological or chemical agent of concern is persistent and/or has potentially deleterious human or ecotoxicological effects. Alternatively, new observations or information (e.g., endocrine disruption) on contaminants (e.g., nonylphenol) can cause the reconsideration of a well known contaminant as a (re)emerging contaminant. Unfortunately, the same analytical advances which bring contaminants to the public's attention do not offer knowledge about whether the newly detected contaminant is of (eco)toxicological interest. Assessing the effects of these contaminants in the environment remains a major time- and resource-intensive challenge. Therefore, it is not surprising that, for the many thousands of chemicals being produced or already on the market and the many new microbes that are being discovered, advances in our understanding of their (eco)toxicological properties are considerably slow and lag significantly behind the public's demand for information. As a result, a contaminant may be considered for several years to be emerging. Regardless of the definition in this chapter we will cover different dimensions of "emerging", including substances that belong to new chemical classes, new types of use, new effects, mechanism of action, source, or exposure route.

## 3 Human vs. Ecological Health Effects

### 3.1 Human Health Effects – Basic Principals

Human health results from complex interactions among genes and the environment. Environmental exposures to chemical, physical, and biological agents may cause or contribute to disease in susceptible individuals. Personal lifestyle factors, such as diet, smoking, alcohol use, level of exercise, and UV exposure, often are a primary focus when considering preventable causes of disease. However, exposures to chemical contaminants at work, at home, in the outdoors, and even in utero, are increasingly recognized as important and preventable contributors to human disease [3].

Toxic effects of chemical agents are often not well understood or appreciated by healthcare providers and the general public. Some chemicals, such as asbestos, vinyl chloride, and lead, are well established as causes of human disease. There is also good evidence available to suggest increases in the incidence of some cancers, asthma, and developmental disorders, can be attributed to chemical exposure, particularly in young children. Other diseases, such as amyotrophic lateral sclerosis (ALS) or Gulf War Syndrome have been hypothesized to be associated with chemical exposures, but the evidence is limited.

The effects of chemical exposures in humans are difficult to study, because controlled human experimentation is not ethically feasible. There is limited human data obtained from accidental exposures, overdoses, or studies of workers exposed occupationally. Environmental exposure studies in the general population also can be useful, though they often have limitations. Many diseases, such as cancer, may not appear until decades after an exposure has occurred, making it difficult for causal associations to be identified. Exposure assessment, a critical step in environmental epidemiologic studies, is difficult. Retrospective exposure assessment usually requires estimates and considerable judgment and is subject to significant error. An individual's exposure may change over time, and exposures to multiple chemicals occur both in the home and work environments. It is difficult for individuals to remember or even know what they have been exposed to. Furthermore, the effects of chemical exposures may vary, depending on the age of exposure (in utero, childhood, adult), the route of exposure (ingestion, inhalation, dermal), amount and duration of exposure, exposures to multiple chemicals simultaneously, and other personal susceptibility factors, including genetic variability.

Because of these challenges, most toxicity research is conducted in animal studies, which contribute important toxicological information and provide strong evidence of disease without human epidemiological studies if the mechanism of action is relevant. Many regulatory decisions to limit or ban the use of a chemical are based on animal data. Furthermore, human epidemiology studies are often conducted after an association has been hypothesized based on animal data. The same is true for most data related to human toxic effects of emerging contaminants described in this chapter.

Although there is a need for much more chemicals to be adequately characterized, a vast amount of data for human acute or chronic toxic effects of various contaminants is already available and published. What is often lacking, both for scientists and regulators, as well as for citizens, is a comprehensive and reliable tool that offers free, scientifically sound, and reliable information about contaminants hazardous to humans. Nevertheless, useful and comprehensive evidence has been recently complied within two independent sources. With the motto: "Mapping the Pollution in People", The Human Toxome Project at the Environmental Working Group in the USA [4] established a web database aimed at collecting and presenting relevant data about health effects of virtually all pollutants that enter the human body. Another source is The Collaborative on Health and the Environment (CHE) Toxicant and Disease Database [5], a searchable database that summarizes links between chemical contaminants and approximately 180 human diseases or conditions.

## 3.2 Ecotoxicological Aspects of Emerging Contaminants

As much as it is difficult to establish clear causal connections between contaminant(s) exposure and human health effects, it is far more difficult to do the same on the ecosystem level, with numerous species involved at different levels of biological organization, and many environmental factors that make the interpretation of field data even more complex. Paradoxically (or not?), knowledge, expertise, and resources being invested in human health issues, outmatch multiple times those invested in the environmental health arena, explaining to a large extent the critical shortage in data needed for a sustainable management of environmental resources.

More specifically, the objective of aquatic toxicity tests with effluents or pure compounds is to estimate the "safe" or "no effect" concentration of these substances, which is defined as the concentration that will permit normal propagation of fish and other aquatic life in the receiving waters. The endpoints which have been considered in tests to determine the adverse effects of toxicants include death and survival, decreased reproduction and growth, locomotor activity, gill ventilation rate, heart rate, blood chemistry, histopathology, enzyme activity, olfactory function, etc. [6]. Since it is not feasible to detect and/or measure all of these (and other possible) effects of toxic substances on a routine basis, observations in toxicity tests generally have been limited to only a few effects, typically including mortality, growth, and reproduction.

Acute lethality is an obvious and easily observed effect which accounts for its wide use in the early period of evaluation of the toxicity of pure compounds and complex effluents. The results of these tests were usually expressed as the concentration lethal to 50% of the test organisms (LC50) over relatively short exposure periods (one-to-four days).

As exposure periods of acute tests were lengthened, the LC50 and lethal threshold concentration were observed to decline for many compounds. By lengthening the tests to include one or more complete life cycles and observing the more subtle effects of the toxicants, such as a reduction in growth and reproduction, more accurate direct estimates of the threshold or safe concentration of the toxicant could be obtained. However, laboratory life-cycle tests may not accurately estimate the "safe" concentration of toxicants, because they are conducted with a limited number of species under highly controlled, steady-state conditions, and the results do not include the effects of the stresses to which the organisms would ordinarily be exposed in the natural environment.

An early published account of a full life-cycle fish toxicity test was that of Mount and Stephan back in 1967 [7]. In this study, fathead minnows, *Pimephales promelas*, were exposed to a graded series of pesticide concentrations throughout their life-cycle, and the effects of the toxicant on survival, growth, and reproduction were measured and evaluated. This work was soon followed by full life-cycle tests using other toxicants and fish species. McKim [8] evaluated the data from 56 full life-cycle tests, 32 of which used the fathead minnow, and concluded that the embryo-larval and early juvenile life-stages were the most sensitive stages. He proposed the use of partial life-cycle toxicity tests with the early life-stages (ELS) of fish to establish water quality criteria. Macek and Sleight [9] found that exposure of critical life-stages of fish to toxicants provides estimates of chronically safe concentrations remarkably similar to those derived from full life-cycle toxicity tests. They reported that for a great majority of toxicants, the concentration which will not be acutely toxic to the most sensitive life stages is the chronically safe concentration for fish, and that the most sensitive life stages are the embryos and fry. Critical life-stage exposure was considered to be exposure of the embryos during most, preferably all, of the embryogenic (incubation) period, and exposure of the fry for 30 days post-hatch for warm water fish with embryogenic periods ranging from 1–14 days, and for 60 days post-hatch for fish with longer embryogenic periods. They concluded that in the majority of cases, the maximum acceptable toxicant concentration (MATC) could be estimated from the results of exposure of the embryos during incubation, and the larvae for 30 days post-hatch.

In a review of the literature on 173 fish full life-cycle and ELS tests performed to determine the chronically safe concentrations of a wide variety of toxicants, such as metals, pesticides, organics, inorganics, detergents, and complex effluents, Woltering [10] found that at the lowest effect concentration, significant reductions were observed in fry survival in 57%, fry growth in 36%, and egg hatchability in 19% of the tests. He also found that fry survival and growth were often equally sensitive, and concluded that the growth response could be deleted from routine application of the ELS tests. The net result would be a significant reduction in the duration and cost of screening tests with no appreciable impact on estimating MATCs for chemical hazard assessments.

Efforts to further reduce the length of partial life-cycle toxicity tests for fish without compromising their predictive value have resulted in the development of an eight-day embryo-larval survival and teratogenicity test for fish and other aquatic vertebrates [11, 12], and a seven-day larval survival and growth test [13]. The similarity of estimates of chronically safe concentrations of toxicants derived from short-term embryo-larval survival and teratogenicity tests to those derived from full life-cycle tests has been firstly demonstrated by Birge et al. [12, 14].

Since that time, most of our knowledge about acute and chronic effects of contaminants originates from the described type of ecotoxicity tests. An overview of the present knowledge related to emerging contaminants/concerns will be presented in the next section.

## 4
## Human and Environmental Health Effects

Among many different categories of emerging contaminants, we will especially take into consideration those which, according to the state-of-the-art litera-

**Table 2** Major human/environmental health concerns and priority status of the most prominent categories of emerging contaminants

| Health concern | Chemical family | | | | | | |
|---|---|---|---|---|---|---|---|
| | Alkyl-phenols | Bisphenol A & BADGE | Brominated dioxins & furans | Per-chlor-ate | Perfluoro-chemicals (PFCs) | Phtal-ates | Polybrom-inated di-phenyl ethers (PBDEs) |
| Birth defects and developmental delays | + | + | | + | + | +++ | ++ |
| Brain and nervous system | | | | | ++ | +++ | +++ |
| Cancer | | + | + | | + | + | + |
| Endocrine system | + | | | +++ | + | | |
| Gastrointestinal (including liver) | | | | | + | | + |
| Hematologic (blood) system | | | | + | | | |
| Hormone activity | + | +++ | | | +++ | +++ | +++ |
| Immune system (including sensi-tization and allergies) | | ++ | + | | +++ | +++ | |
| Kidney and renal system | | + | | | +++ | | |
| Reproduction and fertility | +++ | ++ | ++ | | +++ | +++ | +++ |
| Skin | | + | | | | + | |
| Respiratory system | + | | | | | +++ | |
| Wildlife and environ-mental toxicity | +++ | ++ | | | | | + |
| Persistent, accumulates in wildlife and/or people | ++ | ++ | +++ | | ++ | ++ | ++ |
| OSPAR list | √ | √ | | | | √ | √ |
| Priority substance and/ or banned in the EU, USA or Canada | √ | | | | | √ | |

Weight of evidence: + limited; ++ probable; +++ strong

ture evidence, appear to be of the highest (eco)toxicological relevance and are frequently detected in industrial and/or municipal waste: industrial chemicals (new and recently recognized), personal care products, pharmaceuticals, nonculturable biological pathogens, and, finally, nanomaterials. Instead of referring to numerous studies utilizing various in vivo and in vitro test systems in attempts to characterize toxicity of many different contaminants, what follows in the section(s) below is a brief summary describing relevance and toxic effects reported with a reasonable weight of evidence for the most prominent emerging contaminants. Basic info referring to major human health concerns, wildlife toxicity, bioaccumulation/persistency potential, and the regulatory status of those substances is presented in Table 2.

**Table 2** (continued)

| Health concern | Chemical family<br>Polychlorinated naphthalenes (PCNs) | Fragrances (nitro- and polycyclic musks) | Triclosan | Pharmaceuticals | Non-culturable biological pathogens | Nanomaterials |
|---|---|---|---|---|---|---|
| Birth defects and developmental delays | | | | +++ | | |
| Brain and nervous system | | | | + | | + |
| Cancer | | + | | + | | |
| Endocrine system | | + | + | +++ | | |
| Gastrointestinal (including liver) | +++ | +++ | | + | +++ | |
| Hematologic (blood) system | | | | | | |
| Hormone activity | | | | +++ | | |
| Immune system (including sensitization and allergies) | | + | + | | + | + |
| Kidney and renal system | | | | | | |
| Reproduction and fertility | + | +++ | + | ++ | | |
| Skin | +++ | + | + | + | ++ | |
| Respiratory system | | | | + | +++ | ++ |
| Wildlife and environmental toxicity | ++ | + | + | ++ | | + |
| Persistent, accumulates in wildlife and/or people | ++ | ++ | ++ | + | | |
| OSPAR list | √ | √ | | | | |
| Priority substance and/ or banned in the EU, USA or Canada | √ | | √ | | | |

Weight of evidence: + limited; ++ probable; +++ strong

## 4.1 Industrial Chemicals

### 4.1.1 Alkylphenols

Alkylphenols are widely used industrial chemicals which act as detergents or surfactants. They are added to cosmetics, paints, pesticides, detergents, and cleaning products. Alkylphenols have been recently detected in surface waters contaminated with urban runoff and in wastewater effluents [15, 16] and have been measured in air samples. One study found that newer homes, especially those with poly-vinyl chloride (PVC) materials, have more alkylphenol residues than older houses or outdoor air [17]. As a group they are highly toxic to aquatic organisms. Dozens of recent studies have documented the in vitro and in vivo estrogenic activity of alkylphenols in human cell lines and animals [18–20]. Recent study by McClusky and colleagues [21] revealed harmful effects of *p*-nonylphenol exposure to spermatogenic cycle in male rats. Similar estrogenic activities of alkylphenols have been reported for aquatic organisms, including a recent example of the reduction of reproductive competence of male fathead minnow upon exposure to environmentally relevant mixtures of alkylphenolethoxylates [22]. Further supported by their persistency in aquatic environments and bioaccumulation potential, alkylphenols are put on the OSPAR list of possible substances of concern and included in the list of priority substances in the EU water policy.

### 4.1.2 Bisphenol A and Bisphenol A Diglycidyl Ether

In use since the 1950's, bisphenol A (BPA) is a building block for polycarbonate plastic and epoxy resins. BPA and its derivative, bisphenol A diglycidyl ether (BADGE), are found in many everyday products, such as the lining of metal food and drink cans, plastic baby bottles, pacifiers, and baby toys, dental sealants, computers, cell phones, hard plastic water bottles (such as Nalgene), paints, adhesives, enamels, varnishes, CDs and DVDs, and certain microwavable or reusable food and drink containers. These compounds have been shown to leach into food and water from containers – particularly after heating or as plastic ages.

BPA is a hormone-mimicking chemical that can disrupt the endocrine system at very low concentrations. More than a hundred animal studies have linked low doses of bisphenol A to a variety of adverse health effects, such as reduced sperm count, impaired immune system functioning, increases in prostate tumor proliferation, altered prostate and uterus development, insulin resistance, alteration of brain chemistry, early puberty, and behavioral changes [23–36]. Significantly, many of the studies showing adverse effects are at levels many times lower than what the US Environmental Protection Agency (USEPA) considers safe (50 μg/kg/day).

For BADGE, a bisphenol A derivative used to make epoxy resins and in a variety of industrial, engineering, and construction applications, the major pathway for human exposure is through chemical leaching from the linings of food and drink cans. BADGE is also found in some dental sealants [37].

Some basic toxicological testing has been done on BADGE, but the compound has not been extensively studied. One of the most important toxicological questions is whether BADGE breaks down into bisphenol A in the human body. Based on urinary levels of BPA in workers exposed to BADGE versus unexposed controls, researchers concluded that BADGE breaks down into BPA in the body [38]. However, other research has suggested that there is no such biotransformation [39]. In the human body, BADGE appears in a hydrolysis product known as BADGE 40-H [40]. BADGE is quickly metabolized by the body (within a day or so), therefore body burden levels represent recent exposures.

Considering that its sister chemical, bisphenol A, has a non-monotonic dose response curve, showing nonintuitive patterns of toxicity, it would be difficult to make a final assessment on the toxicity of BADGE without more detailed study. There is some evidence that BADGE is a rodent carcinogen, but data for humans is lacking [41, 42]. Workers using epoxy resin in the construction industry have shown BADGE to be a contact allergen [43]. Males exposed to BADGE through spraying epoxy resin have associated depressed gonadotrophic hormones [38]. A study of BADGE given to pregnant rabbits found that at the lowest dose tested (30 mg/kg/day for days 6 to 18 of gestation) BADGE affected pregnancy ability and the sex ratio of their litters [39]. An in vitro study found that BADGE can induce time and dose-dependent morphological changes and cell detachment from the substratum and can inhibit cell proliferation [44]. Another study found that a BADGE derivative (BADGE.2HCl) can act as an androgen antagonist in in vitro systems [45].

### 4.1.3 Brominated Dioxins and Furans

Brominated dioxins and furans are toxic, persistent, bioaccumulative, and lipophilic ("fat-loving"). Along with dioxins, furans are pollutants produced during PVC plastic production, industrial bleaching, and incineration. They build up in human tissues, are stored in fatty tissues and fluid, such as breast milk, and can be passed on to fetuses and infants during pregnancy and lactation. Brominated dioxins and furans are formed unintentionally, either from incineration of wastes which include consumer products infused with brominated flame retardants, such as polybrominated diphenyl ethers (PBDEs), or as trace contaminants in mixtures of bromine-containing chemicals. Primary (eco)toxicological concern for brominated dioxins and furans is their dioxin-like activity, meaning that they cause birth defects in animals and otherwise disrupt reproductive development and the immune and hormone systems [46–49]. They add to the total dioxin body burden of people, which are near levels where adverse health effects may be occurring in the general population [50].

### 4.1.4 Perchlorate

The vast majority of perchlorate manufactured is used to make solid rocket and missile fuel, while smaller amounts of perchlorate are also used to make firework and road flares. Perchlorate is also a contaminant of certain types of fertilizer which were widely used in the early part of the 20th century, but are in limited use today [51]. According to the analysis of the USEPA's latest data, perchlorate is known to be contaminating at least 160 public drinking water systems in 26 US states [52]. Tests of almost 3000 human urine and breast milk samples, along with tests of more than 1000 fruit, vegetable, cow's milk, beer, and wine samples, reveal that perchlorate exposure in the population is pervasive. Every urine sample tested showed some level of perchlorate contamination, and almost 70% of the fruit and beverage samples tested have had detectable perchlorate [52–60].

Critical toxic effect of perchlorate is inhibition of the thyroid's ability to take up the nutrient iodide, which is a key building block for thyroid hormones. If the thyroid gland does not have enough iodide for a sufficient period of time, body's thyroid hormone levels will eventually drop. Hypothyroidism (low thyroid hormone levels) in adults can cause fatigue, depression, anxiety, unexplained weight gain, hair loss, and low libido. More serious, however, are the effects of thyroid hormone disruption in the developing fetus and child. Small changes in maternal thyroid hormone levels during pregnancy have been associated with reduced IQs in children [61, 62]. A recent epidemiological study by the US Centers for Disease Control (USCDC) shows that perchlorate exposures commonly found in the population can cause significant thyroid hormone disruptions in women – particularly in the population of women with lower iodine intake. Relying on a flawed industry study, the USEPA adopted a water clean-up standard for superfund sites of 24.5 ppb in 2006. Neither the USEPA nor the US Food and Drug Administration (USFDA) have taken any action to address the problem of widespread contamination in food.

Considering animal studies, perchlorate was first discovered to affect the thyroid in the 1950s, but it wasn't until the early 1990s that scientists began to conduct studies that involved feeding low doses of perchlorate to animals and looking for adverse effects. In 1995 the USEPA found that laboratory animals developed thyroid disorders after two weeks of drinking perchlorate-laced water. Subsequent studies found effects on brain and thyroid structure at even lower doses, and noted that rat pups born to exposed mothers were particularly like to show adverse effects [53, 54].

The USCDC conducted the first major epidemiological study on perchlorate exposure in the general population [59]. After testing urine samples of 2299 men and women from around the country for perchlorate, and comparing these findings with the levels of thyroid hormones found in the blood of these same people, the USCDC's researchers discovered that there was a statistically significant relationship between urinary perchlorate and thyroid hormone levels in the 1111 women tested. Furthermore, they found that if low iodine woman started with perchlorate exposure corresponding to 0.19 ppb in urine (the minimum level found), and then ingested enough perchlorate through food and/or drinking water to raise their urinary perchlorate level to 2.9 ppb (the median level found), their T4 thyroid hormone levels would drop by 13 percent. Similarly, if woman's urinary perchlorate level increased to 5.2 ppb (the 75th percentile exposure), their T4 levels would drop by 16 percent. These are significant declines when one considers that recent studies have shown that the cognitive development of the fetus is impaired in mothers with even mild disruptions in thyroid hormone levels [59, 61, 62]. Women with low iodine intake and levels of TSH (a type of thyroid hormone) that were already on the edge of the normal range were found to be even more sensitive to perchlorate exposure. For these women, if they were exposed to 5 parts per billion of perchlorate via food or drinking water, the resulting hormone disruption would push them into sub-clinical hypothyroidism.

### 4.1.5 Perfluorochemicals

The USEPA has described perfluorochemicals (PFCs) as combining "persistence, bioaccumulation, and toxicity properties to an extraordinary degree" [63]. PFCs are industrial chemicals widely used as water, stain, and grease repellants for food wrap, carpet, furniture, and clothing. The family includes such well known name brands as Scotchgard and Teflon.

PFCs are released to the environment in air and water emissions at numerous manufacturing and processing facilities worldwide. PFCs are also likely released to the environment at countless secondary manufacturing facilities, including sites where consumer products are coated for water, stain, and grease repellency. The dominant sources of PFCs in the environment are thought to be fluorotelomer chemicals, the active ingredients in coatings of furniture, clothing, food packaging, and other products. Fluorotelomers break down in the environment and in the body to PFCs differing only in the carbon chain length and end group [64, 65]. Most PFCs are fairly mobile in water, but due to low volatility of the persistent carboxy acids and sulfonates, many do not have the potential to migrate in air far from locations of release as a manufacturing pollutant. In contrast, studies indicate that PFC telomers are relatively volatile and could migrate long distances through the atmosphere.

Fluorotelomers are a likely source of the persistent perfluorochemicals found in newborns, and in wildlife and water in areas remote from manufacturing sites and human populations. Available scientific findings to date show that PFCs widely contaminate human blood [66, 67] and persist in the body for decades [68]. They act through a broad range of toxic mechanisms of action to present potential harm to a wide range of organs (ovaries, liver, kidney, spleen, thymus, thyroid, pituitary, testis), and persist indefinitely in the environment with no known biological or environmental breakdown mechanism [69–71]. Considering their ecotoxicity the newest evidence suggests that PFCc are able to induce and inhibit the activity of xenobiotic efflux transport proteins in marine bivalves [72].

### 4.1.6 Phthalates

Found within many consumer products, phthalates are industrial plasticizers that impart flexibility and resilience to plastic. They are common additives to soft plastic, especially PVC. They are present in clear food wrap, personal care products (detergents and soaps), and pesticides [73].

Phthalates are widely detected in human blood and urine samples. The latest exposure study from USCDC indicates that women are slightly more exposed than men, and younger children (ages 6–11) are more exposed than older children (ages 12–20) [74]. Exposure to phthalates occurs through direct use of cosmetics and other consumer products containing these chemicals, consumption of foods wrapped in products containing these chemicals, and through inhalation of air contaminated with these chemicals [74].

In laboratory animals, fetal exposure to phthalates causes significant developmental toxicity, especially of the male reproductive system. In adult animals, phthalates damage the reproductive organs, adrenal, liver, and kidney [75]. In utero exposure to high levels of phthalate metabolites are associated with marked differences in the reproductive systems of baby boys; the exposure levels associated with these health effects were not extreme, but rather were typical for about one-quarter of all women. Adult men with high levels of phthalates have lower sperm motility and concentration and alterations in hormone levels [76–78]. Concentrations of two phthalates in house dust are associated with asthma and rhinitis in a study of 400 children, half of whom had allergies [79].

### 4.1.7 Polybrominated Diphenyl Ethers

Polybrominated diphenyl ethers (PBDEs) are brominated fire retardants, intentionally added to flexible foam furniture, primarily mattresses, couches, padded chairs, pillows, carpet padding and vehicle upholstery, and to electronic products.

Studies of laboratory animals link PBDE exposure to an array of adverse health effects including thyroid hormone disruption, permanent learning and memory impairment, behavioral changes, hearing deficits, delayed puberty onset, decreased sperm count, and fetal malformations [80–82]. Research in animals shows that exposure to brominated fire retardants in utero or during infancy leads to more significant harm than exposure during adulthood, and at much lower levels [47]. PBDEs are bioaccumulative and lipophilic, and, therefore, are highly persistent in people and the environment. The chemicals build up in the body, are stored in fatty tissues and body fluids, such as blood and breast milk, and can be passed on to fetuses and infants during pregnancy and lactation. People are primarily exposed to PBDEs in their homes, offices, and vehicles. Secondary sources are foods, primarily meat, dairy, fish, and eggs [83].

Some PBDEs were withdrawn from the US market in 2005 due to their toxicity to laboratory animals, and their detection as contaminants in humans, wildlife, house and office buildings, and common foods [84–86]. Deca (PBDE-209), the form used in electronics, continues to be used in televisions, computer monitors and other electronic products. There is widespread concern that Deca breaks down in the environment to more toxic and persistent forms.

### 4.1.8 Polychlorinated Naphthalenes

There are 75 possible chemical variations of polychlorinated naphthalenes (PCNs). They have been used as cable insulation, wood preservatives, engine oil additives, electroplating masking compounds, capacitors, and in dye production. Products are generally mixtures of several different PCNs. The largest source of PCNs believed to be waste incineration and disposal of items containing PCNs, although other potential sources of PCNs to the environment include sewage discharge from municipal and industrial sites leaching from hazardous waste sites. PCNs are also unwanted byproducts formed after the chlorination of drinking water [87]. They have not been used commercially in significant quantities since the 1980s.

PCNs are toxic, persistent and bioaccumulate in people and wildlife. The toxic effects of many PCNs are thought to be similar to dioxin. In humans, severe skin reactions (chloracne) and liver disease have both been reported after occupational exposure to PCNs. Other symptoms found in workers include cirrhosis of the liver, irritation of the eyes, fatigue, headache, anaemia, haematuria, impotentia, anorexia, and nausea. At least ten deaths were reported from liver toxicity. Workers exposed to PCNs also have a slightly higher risk of all cancers combined [88–90].

## 4.2 Personal Care Products (PCPs)

### 4.2.1 Fragrances – Nitromusks and Polycyclic Musks

Nitromusk and polycyclic musks are synthetic fragrances typically used in cosmetics, perfume, air fresheners, cleansing agents, detergents, and soap. Musks are also used as food additives, in cigarettes, and in fish baits. Commonly used musks contaminate lakes and fish in the US and Europe [91–96]. Nitromusk and polycyclic musks tend to accumulate in the fatty tissues of our bodies, and are often detected in breast milk as well as blood [96–98].

In laboratory studies, some nitromusks have been linked to cancer [99, 100]. Studies of nitromusks in people suggest that high levels of some of these chemicals are associated with reproductive and fertility problems in women [101]. Some also produce skin irritation and sensitization [102, 103].

Growing concerns about the health effects of nitromusks have led the EU to ban the use of some of these chemicals in cosmetics and personal care products. As a result, the use of polycyclic musks has increased. However, laboratory studies suggest that polycyclic musks, like nitromusks, may also affect hormone systems [104–109]. Two particular musk chemicals, a nitromusk and a polycyclic musk which both produced neurotoxic effects in laboratory animals, have been removed from the market. In the US, all musk chemicals are unregulated, and safe levels of exposure have not yet been set. Considering their ecotoxic potential, Luckenbah and Epel [110] demonstrated that nitromusk and polycyclic musk compounds act as long-term inhibitors of cellular multixenobiotic resistance (MXR) defense systems mediated in aquatic mollusks by specific transport proteins.

### 4.2.2 Triclosan

Triclosan is essentially a pesticide (antibacterial agent), used in some healthcare facility soaps. It is also the most common antimicrobial agent in household liquid hand soap. It can be found in toothpaste, lip gloss, soap (solid and liquid), plastic products ranging from children's toys to cutting boards, and footwear [111]. It has been detected in human breast milk and serum samples from the general population [98, 112], and in the urine of 61% of 90 girls ages six to eight tested in a recent study spearheaded by Mount Sinai School of Medicine [73].

Triclosan kills microbes by disrupting protein production, binding to the active site of a critical carrier protein reductase essential for fatty acid synthesis, which is present in microbes but not humans. Available studies do not raise major concerns for human health, but some basic questions remain, including the safety of triclosan exposures in utero, and exposures in infancy through contaminated breast milk. Triclosan breaks down in the environment, including in tap water, to chlorinated chemicals that pose both environmental and health concerns [113].

Large quantities of triclosan are washed down drains and into wastewater treatment plants. A fraction is removed during water treatment, but the rest is discharged to lakes and rivers. Studies indicate that its interaction with sunlight results in the formation of methyl triclosan, a chemical that may bioacummulate in wildlife and humans [112, 114], as well as a form of dioxin, which is a chemical linked to a broad range of toxicities including cancer [115]. The Canadian government limits the levels of dioxins and furans allowed as impurities in personal care products that contain triclosan. Triclosan was recently found in 58% of 139 US streams [116], the likely result of its presence in treated discharged wastewater. A safety standard for triclosan has not yet been set, and it does not require testing in tap water. However, it is believed that triclosan likely passes through standard water treatment processes to contaminate treated tap water supplies at low levels. New studies show that triclosan in tap water will readily react with residual chlorine from standard water disinfecting procedures to form a variety of chlorinated byproducts, including chloroform, a suspected human carcinogen [117].

Wildlife species are also contaminated with triclosan and its breakdown products; a recent European study found its breakdown product methyl triclosan in fish, especially concentrated in fatty tissue [113]. Triclosan is known to be acutely toxic to certain types of aquatic organisms, but little is known about its long-term effects on humans [118]. The chemical structure of triclosan is similar to that of diethylstilbestrol (DES), a non-steroidal estrogen, raising concerns about its potential to act as an endocrine disruptor. A recent study showed that triclosan can affect the thyroid gland, significantly altering frog metamorphosis at exposure levels equivalent to those currently found in the environment and human tissues, suggesting that triclosan may represent a potential health risk to human hormone action as well [119]. Studies have also found that triclosan has weakly androgenic effects but no estrogenic effects [120]. In addition, animal studies have shown that prolonged application of triclosan solution to the skin can cause dermal irritation in people with a specific sensitivity. There is no evidence that triclosan is a carcinogen or teratogen [121]. There is concern that the widespread use of antimicrobials such as triclosan in household products may promote antibiotic resistance in bacteria, although the current literature shows a possible association but no definitive link [122].

In addition to the PCPs mentioned above, some other categories like sunscreen agents, preservatives, and nutraceuticals recently got attention as possible emerging contaminants. As for now, however, the weight of evidence does not justify their treatment as immediate hazard to human or wildlife health.

## 4.3 Pharmaceuticals (Human Drugs and Veterinary Medicines)

Recent studies have also identified a number of pharmaceuticals as potential environmental contaminants that may adversely affect reproduction and development of biota in the environment [111, 123]. Some of these substances are not removed in traditional, or even advanced treatment systems, or under best management practices [124, 125]. Several of these substances have recently been detected in well treated effluents and drinking water, showing that sewage treatment frequently does not affect the chemical structure, and, therefore, the toxicity of drugs [126–129]. Emerging data in Europe and North America suggests that these chemicals are widespread in the environment, especially in surface waters exposed to human or agriculture wastes [116, 130]. Consequently, pharmaceuticals often enter the environment at levels similar to better studied agrochemicals.

Traditionally, pharmaceuticals and personal care products have not been viewed as environmental pollutants [131]. However, the potential for these substances to cause a variety of physiological responses in non-target species has raised concerns for possible impacts on the environment. Although these substances are usually found at very low concentrations in the environment, continuous low-dose exposure to these complex mixtures, especially at sensitive life stages, may have significant effects on individuals, populations, or ecosystems. The ecological impact of long-term exposure to large mixtures of those essentially biologically active chemicals is also unknown. Many of these chemicals are known to be persistent in both treatment systems and in the environment. Chemicals found in sewage and manure, such as synthetic estrogens, are known to have biological consequences at extremely low exposures [132]. Exposure of biota to even low doses during critical or sensitive life-stages may have profound effects on development and reproduction for multiple generations.

Due to their intended use in human or veterinary medicine, pharmaceuticals are generally well studied and a large body of toxicological evidence directed to human health issues exists for most of them. Considering their ecotoxicity, however, the available evidence in most cases provides indications of acute effects in vivo for organisms at different trophic levels after short-term exposure, but extremely rarely after long-term chronic exposures. An excellent service called "The Pharmaceuticals in the Environment, Information for Assessing Risk" has been recently developed and is maintained at the National Centers for Coastal Ocean Science (NOAA), Center for Coastal Environmental Health and Biomolecular Research, USA [133]. The database provides information on prescribed amounts, levels detected in aquatic environments, chemical structure, molecular weight, octanol-water partition coefficients, water solubility, environmental persistence, general toxicity information, and specific toxicity levels of pharmaceuticals to five groups of organisms (algae, mollusks, finfish, crustaceans, and select terrestrial animals). Toxicity to terrestrial animals is provided as a general comparison to data found in toxicological literature. All of this information was obtained from available scientific literature and is provided to assist with indentification of locations where risks to aquatic organisms might occur.

Considering the ecotoxicity of human pharmaceuticals, most of the current knowledge is well summarized in several excellent review articles published during the last few years [111, 130, 134–136]. Summarizing the available data, it is clear that there is almost no data about bioaccumulation of pharmaceuticals in biota, and often there is no correlation between the acute toxicity and lipophilicity. Most of pharmaceuticals displayed their LC50 values above 100 mg/L, which classifies them as not being harmful to aquatic organisms. However, variability of data within the same, as well as between different species is considerable, often spanning one or two orders of magnitude. Nevertheless, the overall conclusion is that acute toxicity of pharmaceuticals may be only relevant in case of accidental spills. Chronic toxicity, however, appears to be more relevant to aquatic biota and numerous examples clearly point out that it cannot be derived from acute toxicity data by simple calculations.

Veterinary pharmaceuticals, on the other hand, were traditionally less covered in environmental and human health toxicity studies. Current livestock and aquaculture production practices include the use of a wide variety of pharmaceuticals to enhance animal health and efficient food production, including antimicrobials (antibiotics), growth enhancers, feed supplements, and other medicinal products. Recently, low levels of veterinary medicines were detected in soils, surface waters, and ground waters worldwide [137]. Although the environmental occurrence and associated impacts of some compounds, such as selected antibacterial compounds, have been investigated, the impacts of many other substances found in the environment are not well understood. As a result, questions have arisen about the effects of veterinary medicines on organisms in the environment and on human health.

The interest in veterinary pharmaceuticals as potential emerging contaminants has also stemmed from the proliferation of large-scale animal feeding operations (AFOs) during the last decade. The large number of animals produced creates a proportionately large volume of animal waste and associated emerging contaminants. In a reconnaissance study of liquid waste at swine AFOs in Iowa and North Carolina, US, multiple classes of antibiotics were detected ranging from ppb to ppm concentrations [138]. Compilation of data from liquid waste from swine operations between 1998 and 2002 found one or more antibiotics present in all of the samples. The data from these studies demonstrate that veterinary pharmaceuticals are excreted and frequently occur at detectable levels ranging from ppb to ppm concentrations in liquid and solid waste.

Research to document the presence of antibiotics in fish hatchery recently revealed the occurrence and persistence of antibiotics in medicated feed used in fish hatcheries [139]. It was discovered that ormetoprim and sulfadimethoxine persisted in water for longer periods of time than oxytetracycline in fish hatcheries. Oxytetracycline was detected more frequently in the samples of the intensive hatcheries than samples from the extensive hatcheries. Sulfadimethoxine concentrations were greater in the intensive hatcheries than the extensive hatcheries, but persisted up to 40 days after treatment in both types of fish hatcheries. In addition, antibiotics were detected in untreated hatchery raceways, suggesting that recirculating water within a hatchery can lead to unintentional low-level exposure of antibiotics to healthy fish.

## 4.4
## Nonculturable Biological Pathogens as Emerging Contaminants

Among the viruses infecting humans, many different types are excreted in high concentrations in the feces of patients with gastroenteritis or hepatitis and in lower concentrations in the feces or urine of patients with other viral diseases. Moreover, viruses are also present in healthy individuals, and, thus, high viral loads are detected in urban sewage and are regarded as environmental contaminants [140]. Some viruses, such as humanpolyomaviruses and some adenovirus strains, infect humans during childhood, thereby establishing persistent infections. In the case of many frequent adenoviral respiratory infections, viral particles may continue to be excreted in feces for months or even years afterward. There is available information about some waterborne pathogens, but the improvement in molecular technology for detecting viruses present in water has focused attention on new groups of viruses that could be considered emergent viruses in diverse geographical areas. Technical advances are then most readily associated with the concept of emergent microorganisms, which are defined as newly identified microorganisms, those already existent but characterized by a rapidly increasing incidence and/or geographical ambit, and those for which transmission through food or water has only recently been discovered. Several studies have confirmed that infectious diseases related to water are not only a primordial cause of mortality and morbidity worldwide but also that both the spectrum and incidence of many diseases related to water are increasing. Human polyomaviruses, hepatitis E virus (HEV), and human adenoviruses (HAdV) are three groups of viruses, which are being detected more often in the environment [141]. Adenoviruses, for example, are important human pathogens that are responsible for both enteric illnesses and respiratory and eye infections. Recently, these viruses have been found to be prevalent in rivers, coastal waters, swimming pool waters, and drinking water supplies worldwide. USEPA listed adenovirus as one of nine microorganisms on the Contamination Candidate List for drinking water, because their survival characteristic during water treatment is not yet fully understood. Adenoviruses have been found to be significantly more stable than fecal indicator bacteria and other enteric viruses during UV treatment, and adenovirus infection may be caused by consumption of contaminated water or inhalation of aerosolized droplets during water recreation.

In addition, many species of bacteria pathogenic to humans, such as *Legionella*, are thought to have evolved in association with amoebal hosts. Several novel unculturable bacteria related to *Legionella* have also been found in amoebae, a few of which have been thought to be causes of nosocomial infections in humans [142]. A recent study done by Berk and colleagues in 2006 [143] revealed that it is over 16 times more likely to encounter infected amoebae in cooling towers than in natural environments. Several identified bacteria have novel rRNA sequences, and most strains were not culturable outside of amoebae. Such pathogens of amoebae may spread to the environment via aerosols from cooling towers. Therefore, studies of emerging infectious diseases should strongly consider cooling towers as a source of amoeba-associated pathogens.

Additional example is Campylobacter(s), which are emerging as one of the most significant causes of human infections worldwide, and the role that waterfowl and the aquatic environment have in the spread of disease is beginning to be elucidated [144]. On a world scale, Campylobacters are possibly the major cause of gastrointestinal infections. They are common commensals in the intestinal tract of many species of wild birds, including waterfowl. They are also widely distributed in aquatic environments where their origins may include waterfowl as well as sewage effluents and agricultural runoff. Campylobacters have marked seasonal trends and in temperate aquatic environments they peak during winter, whereas spring-summer is the peak period for human infection. Campylobacter species may survive, and remain potentially pathogenic, for long periods in aquatic environments. The utility of bacterial fecal indicators in predicting the presence of campylobacters in natural waters is questionable. Viable but nonculturable Campylobacter cells may occur, but whether they have any role in the generation of outbreaks of campylobacteriosis is unclear. The routine detection of *Campylobacter* spp. in avian feces and environmental waters largely relies on conventional culture methods, while the recognition of a particular species or strain is based on serotyping and increasingly on molecular methods.

## 4.5
## Antibiotic Resistance Genes

Antibiotic resistance genes (ARGs) are another type of "biological" emerging environmental contaminants. Along with nanoparticles, they may be classical examples of indirect toxicants. The primary health concern in the case of ARGs is related to adverse outcomes of antibiotic's exposures resulting in selection for pathogen resistance or alteration of microbial community structures. The occurrence of ARGs was recently demonstrated in various environmental compartments including river sediments, irrigation ditches, dairy lagoons, and the effluents of wastewater recycling and drinking water treatment plants [145]. Some of ARGs were also present in treated drinking water and recycled wastewater, suggesting that these are potential pathways for the spread of ARGs to and from humans. On the basis of recent studies, there is a need for environmental scientists and engineers to help address the issue of the spread of ARGs in the environment.

## 4.6 Nanomaterials

I close this section with nanomaterials – the concerns of the future and "real" emerging contaminants. Engineered nanomaterials are commonly defined as materials designed and produced to have structural features with at least one dimension of 100 nanometers or less. Such materials typically possess nanostructure-dependent properties (e.g., chemical, mechanical, electrical, optical, magnetic, biological), which make them desirable for commercial or medical applications. However, these same properties potentially may lead to nanostructure-dependent biological activity that differs from and is not directly predicted by the bulk properties of the constituent chemicals and compounds.

The potential for human and ecological toxicity associated with nanomaterials and ultrafine particles is a growing area of investigation as more nanomaterials and products are developed and brought into commercial use. To date, few nanotoxicology studies have addressed the effects of nanomaterials in a variety of organisms and environments. However, the existing research raises some concerns about the safety of nanomaterials and has led to increased interest in studying the toxicity of nanomaterials for use in risk assessment and protection of human health and the environment. A new field of nanotoxicology has been developed to investigate the possibility of harmful effects due to exposure to nanomaterials [146]. Nanotoxicology also encompasses the proper characterization of nanomaterials used in toxicity studies. Characterization has been important in differentiating between naturally occurring forms of nanomaterials, nano-scale byproducts of natural or chemical processes, and manufactured (engineered) nanomaterials. Because of the wide differences in properties among nanomaterials, each of these types of nanoparticles can elicit its own unique biological or ecological responses. As a result, different types of nanomaterials must be categorized, characterized, and studied separately, although certain concepts of nanotoxicology, primarily based on the small size, likely apply to all nanomaterials.

As materials reach the nanoscale, they often no longer display the same reactivity as the bulk compound. For example, even a traditionally inert bulk compound, such as gold, may elicit a biological response when it is introduced as a nanomaterial [147]. The earliest studies investigating the toxicity of nanoparticles focused on atmospheric exposure of humans and environmentally relevant species to heterogeneous mixtures of environmentally produced ultrafine particulate matter (having a diameter $< 100$ nm). These studies examined pulmonary toxicity associated with particulate matter deposition in the respiratory tract of target organisms [148–151]. Epidemiological assessments of the effects of urban air pollution exposure focusing on particulate matter produced as a byproduct of combustion events, such as automobile exhaust and other sources of urban air pollution, showed a link in test populations between morbidity and mortality and the amount of particulate matter [152, 153].

Laboratory-based studies have investigated the effects of a large range of ultrafine materials through in vivo exposures using various animal models as well as cell-culture-based in vitro experiments. To date, animal studies routinely show an increase in pulmonary inflammation, oxidative stress, and distal organ involvement upon respiratory exposure to inhaled or implanted ultrafine particulate matter. Tissue and cell culture analyses have also supported the physiological response seen in whole animal models and yielded data pointing to an increased incidence of oxidative stress, inflammatory cytokine production, and apoptosis in response to exposure to ultrafine particles [154–157]. These studies have also yielded information on gene expression and cell signaling pathways that are activated in response to exposure to a variety of ultrafine particle species ranging from carbon-based combustion products to transition metals. Polytetrafluoroethylene fumes in indoor air pollution are nano-sized highly toxic particles [158]. They elicit a severe inflammatory response at low inhaled particle mass concentrations, suggestive of an oxidative injury.

In contrast to the heterogeneous ultrafine materials produced incidentally by combustion or friction, manufactured nanomaterials can be synthesized in highly homogenous forms of desired sizes and shapes (e.g., spheres, fibers, tubes, rings, planes). Limited research on manufactured nanomaterials has investigated the interrelationship between the size, shape, and dose of a material and its biological effects, and whether a unique toxicological profile may be observed for these different properties within biological models. Typically, the biological activity of particles increases as the particle size decreases. Smaller particles occupy less volume, resulting in a larger number of particles with a greater surface area per unit mass and increased potential for biological interaction [159]. Recent studies have begun to categorize the biological response elicited by various nanomaterials both in the ecosystem and in mammalian systems. Although most current research has focused on the effect of nanomaterials in mammalian systems, some recent studies have shown the potential of nanomaterials to elicit a phytotoxic response in the ecosystem. In the case of alumina nanoparticles, one of the US market leaders for nano-sized materials, 99.6% pure nanoparticles with an average particle size of 13 nm were shown to cause root growth inhibition in five plant species [160].

Charge properties and the ability of carbon nanoparticles to affect the integrity of the blood-brain barrier as well as exhibit chemical effects within the brain have also been studied. Nanoparticles can overcome this physical and electrostatic barrier to the brain. In addition, high concentrations of anionic nanoparticles and cationic nanoparticles are capable of disrupting the integrity of the blood-brain barrier. The brain uptake rates of anionic nanoparticles at lower concentrations were greater than those of neutral or cationic formulations at the same concentrations. This work suggests that neutral nanoparticles and low concentration anionic nanoparticles can serve as carrier molecules providing chemicals direct access to the brain and that cationic nanoparticles have an immediate toxic effect at the blood-brain barrier [161, 162].

Tests with uncoated, water soluble, colloidal C60 fullerenes have shown that redox-active lipophilic carbon nanoparticles are capable of producing oxidative damage in the brains of aquatic species [161]. The bactericidal potential of C60 fullerenes was also observed in these experiments. This property of fullerenes has possible ecological ramifications and is being explored as a potential source of new antimicrobial agents [163]. Oxidative stress as a common mechanism for cell damage induced by nanoparticles and ultrafine particles is well documented; fullerenes are model compounds for producing superoxide. A wide range of nanomaterial species have been shown to create reactive oxygen species both in vivo and in vitro. Species which have been shown to induce free radical damage include the C60 fullerenes, quantum dots, and carbon nanotubes. Nanoparticles of various sizes and chemical compositions are able to preferentially localize in mitochondria where they induce major structural damage and can contribute to oxidative stress [164].

Quantum dots (QDs) such as CdSe QDs have been introduced as new fluorophores for use in bioimaging. When conjugated with antibodies, they are used for immunostaining due to their bright, photostable fluorescence. To date, there is not sufficient analysis of the toxicity of quantum dots in the literature, but some current studies point to issues of concern when these nanomaterials are introduced into biological systems. Recently published research indicates that there is a range of concentrations where quantum dots used in bioimaging have the potential to decrease cell viability, or even cause cell death, thus suggesting that further toxicological evaluation is urgently needed [165, 166]. However, the research also highlights the need to further explore the long-term stability of the coatings used, both in vivo and exposed to environmental conditions.

# 5 Discussion

## 5.1 Regulatory Perspective and Public Concerns

In 2004, the environmental campaign group World Wide Fund (WWF) tested the blood of government ministers from 13 EU Member States for chemicals that can negatively affect human health and wildlife. WWF found on average 37 out of the 103 tested substances in the ministers' blood [167]. Further, it is clear that the EU citizens are concerned. In a recent survey, the impact of chemicals used in everyday products came fifth in a list of 15 environmental issues of concern. When asked about which issue they feel they lack information, citizens cited chemicals first [168]. Do they have reason(s) to be concerned? Undoubtedly, the answer is positive – the overview of the "chemical world", which is in this chapter concentrated only to today's man-made emerging contaminants, clearly suggests that there are real human and environmental health problems that have to be addressed. Considering the issue of chemical contamination, all critical parties – regulators, risk managers, industry sector, politicians, and, finally, scientists – do not offer answers and solutions needed for citizens to be less concerned.

Contamination of water supplies is an evolving problem and will remain an issue as long as technological change continues. Some of the contaminants now being targeted by researchers may come out with a clean slate, while others may require additional scrutiny. One of the hopes of today's researchers is that more sophisticated science will help speed the process of identifying and remedying the problems, before damage to either human health or the environment occurs. In any case, science and regulation must continue to evolve and change, as it has been the case in the past few years, to respond to new needs presented by chemicals and our increasing knowledge of them. At present, however, regulatory communities are placed in a reactive, rather than proactive, position with respect to identifying contaminants and addressing public concern. The current lists of environmental pollutants evolved from those established in 1970s and are mainly focused on conventional "priority pollutants" often referred to as "persistent organic pollutants" (POPs). As was elaborated, these chemicals represent only a tiny part of potential pollutants [1, 2] and biological systems may obviously suffer exposure to many more chemicals stressors, only a small number of which is regulated. Therefore, only a small proportion of potentially hazardous chemicals is toxicologically evaluated, and even smaller number of them is officially regulated.

This position is further emphasized in situations where federal funding is provided only on a short-term basis and only for specifically identified research needs, which by definition are reactionary calls to fill data gaps. Although this approach generates short-term products for stakeholders, it often leads to fragmentary, low profile science. In the long term, such goal-oriented approach to environmental funding does not allow for exploratory research that can be used to anticipate future environmental issues. Unfortunately, in the US, for example, there is no competitive funding scheme for the discovery of new contaminants. In addition, no cohesive plan exists to proactively screen and identify all contaminants of potential concern. On the other hand, both Canada and the EU are actively developing plans that will place them in positions from which they can anticipate future environmental issues. The Registration, Evaluation, and Authorization of Chemicals (REACH) regulation in the EU is a good example [169]. Entered into force in June 2007, it requires that manufacturers of substances and formulators register and provide prescribed (eco)toxicological data for all substances with a volume >1 metric ton per year. In contrast, the USEPA has taken a different tack by sponsoring a voluntary program called the High Production Volume (HPV) Challenge Program [170]. Since the program's inception in 1998, >2200 chemicals have been "adopted" by chemical manufacturers and importers. Unfortunately, this number is small in comparison with the number of chemicals included under REACH, and >200 HPV chemicals are still without the promise of toxicity testing.

## 5.2 (Eco)toxicological Constraints

As may be realized from this overview, (eco)toxicologists often seems to know too little too late, and are far too slow to respond to numerous chemicals that enter the market every day. Moreover, most of (eco)toxicological testing is done using traditional acute toxicity test protocols. As was reliably demonstrated with pharmaceuticals, acute toxicity cannot always serve as a reliable proxy for chronic toxicity effects encountered in real environmental situations. Certain substances may elicit adverse effects weeks, months or years after exposure. Carcinogenicity is a classical example – an ultimate adverse outcome difficult to characterize regarding causal connections. Consequently, chronic exposure assessments cannot be avoided and proper toxicological characterization will probably continue to be a time-consuming process.

The array of chemicals in use will likely continue to diversify and grow with changing use patterns in human populations and animal production facilities. Rapid developments in the pharmaceutical industry will also continue to quickly add to the vast number of chemicals already entering the environment. Due to the ever-increasing potency and specificity of pharmaceuticals, new substances may be of even greater concern for the environment. New approaches for testing and new ways of thinking about new materials are also necessary. The diverse routes of exposure, including inhalation, dermal uptake, ingestion, and injection, can present unique toxicological outcomes that vary with the physicochemical properties of the nanoparticles in question.

The likelihood of constantly introducing new chemicals to commerce pose inevitable doubts as to whether the chemical-by-chemical approach to toxicological testing and regulation of water pollutants will continue to be sustainable. In the past, studies have focused on the effects of single chemicals because chemicals are usually regulated singly. However, chemicals are always present as complex mixtures, thus some might say the regulation approach is naïve. Thus scientists are increasingly focusing on the toxicity of mixtures of chemicals, acknowledging that the toxicity expressed may be a result of additive or multiplicative effects, depending on interactions with other chemicals present in the environment. Furthermore, the issue becomes even more complex taking into account potential toxicity of numerous metabolites being generated from parent compounds.

An alternative approach, formalized as "toxicity apportionment" has been recently proposed [2]. The main principle of this approach would be to assign toxicity according to the total numbers of stressors present, without the need to know their identities in advance. The apportionment approach is especially valuable in accounting for all toxicants sharing the same mechanism of action. As was proposed, water monitoring programs based on that framework should utilize biomarkers and biotests designed around evolutionary biochemical features and mechanisms of action rather than individual chemical entities. This approach may indeed be the best way to simultaneously account for multitude of contaminants having the same mechanism of action, chemicals newly introduced to the market, and pollutants of the future. Looking from the cost-benefit side and trying to obtain relevant toxicological answers in a short time, an efficient screening protocol similar to that shown in Fig. 1, may be based on the extensive use of a series of small scale and in vitro biotests, used to rapidly and sensitively screen for the presence of contaminants of concern, including emerging contaminants, addressing both acute and chronic toxicity and utilizing test species on different levels of biological organizations. It can be used for testing of single chemicals and complex environmental samples. Such a battery of mechanism-based bioassays could be easily incorporated into monitoring efforts.

Nevertheless, whilst they are able to indicate the presence of certain groups of substances in well understood media based on a toxic response, caution is needed in broadening the application of in vitro tests to complex media such as effluents. In vitro tests that typically utilize genetically modified cells, yeast, or bacterial strains, demonstrate promising advantages such as speed, low cost and the ability to give an indication of specific toxicity that usually is not expressed in acute toxicity tests. However, they have to date only been used to a limited extent on effluents, making interpretation of test results difficult or in some cases impossible. Additional experience will be essential to improve the interpretation of test results and their relationship to actual environmental impacts. At present, even the best validated in vitro bioassays are only suitable as an initial screening step to prioritize effluents or effluent fractions for further study. In vivo tests with carefully selected indicator species are more appropriate to assess direct toxicity and should preferably be used for risk assessment purposes. Furthermore, bioassays can give both false negative and false positive results. False negative results may fail to highlight real health or environmental risks; false positives may imply health or environmental risks where, in fact, there are none. Due to the high sensitivity of these tests, false positives are likely when applied to complex mixtures like effluents.

Therefore, methods are now available that detect tiny quantities of chemicals which may potentially be hazardous. However, questions remain about which chemicals are responsible when positive results are obtained from drinking water, wastewater, freshwater and seawater, soil, mud, or any other sample. For effluents, it is a challenge that samples generally contain many compounds,

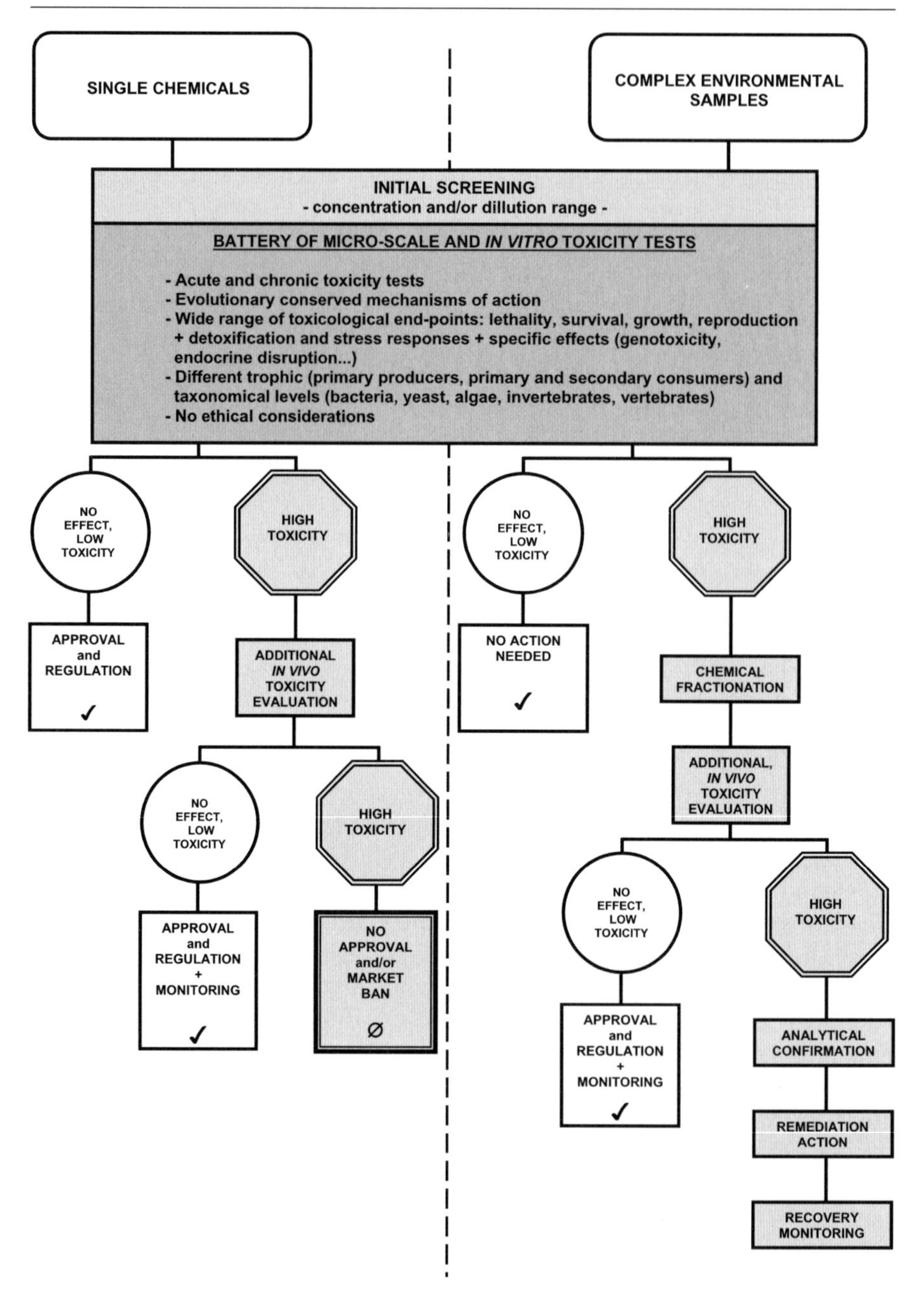


**Fig. 1** Flow chart presentation of the possible (eco)toxicological protocol for rapid screening and characterization of single chemicals and complex environmental samples

resulting in false positives being frequently obtained. In the case of a positive response, the sample may be split up and analytical methods used to try and identify the responsive chemicals. Since these tests are highly sensitive and specific to the cell type used, the relevance of positive results for other species, living animals and longer-term exposures is the subject of ongoing studies. Consequently, a positive assay result should always be complemented with an in vivo assay and analytical detection to confirm the response. Only additional studies – coupled with a proper risk analysis, taking exposure into account – can confirm if the response indicates a genuine environmental risk.

Finally, regardless of the obstacles described, the most important concern regarding the exposure of aquatic and terrestrial organisms to emerging contaminants may be our inability to detect subtle health effects – imperceptible changes ranging from modification or reversal of attraction, behavioral changes related to feeding, matting, predator avoidance, or directional sensing. The changes we may see on the surface would simply be attributed to natural adaptation or any other form of natural changes. This concept of subtle changes, formalized at first by Kurelec in 1993 [171] as the genotoxic disease syndrome (GDS) was described as gradual accumulation of a wide spectrum of toxic events, none of which alone results in an easily detected adverse outcome. However, the final outcome would be an ultimate and often irreversible biological damage – species loss and decrease in biodiversity, unexpected and unexplained due to our inability to detect and act timely. These subtle, cumulative effects could make current toxicity-directed screening strategies largely useless in any effort to test waste effluents for toxic end points. At the moment, unfortunately, in the field of environmental toxicology there is no sound scientific answer to this critical issue. The raise of -omics techniques, however, especially genomics approach based on high-density microarray methodology, may be a future solution theoretically capable of detecting even subtle changes in gene expression patterns.

## 6 Conclusions and Future Directions

In this article, we briefly summarized major human end environmental health effects related to the most prominent categories of emerging contaminants, along with critical (eco)toxicological drawbacks and prerequisites needed for environmentally accountable risk characterization. The most important messages from this chapter, those we want for any reader to take into consideration are:

1. The threat posed by numerous emerging contaminants present in industrial and municipal waste is serious, poorly characterized, and should not be underestimated;

2. The research capacity of (eco)toxicology is at the moment far beyond capacity of analytical chemistry to detect new, emerging contaminants, and even more distant from the capacity of industry sector to design and introduce new chemical entities, likely "emerging" contaminants of the future;
3. Chronic, low-level exposure assessments do not have any scientifically sound alternative and should represent obligatory part of (eco)toxicity characterization of single chemicals and complex environmental mixtures;
4. The necessary improvements in the field of (eco)toxicology will not be possible without major shift in the regulatory arena, including significant changes in the environmental funding schemes.

Countries that adopt proactive approaches, such as the EU REACH initiative, will be afforded distinct environmental, economic, and scientific advantages, because they will be better serving human and nonhuman populations and ecosystems, with tangible savings to the healthcare and environment protection costs. Without the adoption of proactive plans to identify contaminants before they emerge, regulatory communities that remain in reactionary modes will be unable to fully serve the needs of the populations they represent.

**Acknowledgements** Financial support by the EU 6th Framework Specific Targeted Research Project: *Reduction of environmental and health risks, posed by Emerging Contaminants, through advanced treatment of municipal and industrial wastes* (*EMCO*; Contract No. INCO CT 2004-509188) is acknowledged. In addition, this work was partially supported by the Ministry of Science, Education and Sports of the Republic of Croatia, Project No: 098-0982934-2745 and 098-0982934-2712.

## References

1. Daughton CG (2005) Renew Res J 21:6
2. Daughton CG (2004) Environ Imp Assess Rev 24:711
3. Boxall ABA, Sinclair CJ, Fenner K, Kolpin D, Maund SJ (2004) Environ Sci Technol 38:368A
4. The Human Toxome Project, Environmental Working Group, Washington DC, USA. Available at: http://www.ewg.org/sites/humantoxome/about/. Accessed 11 March, 2008
5. The Collaborative on Health and the Environment (CHE) Toxicant and Disease Database. Available et: http://database.healthandenvironment.org/. Accessed 11 March, 2008
6. USEPA (1980) Appendix B – Guidelines for Deriving Water Quality Criteria for the Protection of Aquatic Life and Its Uses. Federal Register, vol 45, No 231, November 28, 1980
7. Mount DI, Stephan CE (1967) Trans Am Fish Soc 96:185
8. McKim JM (1977) J Fish Res Board Can 34:1148
9. Macek KJ, Sleight BH (1977) Utility of toxicity tests with embryos and fry of fish in evaluating hazards associated with the chronic toxicity of chemicals to fishes. In: Mayer FL, Hamelink JL (eds) Aquatic Toxicology and Hazard Evaluation, ASTM STP 634. American Society for Testing and Materials, Philadelphia, p 137

10. Woltering DM (1984) Aquat Toxicol 5:1
11. USEPA (1981) In situ acute/chronic toxicological monitoring of industrial effluents for the NPDES biomonitoring program using fish and amphibian embryo/larval stages as test organisms. OWEP-82-00l. Office of Water Enforcement and Permits, US Environmental Protection Agency, Washington, DC 20460
12. Birge WJ, Black JA, Westerman AG (1985) Environ Toxicol Chem 4:807
13. Norberg TJ, Mount DI (1985) Environ Toxicol Chem 4:711
14. Birge WJ, Black JA, Ramey BA (1981) The reproductive toxicology of aquatic contaminants. In: Saxena J, Fisher F (eds) Hazard Assessments of Chemicals, Current Developments, vol 1. Academic Press, New York, p 59
15. Espejo R (2002) J Chromatogr A 976:335
16. Oros DR, Jarman WM, Lowe T, David N, Lowe S, Davis JA (2003) Mar Pollut Bull 46:1102
17. Saito I, Onuki A, Seto H (2004) Indoor Air 14:325
18. Laws SC, Carey SA, Ferrell JM, Bodman GJ, Cooper RL (2000) Toxicol Sci 54:154
19. Bechi N, Ietta F, Romagnoli R, Focardi S, Corsi I, Buffi C, Paulesu L (2006) Toxicol Sci 93:75
20. Kimura N, Kimura T, Suzuki M, Totsukawa K (2006) J Reprod Dev 52:789
21. McClusky LM, De Jager C, Bornman MS (2006) Toxicol Sci 95:249
22. Bistodeau TJ, Barber LB, Bartell SE, Cediel RA, Grove KJ, Klaustermeier J, Woodard JC, Lee KE, Schoenfuss HL (2006) Aquat Toxicol 79:268
23. Vom Saal F, Hughes C (2005) Environ Health Perspect 113:926
24. Howdeshell KL, Hotchkiss AK, Thayer KA, Vandenbergh JG, Vom Saal FS (1999) Nature 401:763
25. Sakaue LM, Ohsako S, Ishimura R, Kurosawa S, Kurohmaru M, Hayashi Y, Aoki Y (2001) J Occup Health 43:185
26. Al-Hiyasat AS, Darmani H, Elbetieha AM (2002) Eur J Oral Sci 110:163
27. Palanza PL, Howdeshell KL, Parmigiani S, Vom Saal FS (2002) Environ Health Perspect 110:415
28. Schonfelder G, Flick B, Mayr E, Talsness C, Paul M, Chahoud I (2002) Neoplasia 4:98
29. Wetherill YB, Petre CE, Monk KR, Puga A, Knudsen KE (2002) Mol Cancer Ther 1:515
30. Sugita-Konishi Y, Shimurab S, Nishikawab T, Sunagab F, Naitob H, Suzuki Y (2003) Toxicol Lett 136:217
31. Kabuto H, Amakawa M, Shishibori T (2004) Life Sci 74:2931
32. Della Seta D, Minder I, Dessì-Fulgheri F, Farabollini F (2005) Brain Res Bull 65:255
33. Markey CM, Wadia PR, Rubin BS, Sonnenschein C, Soto AM (2005) Biol Reprod 72:1344
34. Porrini S, Bellonia V, Della Seta D, Farabollini F, Giannelli G, Dessì-Fulgheri F (2005) Brain Res Bull 65:261
35. Timms BG, Howdeshell KL, Barton L, Bradley S, Richter CA, Vom Saal FS (2005) Proc Nat Acad Sci USA 102:7014
36. Alonso-Magdalena P, Morimoto S, Ripoll C, Fuentes E, Nadal A (2006) Environ Health Perspect 114:106
37. Olea N, Pulgar R, Pérez P, Olea-Serrano F, Rivas A, Novillo-Fertrell A, Pedraza V, Soto AM, Sonnenschein C (1996) Environ Health Perspect 104:298
38. Hanaoka T, Kawamura N, Hara K, Tsugane S (2002) Occup Environ Med 59:625
39. European Comission (2002) Study on the scientific evaluation of 12 substances in the context of endocrine disrupter priority list of actions – Final Report. WRc-NSF, UK. Available at: http://ec.europa.eu/environment/endocrine/documents/wrc_report.pdf #page=29

40. Inoue K, Yamaguchi A, Wada M, Yoshimura Y, Makino T, Nakazawa H (2001) J Chromatogr B 765:121
41. IARC (1999) Bisphenol A diglycidyl ether. IARC Monogr Eval Carcinog Risks Hum 71:1285
42. Warbrick EV, Dearman RJ, Ashbya J, Schmezer P, Kimber I (2001) Toxicology 163:63
43. Uter W, Rühl R, Pfahlberg A, Geier J, Schnuch A, Gefeller O (2004) Ann Occup Hyg 48:21
44. Ramilo G, Valverde I, Lago J, Vieites J, Cabado A (2006) Arch Toxicol 80:748
45. Satoh K, Ohyama K, Aoki N, Iida M, Nagai F (2004) Food Chem Toxicol 42:983
46. Viberg H, Fredriksson A, Jakobsson E, Örn U, Eriksson P (2003) Toxicol Sci 76:112
47. Viberg H, Johansson N, Fredriksson A, Eriksson J, Marsh G, Eriksson P (2006) Toxicol Sci 92:211
48. Eriksson P, Jakobsson E, Fredriksson A (2001) Environ Health Perspect 109:903
49. Viberg H, Eriksson P (2007) Neurotoxicology 28:136
50. Birnbaum LS, Staskal DF, Diliberto JJ (2003) Environ Int 29:855
51. Dasgupta PK, Dyke JV, Kirk AB, Jackson AW (2006) Environ Sci Technol 40:6608
52. USEPA (2005) Unregulated Contaminant Monitoring Program. US Environmental Protection Agency. Available at: http://www.epa.gov/safewater/ucmr/index.html. Accessed 11 March, 2008
53. EWG (2001) Rocket Science: Perchlorate and the toxic legacy of the cold war. Environmental Working Group, US. Available at: http://www.ewg.org/reports/rocketscience. Accessed 11 March, 2008
54. EWG (2003) Rocket Fuel in Drinking Water: New Studies Show Harm From Much Lower Doses. Environmental Working Group, US. Available at: http://www.ewg.org/node/8445. Accessed 11 March, 2008
55. EWG (2003). Suspect Salads: Toxic rocket fuel found in samples of winter lettuce. Environmental Working Group, US. Available at: http://www.ewg.org/reports/suspectsalads/ . Accessed 11 March, 2008
56. Kirk A, Smith EE, Tian K, Anderson TA, Dasgupta PK (2003) Environ Sci Technol 37:4979
57. Kirk A, Martinelango K, Tian K, Dutta A, Smith EE, Dasgupta PK (2005) Environ Sci Technol 39:2011
58. Sanchez CA, Crump KS, Krieger RI, Khandaker NR, Gibbs JP (2005) Environ Sci Technol 39:9391
59. Blount BC, Pirkle JL, Osterloh JD, Valentin-Blasini L, Caldwell KL (2006) Environ Health Perspect 114:1865
60. El Aribi H, Le Blanc YJC, Antonsen S, Sakuma T (2006) Anal Chim Acta 567:39
61. Haddow JE, Palomaki GE, Allan WC, Williams JR, Knight GJ, Gagnon J, O'Heir CE, Mitchell ML, Hermos RJ, Waisbren SE, Faix JD, Klein RZ (1999) N Engl J Med 341:549
62. Pop VJ, Kuijpens JL, Van Baar AL, Verkerk G, Van Son MM, De Vijlder JJ, Vulsma T, Wiersinga WM, Drexhage HA, Vader HL (1999) Clin Endocrinol 50:149
63. Auer C (2000) May 16, 2000 email message from Charles Auer (EPA) to OECD. EPA administrative record number AR226-0629
64. Hagen DF, Belisle J, Johnson JD, Venkateswarlu P (1981) Anal Biochem 118:336
65. Dinglasan MJ, Ye Y, Edwards EA, Mabury SA (2004) Environ Sci Technol 38:2857
66. Kannan K, Choi J-W, Isekic N, Senthilkumar K, Kima DH, Masunagac S, Giesy JP (2002) Chemosphere 49:225
67. Olsen GW, Burris JM, Lundberg JK, Hansen KJ, Mandel JH, Zobel LR (2002) Final Report: Identification of fluorochemicals in human sera. III. Pediatric participants in a group A streptococci clinical trial investigation US EPA Administrative Record

AR226-1085: Study conducted by Corporate Occupational Medicine. Medical Department, 3M Company, 220-3W-05, St Paul, MN, USA

68. Burris JM, Lundberg JK, Olsen GW, Simpson C, Mandel JH (2002) Determination of serum half-lives of several fluorochemicals. Interim Report No 2, St Paul, MN, 3M Company, US EPA docket AR-226-1086. US Environmental Protection Agency, Washington, DC
69. 3M (2000) Composite analytical laboratory report on the quantitative analysis of fluorochemicals in environmental samples. EPA Administrative Record AR226-0202, 3M
70. 3M (2001) Executive Summary: Environmental monitoring – multi-city study water, sludge, sediment, POTW effluent and landfill leachate samples
71. 3M (2001) Final Report, A longitudinal analysis of serum perfluorooctanesulfonate (PFOS) and perfluorooctanoate (PFOA) in relation to clinical chemistry, thyroid hormone, hematology and urinalysis results from male and female employee participants of the 2000
72. Stevenson CN, MacManus-Spencer LA, Luckenbach T, Luthy RG, Epel D (2006) Environ Sci Technol 40:5580
73. Wolff MS, Teitelbaum SL, Windham G, Pinney SM, Britton JA, Chelimo C, Godbold J, Biro F, Kushi LH, Pfeiffer CM, Calafat AM (2007) Environ Health Perspect 115:116
74. CDC (2005) National Report on Human Exposure to Environmental Chemicals. Centers for Disease Control, USA
75. CERHR (2000) NTP-CERHR expert panel report on di (2-ethylhexyl) phthalate (DEHP). Center for the Evaluation of Risks to Human Reproduction, USA
76. Duty SM, Barr DB, Brock JW, Ryan L, Chen Z, Herrick RF, Christiani DC, Hauser R (2003) Epidemiology 14:269
77. Duty SM, Calafat AM, Silva MJ, Brock JW, Ryan L, Chen Z, Overstreet J, Hauser R (2004) J Androl 25:293
78. Duty SM, Calafat AM, Manori SJ, Ryan L, Hauser R (2005) Hum Reprod 20:604
79. Bornehag C, Sundell J, Weschler CJ (2004) Environ Health Perspect 112:1393
80. Darnerud PO, Eriksen GS, Jóhannesson T, Larsen PB, Viluksela M (2001) Environ Health Perspect 109:49
81. Darnerud PO (2003) Environ Int 29:841
82. Hale RC, Alaee M, Manchester-Neesvig JB, Stapleton HM, Ikonomou MG (2003) Environ Int 29:771
83. Schecter A, Papke O, Tung KC, Joseph J, Harris TR, Dahlgren J (2005) J Occup Environ Med 47:199
84. Sjodin A, Patterson DG Jr, Bergman A (2001) Environ Sci Technol 35:3830
85. Sjodin A, Patterson DG Jr, Bergman A (2003) Environ Int 29:829
86. Sjodin A, McGahee EE III, Zhang Y, Turner WE, Slazyk B, Needham LL, Patterson DG Jr (2004) Environ Health Perspect 112:654
87. Vogelgesang J, Their HP (1986) Z Lebensm-Unters Forsch 182:400
88. Vinitskayaa H, Lachowicz A, Kilanowicz A, Bartkowiak J, Zylinska L (2005) Environ Toxicol Pharmacol 20:450
89. Van de Plassche EJ, Schwegler AM (2002) Polychlorinated naphthalenes. Dossier prepared for the third meeting of the UN-ECE Ad hoc Expert Group on POPs. Royal Haskoning report L0002.A0/R0010/EVDP/TL
90. Fromme H, Otto T, Pilz K (2001) Water Res 35:121
91. Peck AM, Hornbuckle KC (2004) Environ Sci Technol 38:367
92. Duedahl-Olesen L, Cederberga T, Høgsbro Pedersen K, Højgårdc A (2005) Chemosphere 61:422

93. Kannan K, Reiner JL, Yuna S, Perrotta EE, Tao L, Johnson-Restrepo B, Rodan BD (2005) Chemosphere 61:693
94. Peck AM, Linebaugh EK, Hornbuckle KC (2006) Environ Sci Technol 40:5629
95. Rimkus GG, Wolf M (1996) Chemosphere 33:2033
96. Liebl B, Mayer R, Ommer S, Sönnichsen C, Koletzko B (2000) Adv Exp Med Biol 478:289
97. Hutter HP, Wallner P, Moshammer H, Hartl W, Sattelberger R, Lorbeer G, Kundi M (2005) Chemosphere 59:487
98. TNO (2005) Man-made chemicals in maternal and cord blood. TNO Built Environment and Geosciences, Apeldoorn, The Netherlands, p 1
99. Maekawa A, Matsushima Y, Onodera H, Shibutani M, Ogasawara H, Kodama Y, Kurokawa Y, Hayashi Y (1990) Food Chem Toxicol 28:581
100. Apostolidis S, Chandra T, Demirhan I, Cinatl J, Doerr HW, Chandra A (2002) Anticancer Res 22:2657
101. Eisenhardt S, Runnebaum B, Bauer K, Gerhard I (2001) Environ Res 87:123
102. Parker RD, Buehler EV, Newmann EA (1986) Contact Dermatitis 14:103
103. Hayakawa R, Hirose O, Arima Y (1991) J Dermatol 18:420
104. Seinen W, Lemmen JG, Pieters RHH, Verbruggen EMJ, Van der Burg B (1999) Toxicol Lett 111:161
105. Chou YJ, Dietrich DR (1999) Toxicol Lett 111:27
106. Bitsch N, Dudas C, Körner W, Failing K, Biselli S, Rimkus G, Brunn H (2002) Arch Environ Contam Toxicol 43:257
107. Gomez E, Pillon A, Fenet H, Rosain D, Duchesne MJ, Nicolas JC, Balaguer P, Casellas C (2005) J Toxicol Environ Health A 68:239
108. Schreurs RH, Sonneveld E, Jansen JHJ, Seinen W, Van der Burg B (2005) Toxicol Sci 83:264
109. Schreurs RH, Sonneveld E, Van der Saag PT, Van der Burg B, Seinen W (2005) Toxicol Lett 156:261
110. Luckenbach T, Epel D (2005) Environ Health Perspect 113:17
111. Daughton CG, Ternes TA (1999) Environ Health Perspect 107:907
112. Adolfsson-Erici M, Parkkonen J, Sturve J (2002) Chemosphere 46:1485
113. Balmer ME, Poiger T, Droz C, Romanin K, Bergqvist P-A, Muller MD, Buser H-R (2004) Environ Sci Technol 38:390
114. Buser HR, Müller MD, Poiger T, Balmer ME (2002) Environ Sci Technol 36:221
115. Lores M, Llompart M, Sanchez-Prado L, Garcia-Jares C, Cela R (2005) Anal Bioanal Chem 381:1294
116. Kolpin D (2002) Environ Sci Technol 36:1202
117. Rule KL, Ebbet VR, Vikesland P (2005) Environ Sci Technol 39:3176
118. Orvos D, Versteeg VD, Inauen J, Capdevielle M, Rothenstein A, Cunningham V (2002) Environ Toxicol Chem 21:1338
119. Veldhoen N, Skirrow RC, Osachoff H, Wigmore H, Clapson DJ, Gunderson MP, Van Aggelen G, Helbing CC (2006) Aquat Toxicol 80:217
120. Foran CM, Bennett ER, Benson WH (2000) Marine Environ Res 50:153
121. Bhargava HN, Leonard PA (1996) Am J Infect Control 24:209
122. Russell AD (2002) Am J Infect Control 30:495
123. Ternes TA (1998) Water Res 32:3245
124. Ternes TA, Meisenheimer M, McDowell D, Sacher F, Brauch HJ, Haist-Gulde B, Preuss G, Wilme U, Zulei-Seibertet N (2002) Environ Sci Technol 36:3855
125. Webb S, Ternes T, Gibert M, Olejniczaket K (2003) Toxicol Lett 142:157
126. Ternes T, Kreckel P, Muelleret J (1999) Sci Total Environ 225:91

127. Ternes T, Stumpf M, Mueller J, Haberer K, Wilken RD, Servos M (1999) Sci Total Environ 225:81
128. Metcalfe CD, Koenig BG, Bennie DT, Servos M, Ternes TA, Hirschet R (2003) Environ Toxicol Chem 22:2872
129. Metcalfe CD, Miao XS, Koenig BG, Strugeret J (2003) Environ Toxicol Chem 22:2881
130. Halling-Sørensen B, Nors Nielsen S, Lanzky PF, Ingerslev F, Holten Lützhøft HC, Jørgensen SE (1998) Chemosphere 36:357
131. Hewitt LM, Servos MR (2001) Water Qual Res J Can 36:191
132. Metcalfe CD, Metcalfe TL, Kiparissis Y, Koenig BG, Khan C, Hughes RJ, Croley TR March RE, Potteret T (2001) Environ Toxicol Chem 20:297
133. Pharmaceuticals in the Environment, Information for Assessing Risk website. National Centers for Coastal Ocean Science, Center for Coastal Environmental Health and Biomolecular Research. Available et: http://www.chbr.noaa.gov/peiar/default.aspx. Accessed 11 March, 2008
134. Hirsch R, Ternes TA, Haberer K, Kratz KL (1999) Sci Total Environ 225:109
135. Damstra T, Barlow S, Bergman A, Kavlock R, Van der Kraak G (2002) Global assessment of the state of the science of endocrine disruptors. WHO/PCS/EDC/02.2
136. Fent K, Weston AA, Caminada D (2006) Aquat Toxicol 76:122
137. Boxall ABA, Kolpin DW, Halling-Sorensen B, Tolls J (2003) Environ Sci Technol 37:286A
138. Campagnolo ER, Johnson KR, Karpati A, Rubing CS, Kolpin DW, Meyer MT, Esteban JE, Currier RW, Smith K, Thu KM, McGeehin M (2002) Sci Total Environ 299:89
139. Thurman EM, Dietze JE, Scribner EA (2002) Occurrence of antibiotics in water from fish hatcheries. US Geological Survey Fact Sheet 120-02, p 4
140. Albinana-Gimenez N, Clemente-Casares P, Bofill-Mas S, Hundesa A, Ribas F, Girones R (2006) Environ Sci Technol 40:7416
141. Jiang SC (2006) Environ Sci Technol 40:7132
142. Hoge CW, Breiman RF (1991) Epidemiol Rev 13:329
143. Berk SG, Gunderson JH, Newsome AL, Farone AL, Hayes BJ, Redding KS, Uddin N, Williams EL, Johnson RA, Farsian M, Reid A, Skimmyhorn J, Farone MB (2006) Environ Sci Technol 40:7440
144. Abulreesh HH, Paget TA, Goulder R (2006) Environ Sci Technol 40:7122
145. Pruden A, Pei R, Storteboom H, Carlson KH (2006) Environ Sci Technol 40:7445
146. Donaldson K, Stone V, Tran CL, Kreyling W, Borm PJ (2004) Occup Environ Med 61:727
147. Goodman CM, McCusker CD, Yilmaz T, Rotello VM (2004) Bioconjugate Chem 15:897
148. Cheng YS, Hansen GK, Su YF, Yeh HC, Morgan KT (1990) Toxicol Appl Pharmacol 106:222
149. Bermudez E, Mangum JB, Wong BA, Asgharian B, Hext PM, Warheit DB, Everitt JI (2004) Toxicol Sci 77:347
150. Ferin J (1994) Toxicol Lett 72:121
151. Oberdorster G, Oberdorster E, Oberdorster J (2005) Environ Health Perspect 113:823
152. MacNee W, Donaldson K (2000) Monaldi Arch Chest Dis 55:135
153. Oberdorster G, Gelein RM, Ferin J, Weiss B (1995) Inhal Toxicol 7:111
154. Barlow PG, Donaldson K, MacCallum J, Clouter A, Stone V (2005) Toxicol Lett 155:397
155. Brown DM, Donaldson K, Borm PJ, Schins RP, Dehnhardt M, Gilmour P, Jimenez LA, Stone V (2004) Am J Physiol Lung Cell Mol Physiol 286:L344
156. Hetland RB, Cassee FR, Refsnes M, Schwarze PE, Lag M, Boere AJ, Dybing E (2004) Toxicol In Vitro 18:203

157. Stone V, Tuinman M, Vamvakopoulos JE, Shaw J, Brown D, Petterson S, Faux SP, Borm P, MacNee W, Michaelangeli F, Donaldson K (2000) Eur Respir J 15:297
158. De Hartog JJ, Hoek G, Peters A, Timonen KL, Ibald-Mulli A, Brunekreef B, Heinrich J, Tiittanen P, Van Wijnen JH, Kreyling W, Kulmala M, Pekkanen J (2003) Am J Epidemiol 157:613
159. Oberdorster G (1996) Inhal Toxicol 8:73
160. Warheit DB (2004) Mater Today 7:32
161. Oberdorster E (2004) Environ Health Perspect 112:1058
162. Lockman PR, Koziara JM, Mumper RJ, Allen DD (2004) J Drug Target 12:635
163. Yamakoshi YN, Yagami T, Sueyoshi S, Miyata N (1996) J Org Chem 61:7236
164. Li N, Sioutas C, Cho A, Schmitz D, Misra C, Sempf J, Wang M, Oberley T, Froines J, Nel A (2003) Environ Health Perspect 111:455
165. Lovric J, Bazzi HS, Cuie Y, Fortin GR, Winnik FM, Maysinger D (2005) J Mol Med 83:377
166. Shiohara A, Hoshino A, Hanaki K, Suzuki K, Yamamoto K (2004) Microbiol Immunol 48:669
167. WWF (2004) Bad blood? A Survey of chemicals in the blood of European ministers (http://www.worldwildlife.org/toxics/pubs/badblood.pdf). Accessed 11 March, 2008
168. The attitudes of European citizens toward the environment, Special Eurobarometer 217/Wave 62.1, conducted in November 2004, published in April 2005 (http://europa.eu.int/comm/environment/barometer/index.htm).
169. Registration, Evaluation, Authorisation and Restriction of Chemicals (REACH) system in the EU. Available at: http://ecb.jrc.it/reach/
170. High Production Volume (HPV) Challenge Program of the USEPA. Available at: http://www.epa.gov/hpv/. Accessed 11 March, 2008
171. Kurelec B (1993) Mar Environ Res 35:341
172. Chemical Abstracts Service (CAS) of The American Chemical Society. Available at: http://www.cas.org/index.html. Accessed 11 March, 2008

Hdb Env Chem Vol. 5, Part S/1 (2008): 143–168
DOI 10.1007/698_5_101

Published online: 24 January 2008

# Traceability of Emerging Contaminants from Wastewater to Drinking Water

M. Huerta-Fontela[1,2] (✉) · F. Ventura[1]

[1]AGBAR-Aigües de Barcelona, Av. Diagonal 211, 08018 Barcelona, Spain
*mhuerta@agbar.es*

[2]Department of Analytical Chemistry, University of Barcelona, Av. Diagonal 647, 08028 Barcelona, Spain



**Abstract** Due to the incomplete elimination of some human contaminants during wastewater treatment, some of these compounds can be found in surface waters or groundwaters which are used as raw waters for drinking water production. The treatment efficiency to completely eliminate these emerging contaminants or to partially remove them will determine the quality of the final treated water. Up to today, few studies have been performed to evaluate the efficiency of the usual drinking water treatments in eliminating emerging contaminants. Moreover, every day new potential emerging contaminants are discovered and new disinfection by-products are also generated during treatment, with a total ignorance of their potential toxicity or effect on human health. In this chapter, a summary of the state of the art of emerging contaminant occurrence and elimination during drinking water processes at the bench scale or real scale is presented. A study of the presence and elimination of a new group of human contaminants, susceptible to being considered as a new emerging contaminant group, in a real drinking water treatment plant in Spain has also been included.

**Keywords** Carbon · Disinfection by-products · Drinking water · Emerging contaminants · Illicit drugs · Oxidation · Sorption

# 1
# Emerging Contaminants in Drinking Water

The occurrence of emerging contaminants (i.e., human and veterinary drugs, surfactants, textile dyes, algal toxins, etc.) in wastewaters [1–7] and surface waters [1, 8–14] and their removal during conventional treatments has been widely evaluated in recent years. Several organic pollutants, e.g., pharmaceuticals, are not quantitatively eliminated by wastewater treatment and "survive" natural attenuation processes in surface waters. Therefore, the occurrence of these contaminants in these resources can have a negative impact on the quality of drinking water and, perhaps, produce adverse health effects. The incidence of these organic micropollutants in raw water and their elimination during drinking water treatment, as well as the formation of disinfection byproducts (DBPs), are issues related to the quality of raw resources and water supplies. Compared to wastewater treatment plants, much less is known about the behavior of these compounds in drinking water treatment plants (DWTPs). In Table 1, a summary of some of the emerging contaminants detected in drinking water is displayed. The lack of systematic monitoring programs or the fact that they are present at fluctuating concentrations near the analytical method detection limits (some of these compounds usually occur in the low ng/L range) could be some reasons to explain the relatively little knowledge of the occurrence of these compounds in drinking water production [15]. However, several studies have found that the removal of emerging contaminants (mostly polar compounds) during drinking water treatment is incomplete. In 1993, clofibric acid, the active metabolite of some blood lipid regulators such as clofibrate, etofyllin clofibrate, and etofibrate, was found in Berlin tap water at high concentrations above 165 ng/L. Further studies, showed a direct correlation between bank filtration and artificial groundwater enrichment (used by a particular waterworks in drinking water production) and the concentrations of this drug in treated water [16, 17]. The same authors also detected the presence of propylphenazone and diclofenac in finished drinking water. Clofibric acid occurrence was also investigated in drinking waters from southern California [18]. This compound was not found in the samples analyzed; however, ibuprofen, triclosan, several phthalates, and additives were detected in samples of finished drinking water. These authors also performed a seasonal study to evaluate the performance of these compounds through time, concluding that higher concentrations in raw waters were detected between August and November (dry season), probably related to lower flow rates.

Boyd et al. [19] examined the occurrence of nine pharmaceuticals and personal care products (PPCPs) and endocrine disrupting compounds (EDCs), including clofibric acid, anti-inflammatories, analgesics, antibiotics, and hormones, in drinking water from the USA and Canada, and none of them was found in the finished drinking water. The presence of several pharmaceuticals, including lipid regulators, analgesics, anti-inflammatories, and their metabolites, was also evaluated in tap water from Cologne (Germany) [20]. Most of these compounds were found in the rivers and ponds analyzed but none of the eight selected drinking water samples showed the presence of the studied pharmaceuticals. Nevertheless, some hormones and antibiotics were detected in final drinking waters from the USA and Italy in recent years [21, 22]. Thus, McLachlan et al. [21] showed the presence of 17β-estradiol, estriol, and nonylphenol in final drinking waters. Regarding antibiotics, Perret et al. [22] studied the occurrence of 11 sulphonamide compounds (SAs) in mineral and municipal drinking waters from Italy. Concentrations of SAs from 9 to 80 ng/L in four different brands of mineral waters were obtained, while drinking water treatment was shown to be effective in the elimination of these compounds, with concentrations of SAs in municipal waters below the limit of quantification.

MTBE, a gasoline additive used since 1979, has also been detected in finished drinking water from the USA and Europe. Williams [23] reported the occurrence of this contaminant in about 1.3% of the drinking water samples from California (USA) analyzed during a period of 6 years. Concentrations ranged from 5 to 15 μg/L, nevertheless only 27% of the positive samples exceeded California's primary health-based standard of 13 μg/L. MTBE was also found in tap water from Germany; Achten et al. [24] reported maximum concentrations above 71 ng/L in treated water from the Frankfurt area. In 1997, another emerging contaminant, perchlorate, was discovered in water supplies from the USA. Exhaustive surveys were performed in California (USA) and perchlorate was found in 185 out of 2200 drinking water sources analyzed [25].

Algal toxins can also impact humans through drinking water contamination. The most lethal outbreak attributed to the presence of cyanobacteria in drinking water occurred in Brazil, where 88 deaths occurred over a 42-day period [26]. In 1999, toxic cyanobacteria blooms, microcystins, anatoxin-a, and cylindrospermopsin were also found in finished drinking waters from Florida (USA) at levels higher than those proposed in human health guidelines [27].

A more extended study was performed by Stackelberg et al. [28] who evaluated the persistence of 106 organic wastewater-related contaminants through conventional treatment processes and their occurrence in finished treated water. Results showed the presence of 17 of the selected contaminants in final water samples; caffeine (0.119 μg/L), carbamazepine (0.258 μg/L), dehydronifedipine (nifedipine metabolite; 0.004 μg/L), and cotinine (nicotine metabolite; 0.025 μg/L) were detected among the selected prescription and nonprescription drugs. Fragrances such as 7-acetyl-1,1,3,4,4,6-hexamethyl tetrahydronaphthalene (AHTN or Tonalide) and 1,3,4,6,7,8-hexahydro-4,6,6,7,8,8-hexamethylcyclopenta-γ-2-benzopyran (HHCB or Galaxolide), the cosmetic triethyl citrate, and the plasticizer bisphenol A were found at high ng/L concentrations. Some pesticides, flame retardants, and solvents were also detected

**Table 1** Summary of emerging contaminants found in tap water reported in the literature

| Compound | Classification | Concentration (in tap water) | Country | Refs. |
|---|---|---|---|---|
| Bezafibrate | Pharmaceutical | up to 27 ng/L | Germany | [4] |
| | | 0.7 ng/L | Italy | [29] |
| Carbamazepine | Pharmaceutical | up to 30 ng/L | Germany | [4] |
| | | 119 ng/L | USA | [28] |
| | | 43.2 ng/L | France | [30] |
| | | up to 20 ng/L | Germany | [123] |
| | | 5 ng/L | Italy | [29] |
| | | 140 ng/L | USA | [31] |
| Clofibrate | Pharmaceutical | 270 ng/L | Germany | [17] |
| | | 0.58 ng/L | USA | [18] |
| Clofibric acid | Pharmaceutical | 70–7300 ng/L | Germany | [17] |
| | | 3.2–5.3 ng/L | Italy | [32] |
| | | up to 70 ng/L | Germany | [4] |
| | | 0.63 ng/L | USA | [18] |
| Codeine | | 30 ng/L | USA | [31] |
| Diazepam | Pharmaceutical | 19.6–23.5 ng/L | Italy | [32] |
| Diclofenac | Pharmaceutical | up to 6 ng/L | Germany | [4] |
| | | 0.4–0.9 μg/L | Germany | [33] |
| Dilantin | Pharmaceutical | 1.3 ng/L | USA | [34] |
| Fenofibric acid | Pharmaceutical | up to 45 ng/L | Germany | [17] |
| | | up to 42 ng/L | Germany | [4] |
| Gemfibrozil | Pharmaceutical | 0.4 ng/L | Italy | [29] |
| | | 70 ng/L | Canada | [35] |
| Ibuprofen | Pharmaceutical | up to 200 ng/L | Germany | [17] |
| | | up to 3 ng/L | Germany | [4] |
| | | 0.6 ng/L | France | [30] |
| | | 5.85 ng/L | USA | [18] |
| Ibuprofen methyl ester | Metabolite | 9.22 ng/L | USA | [18] |
| Ketoprofen | Pharmaceutical | 3.0 ng/L | France | [30] |
| Meprobamate | Pharmaceutical | 5.9 ng/L | USA | [34] |
| Naproxen | Pharmaceutical | 0.15 ng/L | France | [30] |
| Paracetamol | Pharmaceutical | 211 ng/L | France | [30] |
| Phenazone | Pharmaceutical | up to 1250 ng/L | Germany | [17] |
| | | up to 50 ng/L | Germany | [4] |
| | | 400 ng/L | Germany | [36] |
| | | 250 ng/L | Germany | [37] |
| Primidone | Pharmaceutical | up to 20 ng/L | Germany | [123] |
| Propiphenazone | Pharmaceutical | up to 1465 ng/L | Germany | [17] |
| | | 120 ng/L | Germany | [36] |
| | | 80 ng/L | Germany | [37] |
| Sulfamethizole | Pharmaceutical (veterinary) | 9 ng/L | Italy | [22] |
| Sulfamethoxazole | Pharmaceutical (veterinary) | 8–13 ng/L | Italy | [22] |

**Table 1** (continued)

| Compound | Classification | Concentration (in tap water) | Country | Refs. |
|---|---|---|---|---|
| Sulfadimethoxine | Pharmaceutical (veterinary) | 11 ng/L | Italy | [22] |
| Tylosin | Pharmaceutical | 0.6–1.7 ng/L | Italy | [32] |
| Diatrizoic acid | X-ray contrast | up to 85 ng/L | Germany | [4] |
| Iopamidol | X-ray contrast | up to 79 ng/L | Germany | [4] |
| Iopromid | X-ray contrast | up to 86 ng/L | Germany | [4] |
| Caffeine | Stimulant | 0.119 μg/L | USA | [28] |
| | | 0.237 μg/L | Italy | [29] |
| | | 0.06 μg/L | USA | [31] |
| Cotinine | Nicotine metabolite | 25 ng/L | USA | [28] |
| | | 20 ng/L | USA | [31] |
| 17α-Ethynilestradiol | Hormone | 50 pg/L | Germany | [38] |
| Benzophenone | Sunscreen | 0.13 μg/L | USA | [28] |
| Hydrocinnamic acid | Sunscreen | 12.5 ng/L | USA | [18] |
| Triclosan | Germicide | 0.734 μg/L | USA | [18] |
| | | 0.14 μg/L | USA | [8] |
| AHTN | Fragrance | 0.49 μg/L | USA | [28] |
| | | 0.068 μg/L | USA | [31] |
| Camphor | Fragrance | 0.017 μg/L | USA | [31] |
| HHCB | Fragrance | 0.082 μg/L | USA | [28] |
| Triethyl citrate | Cosmetic | 0.062 μg/L | USA | [28] |
| | | 0.082 μg/L | USA | [31] |
| MTBE | Gasoline additive | <13 μg/L | USA | [23] |
| | | up to 75 ng/L | Germany | [24] |
| Anatoxin-A | Algal toxin | 8.5 μg/L | USA | [27] |
| Cylindrospermopsin | Algal toxin | 97.1 μg/L | USA | [27] |
| Microcystin | Algal toxin | up to 12.5 μg/L | USA | [27] |
| | | up to 1 μg/L | USA | [39] |
| | | <1 μg/L | Germany and Switzerland | [40] |
| Dimethyl phthalate | Plasticizer | 2.36 μg/L | USA | [18] |
| Diethyl phthalate | Plasticizer | 0.16–0.2 μg/L | Germany and Poland | [41] |
| | | 0.3 μg/L | Greece | [42] |
| | | 2.10 μg/L | USA | [18] |
| Dibutyl phthalate | Plasticizer | 0.38–0.64 μg/L | Germany and Poland | [41] |
| | | 0.2–10.4 μg/L | Germany | [43] |
| | | 1.04 μg/L | Greece | [42] |
| | | 3.71 μg/L | USA | [18] |
| Butyl benzyl phthalate | Plasticizer | 0.02–0.05 μg/L | Germany and Poland | [41] |
| | | 0.7 μg/L | Germany | [43] |
| | | 0.651 μg/L | USA | [18] |
| DEHP | Plasticizer | 0.05–0.06 μg/L | Germany and Poland | [41] |
| | | 0.93 μg/L | Greece | [42] |

but their concentrations did not exceed the maximum concentration levels established by the US Environmental Protection Agency.

Recently, Loos et al. [29] performed a survey of the anthropogenic environmental pollutants in surface and drinking waters from Italy. Fifty-one contaminants including pharmaceuticals, hormones, phthalates, surfactants, and herbicides were analyzed in both water matrices. Results achieved from surface waters coming from a lake showed the presence of 28 contaminants in the ng/L concentration range and similar concentration levels were obtained in tap water for the 23 detected compounds. For instance, pharmaceuticals such as carbamazepine, gemfibrozil, and bezafibrate were found at 5, 0.4, and 0.7 ng/L concentration levels in the tap water samples analyzed.

## 2 Emerging Contaminants During Drinking Water Treatment

### 2.1 Activated Carbon Adsorption

Activated carbon is a commonly used adsorbent for the removal of organic compounds such as pesticides, pharmaceuticals, and odor and taste compounds [44–46]. Adsorption on activated carbon depends on the intrinsic properties of the activated carbon sorbent (surface area and charge, pore size distribution, oxygen content) and on the solute properties (shape, size, charge, and hydrophobicity). Removal of such organic compounds is mainly controlled by hydrophobic interactions.

Powdered activated carbon (PAC) was evaluated for the elimination of selected PPCPs and endocrine disruptors during simulated drinking water treatment processes in the laboratory [47]. Octanol–water partition coefficients were shown to be a reasonable indicator of compound removal in PAC test conditions. Therefore, compounds with $\log K_{ow}$ values higher than 3 (i.e., sulfamethoxazole or carbamazepine) showed elimination percentages higher than 70% (5 mg/L; 4-h contact time) except for compounds with deprotonated acid functional groups (i.e., naproxen or ibuprofen), which seemed the most difficult to remove with PAC. Deviations from this correlation were also detected for N-heterocyclic compounds (i.e., caffeine or trimethoprim) or protonated bases (i.e., acetaminophen) with low $K_{ow}$, which showed higher removal percentages than expected.

Granular activated carbon (GAC) was also evaluated for the elimination of pharmaceuticals (bezafibrate, clofibric acid, diclofenac, and carbamazepine) under laboratory, pilot, and waterworks conditions in Germany [33]. Pilot experiments showed high adsorption capacities for all the compounds except for clofibric acid, which due to its acidic properties had a low breakthrough volume (17 $m^3$/kg in a 160-cm carbon layer). In waterworks, GAC filtration was also shown to be a very effective removal process, even at high concentrations of pharmaceuticals. They were almost completely removed at throughputs over 70 $m^3$ $kg^{-1}$ except for clofibric acid, which could be removed completely at 15–20 $m^3$ $kg^{-1}$. Nevertheless, the results obtained for carbamazepine were contradicted by a subsequent study performed in a DWTP in the USA [28]. In this work, GAC efficiency was evaluated for the elimination of prescription and nonprescription drugs, fragrance compounds, PPCPs, and other organic contaminants. These studies indicated that carbamazepine and other hydrophobic compounds, such as fragrances HHCB (Galaxolide) and AHTN (Tonalide), persisted through DWTPs including filtration with GAC. The authors suggested that different sorption efficiencies depend on competition with other organic compounds; therefore, the adsorption capacities for these compounds result in smaller values in a DWTP that contains amounts of organic compounds rather than in a laboratory or pilot-scale experiment.

## 2.2 Oxidation Processes

In drinking water treatment systems, the oxidants commonly used are chlorine, chlorine dioxide, and ozone. Ozone is widely used in Europe for the treatment of surface waters while free chlorine is preferred in the USA, although in recent years ozone use has experienced an increase. All three oxidants are strong electrophiles that exhibit selective reactivity with organic compounds. Among them, ozone tends to be more reactive, following the order $O_3 > ClO_2 > HOCl$. One exception is waters with high ammonia content where chlorine has the highest reactivity.

Oxidation processes have to deal with one major drawback, the formation of undesirable DBPs which in some cases can exhibit higher toxicity than the precursors. A summary of some DBPs from pharmaceuticals produced during oxidation processes described in the literature is displayed in Table 2.

### 2.2.1 Ozonation

Ozone is used in water treatment as both disinfectant and oxidant and reacts with a large number of organic and inorganic compounds [48–50]. Rate constants for the reaction with ozone range several orders of magnitude, showing that ozone is a very selective oxidant. Regarding organic compounds, ozone is particularly reactive toward amines, phenols, and compounds with double bonds, especially in aliphatic compounds. In addition, ozone is unstable in water (from seconds to hours) and its decomposition leads to a major secondary oxidant, the hydroxyl radical [50, 51]. The OH radical is a powerful but less selective oxidant; it reacts with high rate constants with most organic compounds but these reactions are less efficient because a large fraction is scavenged by

**Table 2** Summary of DBPs described in the literature and produced from pharmaceutical precursors during oxidation processes

| Compound | Class | Process | DBPs/Intermediates | Refs. |
|---|---|---|---|---|
| Amoxicillin | Antibiotic | Ozonation | Elemental sulfur | [83] |
| | | Ozonation | Hydroxylation of phenol ring | [84] |
| MMTD | Antibiotic | $H_2O_2$/UV | Degradation pathway | [85] |
| | | Photolysis | One DBP | [85] |
| MMTD-Me | Antibiotic | $H_2O_2$/UV | S oxidation | [86] |
| | | Photolysis | Two DBPs | [86] |
| Lincomycin | Antibiotic | $TiO_2/h\nu$ | Mineralization | [87] |
| Sulfadiazine | Antibiotic | $TiO_2/h\nu$ | 4-Methyl-2-aminopyrimidine | [88] |
| Sulfamethoxazole | Antibiotic | Ozonation | Hydroxylamine formation | [60] |
| | | Chlorination | 3-Amino-5methylisoxazole; *N*-chloro-*p*-benzoquinoneimine | [75] |
| Sulfamethoxine | Antibiotic | $TiO_2/h\nu$ | 2,6-Dimethoxy-4-aminopyrimidine; 2-aminothiazole | [88] |
| Sulfamerazine | Antibiotic | $TiO_2/h\nu$ | 4-Methyl-2-aminopyrimidine | [88] |
| Sulfathiazole | Antibiotic | $TiO_2/h\nu$ | 2,6-Dimethoxy-4-aminopyrimidine; 2-aminothiazole | [88] |
| Trimethoprim | Antibiotic | Ozonation/chlorination | Degradation pathway | [89] |
| | | Chlorination | Chlorinated and hydroxylated products from TMP's 3,4,5-trimethoxybenzyl moiety | [90] |
| Busperidone | Antianxiety | $TiO_2/h\nu$ | Hydroxybusperidone; dihydroxybusperidone, dipyrimidinylbusperidone; 1-pyrimidinyl piperazine | [91] |
| Carbamazepine | Anticonvulsant | Ozonation | Degradation pathway | [92] |
| | | Ozonation | 1-(2-Benzaldehyde)-4-hydro-(1*H*,3*H*)-quinazoline-2-one (BQM) 1-(2-Benzaldehyde)-(1*H*,3*H*)-quinazoline-2,4-dione (BQD) 1-(2-Benzoic acid)-(1*H*,3*H*)-quinazoline-2,4-dione (BaQD) | [57] |
| | | $H_2O_2$/UV | Acridine, salicylic acid, catechol, and anthranilic acid | [93] |
| | | $TiO_2/h\nu$ | Degradation pathway | [94] |

**Table 2** (continued)

| Compound | Class | Process | DBPs/Intermediates | Refs. |
|---|---|---|---|---|
| Acetaminophen | Analgesic | Chlorination | Chloro-4-acetamidophenol; dichloro-4-acetaminophenol, 1,4-benzoquinone; *N*-acetyl-*p*-benzoquinone imine | [72] |
| Diclofenac | Anti-inflammatory | Ozonation | Degradation pathway | [95] |
| | | $H_2O_2$/UV | Degradation pathway | [95] |
| | | Photo-Fenton | Degradation pathway | [96] |
| Paracetamol | Anti-inflammatory | Ozonation | *N*-(4-hydroxyphenyl) acetamide | [92] |
| | | Ozonation | *N*-(4-hydroxyphenyl) acetamide | [57] |
| | | $H_2O_2$/UV | 2-[(2,6-Dichlorophenyl)amino]-5-hydroxyphenylacetic acid); 2,5-dihydroxyphenylacetic acid | [97, 98] |
| | | Anodic oxidation | Oxalic acid, oxamic acid | [99] |
| Cimetidine | Histamine receptor | Fenton | Cimetidine sulfoxide, *N*-desmethylcimetidine, *N*-desmethylcimetidine sulfoxide, cimetidine guanylurea, and 5-hydroxymethylimidazole | [100] |
| Ranitidine | Histamine receptor | $TiO_2/h\nu$ | Mineralization | [101] |
| 17β-Estradiol | Hormone | Ozonation | Degradation pathway | [71] |
| | | Ozonation | Oxidized and chlorine substituted compounds | [72] |
| | | $TiO_2/h\nu$ | 10ε-17β-Dihydroxy-1,4-estradien-3-one and testosterone-like species | [102] |
| Estrone | Hormone | Ozonation | Degradation pathway | [70] |
| | | Photo-Fenton | Six intermediates | [102] |
| | | Chlorination | 2-Chloroestrone, 4-chloroestrone, 2,4-dichloroestrone, and 1,4-estradiene-3,17-dione | [103] |
| 17α-Ethinylestradiol | Hormone | Ozonation | Dehydrated and decarboxylated compounds (five products) | [71] |
| | | Ozonation | Oxidized and chlorine substituted compounds | [72] |
| Clofibric acid | Lipid regulator | $TiO_2/h\nu$ | Degradation pathway | [94] |
| Iomeprol | X-ray contrast | $TiO_2/h\nu$ | By-products unidentified | [94] |

the water matrix. Additional oxidation processes are the advanced oxidation processes (AOPs) which use OH radicals as the main oxidants. These processes accelerate the formation of radicals by increasing the pH in water, by adding hydrogen peroxide, or by applying UV radiation [50, 52, 53].

Several experiments have been developed in the laboratory in order to evaluate the oxidation of organic compounds with ozone during drinking water treatment [54]. These experiments showed that certain pharmaceuticals react quickly with ozone while others show no reaction, depending on their structural characteristics. Diclofenac, tetracyclines, carbamazepine, 17α-ethinylestradiol, and estradiol showed rate constants higher than $10^6\ M^{-1}\ s^{-1}$ (pH 7 at 20 °C). For water treatment conditions (pH 7–8; $[O_3] = 1$ mg/L) half-lives for these compounds are lower than 1 s, indicating the complete transformation of the parent compound during the ozonation process. Compounds with no reactive sites for ozone reaction, with lower rate constants, were more efficiently removed by reaction with OH radicals when AOPs were used, with rate constants about two or three times faster. For instance, iopromine, with an ozonation constant of $<0.8\ M^{-1}\ s^{-1}$, showed a $K_{OH}$ of $3.3 \times 10^9\ M^{-1}\ s^{-1}$ and ibuprofen, which was only oxidized above 31%, increased this percentage to 84% when OH radicals were formed.

Ternes et al. [33] evaluated the elimination of bezafibrate, clofibric acid, diclofenac, carbamazepine, and pirimidone under laboratory and full-scale DWTP conditions. Ozone was shown to be effective in eliminating carbamazepine and diclofenac (97%, 0.5 mg/L ozone dose), bezafibrate and pirimidone were appreciably removed with percentages above 50% (1 mg/L ozone doses), while clofibric acid was poorly removed even at high ozone doses (<40%, 2.5–3.0 mg/L ozone doses).

Oxidation of EDCs by reaction with ozone has also been experimentally evaluated. Estrogen steroids and nonylphenols reacted with ozone under similar conditions to those applied in water treatment systems [46]. Petrovic et al. evaluated the elimination of neutral and acidic nonylphenols in a Spanish DWTP [55, 56]. An efficiency of 87% in the elimination of these compounds and their halogenated by-products under ozone treatment was obtained.

More recently, bench-, pilot-, and full-scale studies have been performed to evaluate the ozone efficiency in the elimination of 36 diverse contaminants, including PPCPs, hormones, and pesticides in the USA [57]. Results showed that all the compounds were removed with percentages higher than 50% except for TCEP, lindane, and musk ketone, which were eliminated with percentages lower than 20%, and atrazine, iopromide, and meprobamate with removal percentages between 20 and 50%.

Regarding the transformation products generated from emerging contaminant precursors during ozonation, little information is found [58]. Ozonation of carbamazepine was studied in a German waterworks [59], with the conclusion that when this compound was present in raw water, two main products were formed, BQM (benzaldehydehydroquinazolineone) and BQD (benzaldehydequinazolinedione). Additionally, some transformation products could be predicted on the basis of known reaction pathways for specific functional groups [50]. For instance, it is known that secondary and tertiary amines react with ozone giving hydroxylamines and amine oxides, respectively [60]. Formation of these hydroxylamines could be problematic, for example in the case of sulfonamides, of which hydroxylamines are related to hypersensitivity reactions [61].

### 2.2.2 Chlorination

Chlorine is an oxidant used for disinfection of water supplies. Free chlorine (i.e., HOCl and $OCl^-$) is commonly used in the USA for disinfection and oxidation of inorganic species. One major drawback in chlorination use is the formation of chlorinated organic compounds (mainly trihalomethanes and haloacetic acids) as DBPs, which are classified as carcinogenic and/or mutagenic compounds [62, 63]. Although the oxidation kinetics for organic compounds are lower than those for ozone or chlorine dioxide, it reacts rapidly with phenolic compounds, mainly through the reaction between HOCl and the deprotonated phenolate anion [64]. The sequential addition to the aromatic ring leads to ring cleavage. The reactivity with phenol moieties could explain the transformation of hormones (estradiol, ethynylestradiol, estriol, estrone) and nonylphenol by chlorine, evaluated in laboratory experiments [46].

Some experiments have been performed in order to assess chlorination effects over several emerging compounds at the laboratory scale [65–68]. The fate and occurrence of PPCPs (including musk fragrances), endocrine disruptors, and other organic contaminants were evaluated during simulated drinking water treatment (25 mg/L of $Cl_2$; contact time 24 h) [47]. Under these conditions gemfibrocil, hydrocodone, carbamazepine, compounds with primary or secondary amines (diclofenac, sulfamethoxazole, trimethoprim), and compounds with phenolic moieties (estradiol, estrone, ethynylestradiol, acetaminophen; oxybenzone, triclosan; bisphenol A) showed high reactivity with chlorine. On the other hand, the least reactive compounds were those that have electron-withdrawing functional groups or no conjugated carbon bonds (atrazine, BHC, DEET, fluoxetine, iopromide, meprobamate, and TCEP). The chlorination efficiency to eliminate ten antibiotics (carbadox, erythromycin, roxithromycin, sulfadimethoxine, sulfamerazine, sulfamethazine, sulfamethizole, sulfamethoxazole, sulfathiozole, and tylosin) was also evaluated on the laboratory scale and in surface waters [69]. The results obtained showed that a significant removal of all these compounds could be expected during free chlorination in most water treatment utilities. For instance, carbadox was completely removed within 1 min of contact time and at a chlorine concentration of 0.1 mg/L, while macrolides were removed above 85% with 2 h of contact time and 1 mg/L of chlorine.

Oxidation of organic contaminants has been also evaluated in full-scale treatments. Chlorination studies performed in different DWTPs in the USA and Canada, to assess the elimination of pharmaceuticals (i.e., clofibric acid, naproxen, ibuprofen, acetaminophen, fluoxetine), steroids (i.e., estrone, 17β-estradiol), and plasticizers (bisphenol A), showed nondetectable concentrations of the target compounds after the chlorination step [19].

More recently, the effect of chlorine residual to eliminate several pharmaceuticals and other organic compounds (POOCs) has been evaluated in drinking waters from the USA. The addition of free chlorine to finished drinking water is a common practice as a distribution system disinfectant residual. Gibs et al. [70] have evaluated the effect of the addition of 1.2 mg/L of free chlorine in a finished drinking water with 98 POOCs. Results showed that 52 POOCs would remain after 10 days, with an unremarkable reduction in their concentrations.

As previously described, chlorine usually produces undesirable chlorination by-products to some extent. The formation, fate, and toxicity of oxidative by-products from pesticides and EDCs/PPCPs has been studied and assessed as of potential concern [53, 71]. The E-screen performed after chlorination of bisphenol A, 17α-estradiol, and 17α-ethynylestradiol showed a reduction in estrogenic activity after extended exposure time (120 min). Nevertheless, all compounds showed a similar estrogenicity trend, with a higher estrogenicity activity registered during the first phases of oxidation probably related to the formation of chlorination by-products [72]. Chlorination of acetaminophen has also been studied showing the formation of two chlorination ring products, chloro-4-acetamidophenol and dichloro-4-acetamidophenol, and two quinoidal oxidation by-products, 1,4-benzoquinone and *N*-acetyl-*p*-benzoquinone imine (NAPQI). These toxic compounds are associated with acetaminophen overdoses in humans with lethal effects [73].

Chlorination of sulfamethoxazole (SMX), a member of the sulfonamide antibacterial class, has also been studied in wastewater and drinking water matrices. Chlorine reacted with the aniline nitrogen giving the halogenation of the aniline moiety, yielding a ring chlorinated product, and with the SMX sulfonamide moiety to yield the formation of 3-amino-5-methylisoxazole and *N*-chloro-*p*-benzoquinoneimine subproducts [74].

### 2.2.3 Chlorine Dioxide

Chlorine dioxide ($ClO_2$) is an oxidant used for disinfection of high quality water, such as groundwater or treated surface water. In Europe, it is also used to protect drinking water distribution at residual concentrations (0.05 to 0.1 mg/L), while in the USA it is mainly used for the preoxidation of surface waters. Compared to chlorine, $ClO_2$ is generally a stronger and faster oxidant, [75] and is more effective for the inactivation of viruses, bacteria, and protozoa (including the cyst of Giardia and the oocysts of Cryptosporidium). Chemically, $ClO_2$ has been demonstrated to be a very selective oxidant of specific functional groups of organic compounds, such as phenolic moieties, tertiary amino groups, or thiol groups [76]. Additionally, halogenated DBPs are not formed even under suitable conditions [77]. Nevertheless, other DBPs are formed during $ClO_2$ reaction. Therefore, chlorite is the major reduction product of $ClO_2$, considered to be a blood poison [61, 78] and regulated by the USEPA at the 1 mg/L level [79].

Due to the oxidant doses used of $ClO_2$ in drinking water treatment and its specific reactivity, a complete elimination of parent contaminants is not expected. Nevertheless, this treatment could lead to the deactivation of specific functional groups responsible for parent activity. Chlorine dioxide has demonstrated cleavage of one of the N – C bonds of tertiary amines [80], which would mean the loss of a methyl or amino group in macrolide antibiotics leading to a related and expected decrease in pharmacological activity [81].

Oxidation of several pharmaceuticals by $ClO_2$ was evaluated in samples from a German DWTP [82]. Water samples were collected before $ClO_2$ treatment and spiked with the selected pharmaceuticals. Then, $ClO_2$ doses of 0.95 and 11.5 mg/L were added and samples were analyzed after 30 min of contact time. Under these experimental conditions, bezafibrate, carbamazepine, diazepam, and ibuprofen showed no reactivity while diclofenac was completely oxidized and phenazone derivatives and naproxen showed an appreciable reactivity.

## 2.3 Membrane Separation

In membrane processes, a semipermeable membrane separates contaminants from the water by a process known as crossflow filtration (also called tangential flow filtration). The bulk solution flows over, and parallel to, the filter surface while, under pressure, a portion of the water is forced through the membrane to produce a permeate stream. The turbulent flow of the feedwater over the membrane surface minimizes accumulation of particulate matter there, and facilitates continuous operation.

Different types of membranes are applied to drinking water treatment with different characteristic separations depending on their composition and pores. Several classifications can be made to characterize membranes; size exclusion is one of the most significant mechanisms to separate contaminants.

### 2.3.1 Ultrafiltration

Ultrafiltration (UF) allows the removal of turbidity, microorganisms, and many hydrophobic macromolecules (0.001–0.1 μm) with $\log K_{ow} > 4$. The removal properties of UF membranes are usually expressed in terms of molecular weight cutoff (MWCO) which ranges from 1000 up to 50 000 Da. Nevertheless, most organic EDC/PPCP compounds range from 150 to 500 Da, and only those associated with particles or colloidal organic matter are removed.

An investigation on the removal of 52 EDCs and PPCPs with different physicochemical properties such as size, hydrophobicity, and acidity by UF and nanofiltration (NF) has been carried out in model and natural waters [104, 105]. The results showed that the UF membrane retained hydrophobic EDCs mainly by adsorption processes. UF membranes showed retention percentages lower than 40% for all compounds except triclosan (87%), oxybenzone (77%), and progesterone (56%). In most cases, the concentration of EDCs and PPCPs was feed > retentate > permeate except for few compounds (i.e., diclofenac, erythromycin, estriol, gemfibrozil, ibuprofen, chlordane, dieldrin) that showed lower concentrations in retentate than initial ones. These compounds were probably adsorbed onto the membrane and into the membrane pores. It has been reported that retention of relatively hydrophobic compounds and hormones (i.e., $\log K_{ow} > 3$) by UF, reverse osmosis (RO), and NF membranes is mainly due to adsorption [106, 107]. Yoon et al. [105] stated that compounds highly retained by UF (30–80%) have common structural properties including aromatic ring structures, high $pK_a$, and/or high $\log K_{ow}$ values, whereas poorly retained compounds include those with low $\log K_{ow}$ due to aliphatic, aromatic, nitrogen, carbonyl, phosphate, amine, or hydroxyl functional groups.

### 2.3.2 Nanofiltration/Reverse Osmosis

NF and RO are effective physical diffusion-controlled and size-exclusion processes which have been demonstrated to effectively remove pathogens and organic contaminants. However, the rejection efficiency correlates to different parameters affecting the solute, the membrane, and the feed water composition; moreover, it is also correlated with the concentration of the organic contaminant and its effective charge state. Both processes have the broadest duration of treatment capability but require a great degree of pretreatment, and in addition RO has a high relative cost compared with other technologies.

Bench-scale tests have been performed in order to evaluate the removal of several emerging contaminants by NF and/or RO. A pilot system with RO membranes was used to evaluate the elimination ratio of several pharmaceuticals, pesticides, and PPCPs. The system evaluated both virgin and fouled membranes, showing that target analytes were well-rejected and no effect of membrane fouling was detected [108]. Another study evaluated the elimination of steroid hormones by RO in wastewater matrices. Results showed removals greater than 90% for 17β-estradiol and 17α-ethinylestradiol [66].

NF membranes have also been evaluated by bench-scale tests for the analysis of EDCs and PPCPs [109]. Results showed that NF membranes had a low adsorption capacity for the less volatile and less hydrophobic compounds. Average retention percentages were 30–90% depending on their properties, except for naproxen which showed poor retention lower than 10%. In these tests, hydrophobicity led to adsorption and polarity to charge repulsions that were more important than molecular weight in removing EDCs and PPCPs.

A study of the removal of pesticides [110] and pharmaceuticals [111] by NF and RO membranes in a real DWTP has been performed. The DWTP supplies treated water to 20 000 inhabitants and uses one NF line and two parallel RO lines with a final mixing of the three permeates to obtain treated water. Triazines (i.e., simazine, atrazine, terbutylazine, and terbutryn) and metabolites (DIA, DEA) were fully eliminated in both NF and RO lines. On the other hand, removal of pharmaceuticals showed very similar percentages to those obtained for triazines, and high values above 80% were obtained in both NF and RO lines for most of the selected compounds (i.e., hydrochlorothiazide, ketoprofen, gemfibrozil, diclofenac, sulfamethoxazole, sotalol, metoprolol, propylphenazone, and carbamazepine). However, strong fluctuations in the permeate concentrations for some compounds, such as acetaminophen and mefenamic acid, were measured.

An assessment of removal possibilities with NF of priority pollutants in water sources of Flanders and The Netherlands has been recently reviewed [112]. The authors suggested that rejection of organic pollutants in NF could be qualitatively predicted as a function of a limited set of solute parameters, such as $\log K_{ow}$, $pK_a$, and molar mass. The prediction was based on the scheme proposed by Bellona et al. [113] but using hydrophobicity as the primary solute parameter. Their qualitative predictions for target compounds (hormones, industrial chemicals, pesticides, and pharmaceuticals) roughly correlated with values from the literature. The authors stated that the solute parameters together with a knowledge of the membrane material can give real estimations of the rejection of organic micropollutants and can provide feasible evaluations of NF in drinking water plant designs.

## 3 Emerging Disinfection By-Products

A widely known group of drinking water contaminants are DBPs which are generated during the treatment process. Some of these compounds, such as trihalomethanes, haloacetic acids, bromates, or chlorites, are widely known and they have been studied and regulated for the last 30 years. However, emerging contaminants in raw waters and new alternative disinfectants and treatments for drinking water production, implemented by the DWTPs, could lead to the formation of new DBPs. In Sect. 2, DBP formation from pharmaceuticals and hormones was examined. In this section, the emerging DBPs generated during water treatment due to alternative disinfectants from chlorine (i.e., ozone, chlorine dioxide, and chloramines) will be discussed. Up to now, scarce information about the potential toxicity of DBPs generated from these alternative disinfectants can be found. New DBPs identified include iodo-acids, bromonitromethanes, iodo-trihalomethanes, brominated forms of MX, bromoamines and bromopyrrole [114], nitrosodimethylamine (NDMA), and other nitrosamines. Recent studies [115] of their toxicity have demonstrated that some of these compounds are more genotoxic than many of the DBPs regulated, and are present at similar concentration levels to those regulated.

Among the emerging DBPs investigated, one remarkable compound is NDMA [116–118] which is generated from chloramines or chlorine disinfection (Fig. 1) [128–130]. This compound belongs to the chemical class of the *N*-nitrosoamines and its importance remains, as it is considered a potential human carcinogen with more cancer potencies than those reported for trihalomethanes [119, 120]. In 1989, NDMA was first detected in treated drinking water from Ohsweken (Ontario, Canada) at elevated concentrations (up to 0.3 μg/L). This finding prompted a survey of 145 Ontario DWTPs [116, 121, 122] and the concentrations of NDMA detected in the treated water were lower than 5 ng/L (except for some samples exceeding 9 ng/L). More recently, similar results were obtained for NDMA concentrations in drinking water systems from the USA. Results showed that NDMA was detected at concentra-

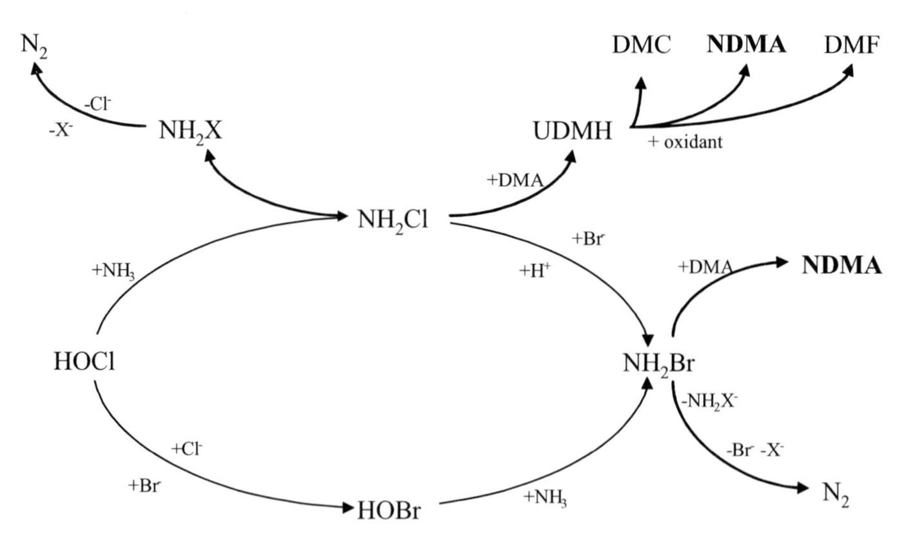


**Fig. 1** NDMA formation mechanism for the chloramine/bromamine pathway [128–130]. X: Cl/Br; UDMH: unsymmetrical dimethylhydrazine; DMC: dimethylcyanamide; DMF: dimethylformamide

tion levels lower than 5 ng/L in water supplies which used only free chlorine, while 3 out of 20 chloraminated supplies contained concentrations higher than 10 ng/L [99]. Nevertheless, a more extended survey performed from 2001 to 2002 in 21 USA water systems indicated median concentrations of NDMA in chlorinated or chloraminated waters lower than 1 ng $L^{-1}$ [123]. Regarding legislation, although NDMA is listed as a priority pollutant [124], a maximum contaminant level (MCL) has not been established and it has not yet been included on the candidate contaminant list (CCL), which is a list of unregulated contaminants for monitoring in the USA [125]. Nevertheless, some regulatory agencies have established guidelines for maximum concentrations of NDMA; the Ontario Ministry of the Environment and Energy has fixed a value of 9 ng/L [126], while the California Department of Health Services has suggested a value of 10 ng/L [127].

With regard to other emerging DBPs, Richardson et al. [131] studied the formation of DBPs when alternative disinfectants were used. Over 200 DBPs were identified and a comparison between by-products formed from different treatments was also performed. The effect of high concentrations of bromide on the formation of chlorine dioxide DBPs was also evaluated by selecting natural waters from Israel (Sea of Galilee) with high natural levels of this compound. The DBP structures identified showed high degrees of bromide, such as 1,1,3,3-tetrabromopropane.

Finally, new alternative routes of exposure to drinking water DBPs are now being recognized. Inhalation or dermal absorption during bathing or showering can be translated into high exposure to toxic/carcinogenic compounds [132]. A recent study performed by Villanueva et al. [133] revealed a correlation between these activities and a higher risk of bladder cancer. An additional new route of exposure to DBPs is swimming pools. Zwiener et al. [134] published a review article on the formation of DBPs in swimming pool waters and the adverse health effects that could be related to them.

# 4 Removal of New Emerging Contaminants in a Drinking Water Treatment Plant (DWTP)

Human habits and activities have been widely demonstrated to impact the environment in many ways. Recently, a new group of human-use contaminants, illicit drugs, have been detected in aquatic media from the USA [135], Italy [136, 137], Germany [138], Spain [139, 140], and Ireland [141]. Due to the high consumption rates—around 200 million people have consumed illicit drugs in the last year—the determination of these compounds has become an important issue, not only for forensic sciences but also in environmental studies [135]. Some of these drugs are released unaltered or as slightly transformed metabolites. Therefore, they reach municipal wastewater treat-

**Table 3** Drug concentrations in WWTP samples (NE Spain) (April to September, 2006) and in water from the Llobregat river (NE Spain) (September 2006) [139]

| Compound | WWTP influent ($n:16$) Samples (>LOQ[a]) | C max ng/L | C mean ng/L | WWTP effluent ($n:16$) Samples (>LOQ[a]) | C max ng/L | C mean ng/L | River ($n:6$) Samples (>LOQ[a]) | C max ng/L | C mean ng/L |
|---|---|---|---|---|---|---|---|---|---|
| Nicotine | 10 | 56053 | 13082 | 6 | 4775 | 2669 | 5 | 815 | 595 |
| Cotinine | 16 | 6820 | 2732 | 12 | 2726 | 1419 | 5 | 516 | 331 |
| Caffeine | 15 | 61638 | 23134 | 14 | 22848 | 4356 | 6 | 2991 | 1926 |
| Paraxanthine | 14 | 54220 | 14240 | 12 | 45681 | 5932 | 5 | 2709 | 1756 |
| Amphetamine | 1 | 15 | 15 | 0 | <LOQ[b] | <LOQ[b] | 0 | <LOQ[b] | <LOQ[b] |
| MDMA | 5 | 91 | 49 | 4 | 67 | 41 | 2 | 3.5 | 3 |
| MDEA | 1 | 27 | 28 | 0 | <LOQ[b] | <LOQ[b] | 0 | <LOD | <LOD |
| Ketamine | 9 | 50 | 41 | 2 | 49 | 19 | 0 | <LOD | <LOD |
| Cocaine | 14 | 225 | 79 | 6 | 47 | 17 | 2 | 10 | 6 |
| Benzoylecgonine | 14 | 2307 | 810 | 11 | 928 | 216 | 4 | 111 | 77 |

[a] Number of samples with concentrations higher than LOQ value

[b] Concentrations between LOQ and LOD. LOQs (wastewater): nicotine (800 ng/L); cotinine (500 ng/L); caffeine (5 ng/L); paraxanthine (850 ng/L); amphetamine and MDA (1 ng/L); METH (0.9 ng/L); MDMA (1.5 ng/L); MDEA (2.5 ng/L); ketamine (5 ng/L); cocaine and BE (0.2 ng/L). LOQs (surface water): nicotine, cotinine, and paraxanthine (200 ng/L); caffeine (1.5 ng/L); amphetamine, MDA, and MDEA (0.8 ng/L); METH (0.7 ng/L); MDMA (0.3 ng/L); ketamine (3.1 ng/L); cocaine (0.15 ng/L); BE (0.1 ng/L)

ment plants (WWTPs) where, depending on the efficiency of the treatment, they are totally removed or, on the contrary, persist during the treatment and can be detected in receiving waters. The effectiveness of the water treatment processes and the impact of most of these compounds on the aquatic environment are still unknown.

An UPLC-MS/MS method was developed for the analysis of caffeine, nicotine, cocaine, amphetamine related compounds, and other synthetic controlled drugs, and their metabolites, in waste and surface waters [139]. Once the method was optimized and the quality parameters were established, the method was applied to the estimation of the occurrence of these substances in water samples from Catalonia (NE Spain) (Table 3). Results displayed in this table have been already submitted for publication. The analysis of several samples from WWTPs revealed the presence of drugs, such as cocaine and amphetamine related compounds, in both influent and effluent samples. Several illicit drugs, such as cocaine or MDMA (ecstasy), were also found in surface waters while nicotine and caffeine were detected in all the analyzed samples. The results obtained demonstrate that the presence of these drugs in aquatic media must be considered a matter of environmental concern [139].

The incidence of these illicit drugs in surface waters posed the need to investigate the elimination of these compounds during drinking water treatment and their presence in final treated water. The treatment in the DWTP investigated consisted in prechlorination (with chlorine or chlorine dioxide), sand filtration, flocculation and sedimentation, ozonation, GAC filtration, and final postchlorination.

**Table 4** Drug concentrations of raw water, treated water, and elimination percentages in a DWTP (Spain)

| | Intake[a] ng/L | Treated[a] ng/L | Elimination (%) |
|---|---|---|---|
| Nicotine | nd–1047 | <LOQ | >99.9 |
| Cotinine | nd–516 | nd–276 | 74 |
| Caffeine | nd–2991 | nd–126 | 93 |
| Paraxanthine | nd–2709 | <LOQ | >99.9 |
| Amphetamine | nd–165 | <LOQ | >99.9 |
| MDA | nd–6 | <LOQ | >99.9 |
| MDMA | nd–123 | <LOQ | >99.9 |
| MDEA | nd–54 | <LOQ | >99.9 |
| Ketamine | nd–61 | <LOQ | >99.9 |
| Cocaine | nd–411 | <LOQ | >99.9 |
| Benzoylecgonine | nd–1047 | nd–24 | 89 |

nd: non detected
[a] $n = 24$

Several controlled drugs, such as cocaine, benzoylecgonine (cocaine metabolite) and some amphetamine type stimulants (i.e., amphetamine or ecstasy), were detected with concentrations higher than their limit of quantitation (LOQ) at the intake of the selected DWTP [141]. For instance, maximum concentrations of 22 ng/L were obtained for cocaine and up to 37 ng/L for ecstasy. The removal efficiency during treatment was also evaluated and the results (Table 4) showed that removal percentages higher than 99.9% were obtained for most of the compounds found at the intake, including cocaine and ecstasy.

Only three of the studied compounds were detected in some samples with concentrations higher than the LOQs. Cotinine and caffeine among the controlled drugs were found in treated water with removal percentages of about 74 and 93%, respectively, and among the illicit drugs only the biologically inactive metabolite of cocaine was found in treated water at low ng/L levels with a removal of 89%. The analyses were performed by using an UPLC system coupled to tandem mass spectrometry (MS/MS) and the quality parameters were already established [139]. An extracted chromatogram from a treated water sample is displayed in Fig. 2. Two transitions were acquired for each compound in order to obtain four identification points, fulfilling the

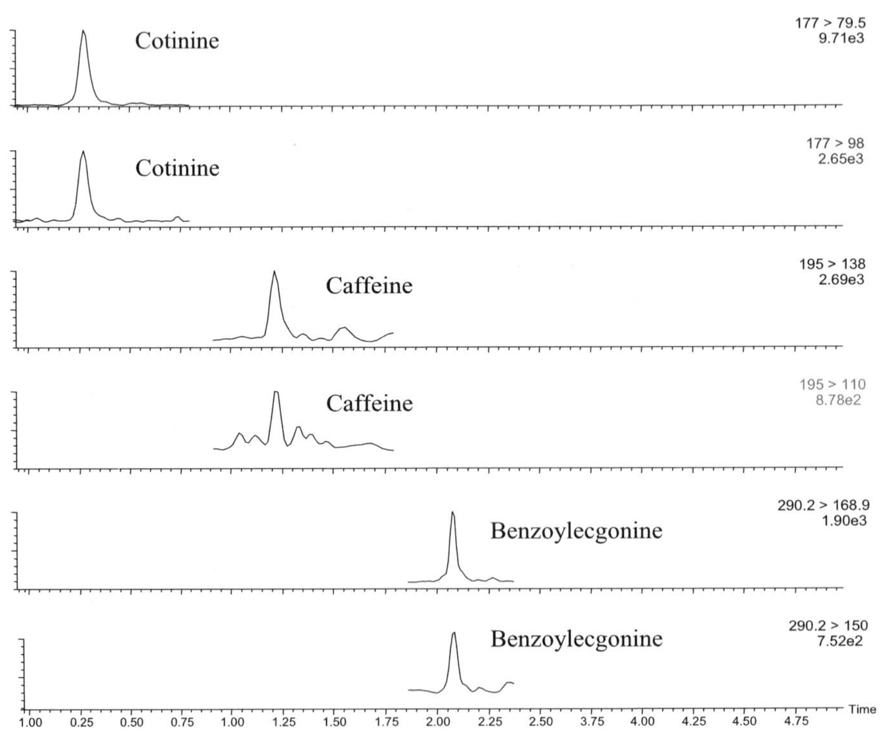


**Fig. 2** Extracted ion chromatogram obtained at the intake of a Spanish DWTP. SRM acquisition mode

European Council directives (96/23/EC) regarding mass spectrometric detection [142] and the general criteria for forensic analysis [143].

## 5
## Concluding Remarks

The occurrence of emerging contaminants in aquatic media has been widely assessed in the last decade. Nevertheless, much more data are needed in order to improve the knowledge of the behavior/removal of these compounds in wastewaters and surface waters, as well as their toxicological impact on both aquatic life and human beings and to establish safe guideline values. Moreover, the occurrence of these contaminants in drinking water, just like the removal efficiency of the treatment processes, is still relatively unknown.

In this chapter, a summary of the works published regarding the elimination of emerging contaminants through conventional drinking water treatments and the persistence of some of them through treatments has been presented. Activated carbon adsorption (PAC or GAC) has been shown to be effective to remove nonionic compounds with log $K_{ow}$ higher than 3. Nevertheless, some pharmaceuticals such as carbamazepine and some fragrances such as HHCB (Galaxolide) persisted throughout treatment. NF and RO membranes were also found to remove organic contaminants to a very high extent. Oxidation processes such as ozonation and chlorination have also been evaluated in the elimination of emerging contaminants. Ozone was shown to be very effective in eliminating several pharmaceuticals, hormones, and nonylphenols with percentages higher than 50%, while poorer elimination rates were found for some pesticides (i.e., lindane, atrazine), fragrances (i.e., musk ketone), and pharmaceuticals (i.e., clofibric acid, meprobamate). Oxidation with chlorine or chlorine dioxide was shown to be less efficient but high reactivities were obtained when contaminants contained phenolic or amino moieties (i.e., hormones, nonylphenols, sulfonamides). One major drawback of the oxidation processes is the formation of undesirable DBPs which could have toxic effects. The formation of DBPs from these emerging contaminants together with new disinfection treatments could lead to emerging DBPs. Up to now some new DBPs, such as NDMA, bromonitromethanes, or iodo-trihalomethanes, have already been identified.

Finally, it must be emphasized that the emerging contaminants field is still growing. New human habits or activities could cause the appearance of novel contaminants in aquatic media that may become emerging contaminants. One example of new contaminants derived from human activities and detected in water sources are illicit drugs. These contaminants have recently been detected in aquatic media from the USA and Europe, thus demonstrating once more the cause–effect relationship between human activities and environmental contamination.

## References


1. Ternes T (1998) Water Res 32:3245
2. Ternes TA, Kreckel P, Mueller J (1999) Sci Total Environ 225:91
3. Johnson AC, Sumter JP (2001) Environ Sci Technol 35:4697
4. Ternes T (2001) Pharmaceuticals and metabolites as contaminants of the aquatic environment. In: Daughton C, Jones-Lepp T (eds) Pharmaceuticals and personal care products in the environment: scientific and regulatory issues. American Chemical Society, Washington, p 39
5. Ternes T, Stuber J, Herrmann N, McDowell D, Ried A, Kampmann M, Teiser B (2003) Water Res 37:1976
6. Joss A, Keller E, Alder A, Gobel A, McArdell C, Ternes T, Siegrist H (2005) Water Res 39:3139
7. Thomas P, Foster G (2005) Environ Toxicol Chem 24:25
8. Kolpin DW, Furlong ET, Meyer MT, Thurman EM, Zaugg SD, Barber LB, Buxton HT (2002) Environ Sci Technol 36:1202
9. Stumpf M, Ternes T, Wilken R-D, Rodriques S, Baumann W (1999) Sci Total Environ 225:135
10. Snyder S, Kelly K, Grange A, Sovocool GW, Snyder E, Giesy J (2001) Pharmaceuticals and personal care products in the waters of Lake Mead Nevada. In: Daughton C, Jones-Lepp T (eds) Pharmaceuticals and personal care products in the environment: scientific and regulatory issues. American Chemical Society Washington, p 117
11. Barber LB, Leenheer JA, Pereira W, Noyes T, Brown G, Tabor C, Writer J (1995) Contaminants in the Mississippi River from municipal and industrial wastewater. In: Us Geological survey circular 1133, Virginia
12. Singer H, Muler S, Tixier C, Pillonel L (2002) Environ Sci Technol 36:4998
13. Calamari D, Zuccato E, Castiglioni S, Bagnati R, Fanelli R (2003) Environ Sci Technol 37:1241
14. Daughton C (2004) Environ Impact Assess Rev 24:711
15. Zwiener C (2007) Anal Chem 387:1159
16. Heberer T, Stan H-J (1997) Int J Environ Anal Chem 67:113
17. Heberer T, Schmidt-Bäumler K, Stan H-J (1998) Acta Hydrochim Hydrobiol 26:272
18. Loraine G, Pettigrove M (2006) Environ Sci Technol 40:687
19. Boyd GR, Reemstsma H, Grimm DA, Mitra S (2003) Sci Total Environ 311:135
20. Jux U, Baginski RM, Arnold H, Kronke M, Seng PN (2002) Int J Hyg Environ Health 205:393
21. McLachlan JA, Guillette LJ, Iguchi T Jr, Toscano WA Jr (2001) Ann NY Acad Sci 948:153
22. Perret D, Gentili A, Marchese S, Greco A, Curini R (2006) Chromatographia 63:225
23. Williams PRD (2001) Environ Forensics 2:75
24. Achten C, Kolb A, Puttmann W (2002) Environ Sci Technol 36:3662
25. California Department of Health Services. California's experience with perchlorate in drinking water. http://www.drs.ca.gov/ps/ddwem/chemicals/perchl/perchlindx.htm
26. Teixeira G, Costa C, de Carvalho VL, Pereira S, Hage E (1993) Bull Pan Am Health Organ 27:244
27. Burns J (2003) Report to the Florida Department of Health. Tallahassee
28. Stackelberg PE, Furlong ET, Meyer MT, Zaugg SD, Hendersond AK, Reissmand DB (2004) Sci Total Environ 329:99

29. Loos R, Wollgast J, Huber T, Hanke G (2007) Anal Bioanal Chem 387:1469
30. Rabiet M, Togola A, Brissaud F, Seidel JL, Budzinski H, Elbaz-Poulichet F (2006) Environ Sci Technol 40:5282
31. Stackelberg PE, Gibs J, Furlong ET, Meyer MT, Zaugg SD, Lippincott L (2007) Sci Total Environ 377:255
32. Zuccato E, Calamari D, Natangelo M, Fanelli R (2000) Lancet 355:1789
33. Ternes TA, Meisenheimer M, McDowell D, Sacher H, Brauch J, Gulde BH, Preuss G, Wilme U, Seirbet NZ (2002) Sci Total Environ 36:3855
34. Vanderford BJ, Snyder SA (2006) Environ Sci Technol 40:7312
35. Tauber R (2003) Quantitative analysis of pharmaceuticals in drinking water from ten Canadian cities. Enviro-Test Laboratories, Xenos Division, Ontario
36. Reddersen K, Heberer T, Dunnbier U (2002) Chemosphere 49:539
37. Zuehlke S, Duennbier U, Heberer T (2004) J Chromatogr A 1050:201
38. Holger MK, Ballschmiter K (2001) Environ Sci Technol 35:3201
39. Carmichael WW (2001) American Water Works Association Research Foundation. Denver
40. Hoeger SJ, Hitzfeld BC, Dietrich DR (2005) Toxicol Appl Pharmacol 203:231
41. Luks-Betlej K, Popp P, Janoszka B, Paschke H (2001) J Chromatogr A 938:93
42. Psillakis E, Kalogerakis N (2003) J Chromatogr A 999:145
43. Fromme H, Kuchler T, Otto T, Pilz K, Muller J, Wenzel A (2002) Water Res 36:1429
44. Robeck GG, Dostal KA, Cohen JM, Kriessl JF (1965) J Am Water Works Assoc 57:181
45. Sacher F, Haist-Gulde B, Brauch H-J, Preuss G, Wilme U, Zullei-Seibert N, Meisenheimer M, Welsch H, Ternes TA (2000) 219th ACS national meeting, San Francisco, p 116
46. West P (2000) AWWA Annual Conference. Denver
47. Westerhoff P, Yoon Y, Snyder S, Wert E (2005) Environ Sci Technol 36:6649
48. Hoigne J, Bader H (1983) Water Res 17:185
49. Hoigne J, Bader H (1985) Water Res 19:993
50. von Gunten U (2002) Water Res 37:1443
51. Haag WR, Yao CCD (1992) Environ Sci Technol 26:1005
52. Hoigne J (1998) In: Hubrec J (ed) Handbook of environmental chemistry. Springer, Berlin, p 83
53. Acero JL, von Gunten U (2001) J Am Water Works Assoc 93:90
54. Huber MM, Canonica S, Park G-Y, von Gunten U (2003) Environ Sci Technol 37:1016
55. Petrovic M, Diaz A, Ventura F, Barceló D (2001) Anal Chem 73:5886
56. Petrovic M, Diaz A, Ventura F, Barceló D (2003) Environ Sci Technol 37:4442
57. Snyder SA, Wert EC, Rexing DJ, Zegers RE, Drury DD (2006) Ozone Sci Eng 28:445
58. Ikehata K, Naghashkar NJ, El-Din MG (2006) Ozone Sci Eng 28:353
59. McDowell DC (2005) Environ Sci Technol 39:8014
60. Muñoz F, von Sonntag C (2000) Chem Soc Perkin Trans 2:2029
61. Sisson ME, Rieder MJ, Bird IA, Almawi WY (1997) Int J Immunopharmacol 19:299
62. WHO guidelines for drinking-water quality (2004) WHO, Geneva, p 451
63. Miles AM, Singer PC, Ashley DL, Lynberg MC, Langlois PH, Nuckols JR (2002) Environ Sci Technol 36:1692
64. Faust BC, Hoigne J (1987) Environ Sci Technol 21:957
65. Adams SM, Greeley MS (2000) Water Air Soil Pollut 123:103
66. Huang C, Sedlak DL (2001) Environ Toxicol Chem 20:133
67. Sedlak DL, Pinkston KE (2001) Water Res Update 120:56
68. Gould JP, Richards JT (1984) Water Res 18:1001

69. Chamberlain E, Adams C (2006) Water Res 40:2517
70. Gibs J, Stackelberg PE, Furlong ET, Meyer MT, Zaugg SD, Lippincott RL (2007) Sci Total Environ 373:240
71. Huber MM, Ternes TA, von Gunten U (2004) Environ Sci Technol 38:5177
72. Alum A, Yoon Y, Westerhoff P, Abbaszadegan M (2004) Environ Toxicol 19:257
73. Bredner M, MacCrehan WA (2006) Environ Sci Technol 40:516
74. Dodd M, Huang C (2004) Environ Sci Technol 38:5607
75. Ravacha C, Blits R (1984) Water Res 19:1273
76. Hoigne J, Bader H (1994) Water Res 28:45
77. USEPA: Stage 1 disinfectants and disinfection by-product rule (1998). EPA 63-FR 69390–69476
78. Condie LW (1986) J Am Water Works Assoc 73:78
79. USEPA: Alternative disinfectants and oxidants guidance manual (1999). EPA 815-R-99–014
80. Rosenblatt DH, Hull LA, De Luca DC, Davis GT, Weglein RC, Williams HKR (1967) J Am Chem Soc 89:1158
81. Li XQ, Zhong DF, Huang HH, Wu SD (2001) Acta Pharmacol Sin 22:469
82. Huber MM, Korhonen S, Ternes TA, von Gunten U (2005) Water Res 39:3607
83. Arslan-Alaton I, Dogruel S (2004) J Hazard Mater 112:105
84. Andreozzi R, Canterino M, Marotta R, Paxeus N (2005) J Hazard Mater 122:243
85. Lopez A, Bozzi A, Mascolo G, Ciannarella R, Passino R (2002) Ann Chim 92:41
86. Bozzi A, Lopez A, Mascolo G, Tiravanti G (2002) Water Sci Technol Water Suppl 2:19
87. Addamo M, Augugliaro V, Di Paola A, García-López E, Loddo V, Marcí G, Palmisano L (2005) J Appl Electrochem 35:765
88. Calza P, Pazzi M, Medana C, Baiocchi C, Pelizzetti E (2004) J Pharm Biomed Anal 35:9
89. Adams C, Wang Y, Loftin K, Meyer M (2002) J Environ Eng ASCE 128:253
90. Dodd MC, Huang CH (2007) Water Res 41:647
91. Calza P, Medana C, Pazzi M, Baiocchi C, Pelizzetti E (2004) Appl Catal B Environ 53:63
92. Andreozzi R, Caprio V, Marotta R, Vogna D (2003) Water Res 37:993
93. Vogna D, Marotta R, Napolitano A, Andreozzi R, d'Ischia M (2004) Water Res 38:414
94. Doll TE, Frimmel FH (2005) Water Res 39:847
95. Vogna D, Marotta R, Andreozzi R, Napolitano A, d'Ischia M (2004) Chemosphere 54:497
96. Pérez-Estrada LA, Maldonado MI, Gernjak W, Agüera A, Fernández-Alba AR, Ballesteros MM, Malato S (2005) Catal Today 101:219
97. Vogna D, Marotta R, Napolitano A, d'Ischia M (2002) J Org Chem 67:6143
98. Andreozzi R, Caprio V, Marotta R, Vogna D (2003) Water Res 37:993
99. Brillas E, Sirés I, Arias C, Cabot PL, Centellas F, Rodríguez RM, Garrido JA (2005) Chemosphere 58:399
100. Zbaida S, Kariv R, Fischer P, Silmangreenspan J, Tashma Z (1986) Eur J Biochem 154:603
101. Addamo M, Augugliaro V, Di Paola A, García-López E, Loddo V, Marcí G, Palmisano L (2005) J Appl Electrochem 35:765
102. Ohko Y, Iuchi KI, Niwa C, Tatsuma T, Nakashima T, Iguchi T, Kubota Y, Fujishima A (2002) Environ Sci Technol 36:4175
103. Nakamura H, Kuruto-Niwa R, Uchida M, Terao Y (2007) Chemosphere 66:144
104. Feng XH, Ding SM, Tu JF, Wu F, Deng NS (2005) Sci Total Environ 345:229

105. Yoon Y, Westerhoff P, Snyder SA, Wert EC (2006) J Memb Sci 270:88
106. Kimura K, Amy G, Drewes J, Watanabe Y (2003) J Memb Sci 221:89
107. Nghiem LD, Schaeffer AI, Elimelech M (2004) Environ Sci Technol 38:1888
108. Snyder SA, Adham S, Redding AM, Cannon FS, Decarolish J, Oppenheimer J, Wert EC, Yoon Y (2006) Desalination 202:156
109. Yoon Y, Westerhoff P, Snyder SA, Wert EC, Yoon J (2006) Desalination 202:16
110. Quintana J, Ventura F, Martí I, Luque F (2005) EMCO workshop, Dubrovnik
111. Radjenovic J, Petrovic M, Ventura F, Barceló D (2007) EMCO workshop, Belgrade
112. Verliefde A, Cornelissen E, Amy G, Van der Bruggen B, van Dijk H (2007) Environ Pollut 146:281
113. Bellona C, Drewes JE, Xu P, Amy G (2004) Water Res 38:2795
114. Krasner SW, Weinberg HS, Richardson SD, Pastor SJ, Chinn R, Sclimenti MJ, Onstad GD, Thruston AD Jr (2006) Environ Sci Technol 40:7175
115. Richardson SD, Thruston AD Jr, Rav-Acha C, Groisman L, Popilevsky I, Juraev O, Glezer V, McKague AB, Plewa MJ, Wagner ED (2003) Environ Sci Technol 37:3782
116. Jobb DB, Hunsinger RB, Meresz O, Taguchi VY (1995) Proc Am Water Works Assoc, water quality technology conference, Denver
117. Graham JE, Meresz O, Farquhar GJ, Andrews SA (1995) Proc Am Water Works Assoc, water quality technology conference, Denver
118. Najm I, Trussell RR (2001) J Am Water Works Assoc 93:92
119. Mitch WA, Sharp JO, Trussell RR, Valentine RL, Alvarez-Cohen L, Sedlak DL (2003) Environ Eng Sci 20:389
120. US EPA (2002) Integrated risk information system. Office of Research and Development (ORD), National Center for Environmental Assessment, www.epa.gov/ngispgm3/iris/search.htm
121. MOE (1998) Ontario Ministry of the Environment. Drinking Water Surveillance Program, 1996–1997. Executive Summary Report, www.ene.gov.on.ca/envision/dwsp/index96_97.htm
122. DHS (2002) California Department of Health Services; NDMA in California drinking water. March 15, www.dhs.ca.gov/ps/ddwem/chemicals/NDMA/history.htm
123. Barrett S, Hwang C, Guo Y, Andrews SA, Valentine R (2003) Proceedings of the 2003 AWWA annual conference, Anaheim
124. CFR (2001) Code of Federal Regulations, Title 40, Chapter 1, Part 131.36
125. US EPA (1998) Announcement of drinking water candidate contaminant list. Fed Reg 63(40):10273
126. MOE (2000) Ontario Ministry of the Environment and Energy. Regulation made under the Ontario Water Resources Act: Drinking Water Protection—Larger Water Works, www.ene.gov.on.ca/envision/WaterReg/Reg-final.pdf
127. www.epa.gov/safewater/mdbp/dbp1.html
128. www.valleywater.org/media/pdf/SFPUC_NDMA_White_Paper.pdf
129. Mitch WA, Sedlak DL (2002) Environ Sci Technol 36:588
130. Mitch WA, Sharp JO, Trussell RR, Valentine RL, Alvarez-Cohen L, Sedlak DL (2003) Environ Eng Sci 20:389
131. Richardson SD, Thruston AD, Caughran T, Chen PH, Collette TW, Schenck KM, Lykins BW Jr, Rav-Acha C, Glezer V (2000) Water Air Soil Pollut 123:95
132. Richardson S (2007) Anal Chem 79:4295
133. Villanueva CM, Cantor KP, Grimalt JO, Malats N, Silverman D, Tardon A, Garcia-Closas R, Serra C, Carrato A, Castano-Vinyals G, Marcos R, Rothman N, Real FX, Dosemeci M, Kogevinas M (2007) Am J Epidemiol 165:148

134. Zwiener C, Richardson SD, DeMarini DM, Grummt T, Glauner T, Frimmel FH (2007) Environ Sci Technol 41:363
135. Jones-Lepp TL, Alvarez DA, Petty JD, Huckins JN (2004) Arch Environ Contam Toxicol 47:427
136. Zuccato E, Chiabrando C, Castiglioni S, Calamari D, Bagnati R, Schiarea S, Fanelli R (2005) Environ Health 4:1
137. Castiglioni S, Zuccato E, Crisci E, Chiabrando C, Fanelli R, Bagnati R (2006) Anal Chem 78:8421
138. Hummel D, Loffer D, Fink G, Ternes TA (2006) Environ Sci Technol 40:7321
139. Huerta-Fontela M, Galceran MT, Ventura F (2007) Anal Chem 79:3821
140. Boleda MT, Galceran MT, Ventura F (2007) J Chromatogr A 115:38
141. Bones J, Thomas KV, Pull B (2007) J Environ Monit 9:701
142. Commission of the European Communities Official Journal of the European Communities (2002) p 221
143. Rivier L (2003) Anal Chim Acta 492:69

Hdb Env Chem Vol. 5, Part S/1 (2008): 169–188
DOI 10.1007/698_5_107

Published online: 5 April 2008

# Impact of Emergent Contaminants in the Environment: Environmental Risk Assessment

Julián Blasco[1] (✉) · Angel DelValls[2]

[1]Instituto de Ciencias Marinas de Andalucía (CSIC), Campus Río San Pedro, 11510 Puerto Real (Cádiz), Spain
*julian.blasco@icman.csic.es*

[2]Departamento Química-Física, Facultad de Ciencias del Mar y Ambientales, Universidad de Cádiz, Campus Río San Pedro, 11510 Puerto Real (Cádiz), Spain



**Abstract** Human pharmaceuticals enter the environment mainly through regular domestic use. Their presence in the aquatic environment has been recorded in the range ng $L^{-1}$ to μg $L^{-1}$. Knowledge of the risk associated with the use of pharmaceuticals involves establishing the ratio between predicted environmental concentrations (PECs) and predicted no effect concentration (PNECs). The European Union (EMEA) and USA (FDA) have implemented two-tiered strategies for environmental risk assessment (ERA) of pharmaceuticals. Advances in analytical techniques have allowed us to measure pharmaceuticals in the environmental compartment and the refinement of ERA. On the other hand, for calculation of PNECs, acute and chronic toxicity tests are employed; a critical analysis of the available information was carried out, indicating that acute toxicity was only likely for spills, although an exception to this general behavior is shown by endocrine-active substances. Studies including mixtures of pharmaceuticals are not common in the study of pharmaceutical effects. Only for a limited number of drugs, are the ecotoxicity data available adequate for risk assessment. Selection of model compounds with a priori knowledge about the target biological compounds, and the selection of species, life stages and endpoints would be helpful. New technologies such as proteomics and genomics could be valuable resources to be included in the framework of pharmaceutical environmental risk assessment.

**Keywords** Ecotoxicology · Environmental concentration · Pharmaceuticals · Risk assessment

**Abbreviations**

| | |
|---|---|
| AF | Assessment factor |
| BAF | Bioaccumulation factor |
| CPMP | Committee for Proprietary Medicinal Products |
| $EC_{50}$ | Effect concentration 50% |
| EE2 | Ethynilestradiol |
| EEC | Expected environmental concentration |
| EIC | Expected introduction concentration |
| EMC | Endocrine modulating chemicals |
| EMEA | European Medicines Agency |
| ERMS | European Risk Management Strategy |
| FDA | Food and Drug Administration |
| GMOs | Genetically modified organisms |
| ICH | International Conference on Harmonization of Pharmaceuticals for Human Use |
| ISO | International Organization for Standardization |
| $LC_{50}$ | Lethal concentration 50% |
| LC-MS | Liquid chromatography tandem mass spectrometry |
| LOQ | Limit of quantification |
| NOEC | No observed effect concentration |
| OA | Oxolonic acid |
| OECD | Organization for Economic and Cooperation Development |
| OTC | Oxytetracycline |
| PBDEs | Polybromated diphenylethers |
| PEC | Predicted environmental concentration |
| PNEC | Predicted no effect concentration |
| PPCPs | Pharmaceutical and personal care products |
| QSARs | Quantitative structure—activity relationships |
| SSRI | Selective serotonin re-uptake inhibitors |
| STP | Sewage treatment plant |
| TGD | Technical Guide Document in Support of Commission Directive 93/67/EEC |

# 1 Introduction

Emergent contaminants are not easy to define because they represent a changing reality, dependent on perspective and timing [1]. The permanence in this status is dependent on its persistence in the environment, effects on humans and ecotoxicity. In this sense, knowledge of new properties of chemicals that are well known can re-introduce them as emergent contaminants. Recently, an editorial of Environmental Toxicology and Chemistry [2] pointed out that the level of concern about the new emergent contaminants is unknown and it is necessary to evaluate their significance for human and ecological health.

Four broad categories have been established for emergent contaminants: (a) pharmaceuticals and personal-care products (PPCPs); (b) polybromated diphenylethers (PBDEs) and other persistent organic contaminants; (c) endocrine modulating chemicals (EMCs) and (d) nanotechnology products. These categories are not totally separated because a compound could be at the same time a PPCP and an EMC.

Herein we will focus on the environmental risk assessment of human pharmaceuticals because the ERA of the different types of emergent contaminants pointed out above is beyond the scope of this work.

Entry of human pharmaceuticals and PCCPs to the environment is mainly via regular domestic use [3]. After their use, pharmaceuticals are excreted, some of them are partially metabolized (slightly transformed or conjugated to polar molecules) and released into the aquatic environment via wastewater effluent. Unused drugs are stored until the expiration date and finally exposed of down drains reaching the aquatic environment. Consequently, they can potentially affect drinking water quality. The entry path scenarios for human pharmaceutical products have been summarized by the Committee for Proprietary Medicinal Products (CPMP) (Fig. 1) [4].

Variable quantities of pharmaceuticals are present in surface waters, ground waters, and sediment, ranging in concentrations between ng $L^{-1}$ to µg $L^{-1}$ [5, 6]. Knowledge of pharmaceuticals in environmental compartments has been supported by the great advance in analytical techniques, which has improved detection levels of these compounds in the environment. New chemical methods, such as liquid chromatography tandem mass

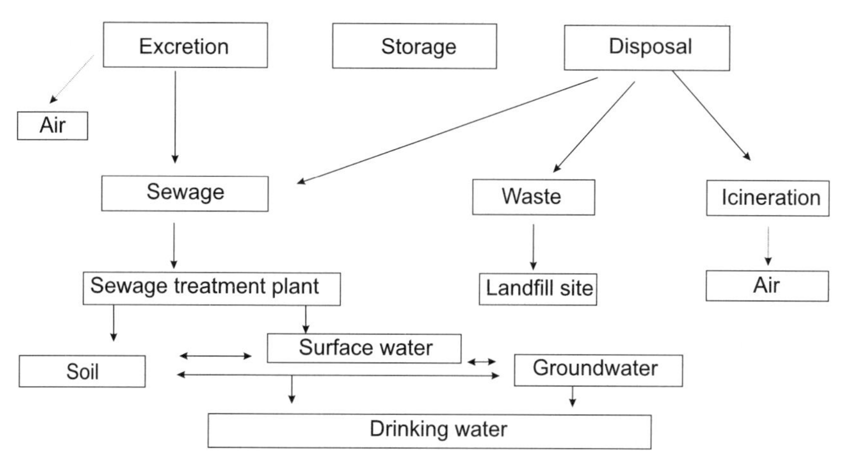


**Fig. 1** Routes of entry to the environment for human pharmaceuticals [4]

spectrometry (LC-MS), are able to determine more organic polar compounds without derivatization [7–9]. As a consequence, several monitoring programs have been carried out in different countries that have demonstrated the presence of drug residues to be widely distributed.

On the other hand, knowledge concerning the ecotoxicological effects of pharmaceuticals on aquatic and terrestrial organisms and wildlife is scarce, especially the aspects related to chronic toxicity and more-subtle effects [10]. Most of the published aquatic toxicity data and risk assessments for human pharmaceuticals are based on short-term acute studies [5, 11, 12]. Nevertheless, information about the chronic effects of human pharmaceuticals on aquatic organisms has been recently reviewed by Crane et al. [13].

Although the amounts of human drugs released to the environment are quite high, only recently have detailed guidelines been developed about how pharmaceuticals should be assessed.

## 2 Environmental Risk Assessment Regulations

Environmental risk assessment is a process that evaluates the likelihood that adverse effects may occur as a result of exposure to one or more stressors [15]. The characterization of the risk involves knowing the ratio between predicted environmental concentration (PEC) and predicted no effect concentration (PNEC); if this value is less than 1 there is no risk to the ecosystem, but if the value is equal to or higher than 1 there is a risk and regulation activities will be needed.

Although the market for pharmaceuticals is highly globalized, and harmonization for testing guidelines have been supported by the International Conference on Harmonization of Pharmaceuticals for Human Use (ICH), for the ERA of human pharmaceuticals different strategies have been followed in different countries according to specific regulations.

### 2.1 Regulations in the EU

The European Commission has released a guideline about the environmental risk assessment of medicinal products for human use, in accordance with Article 8(3) of Directive 2001/83/EC, as amended, the evaluation of the potential environmental risks posed by medicinal products, their environmental impact should be assessed and, on a case-by-case basis, specific arrangements to limit the impact should be considered [14]. The ERA should accompany any application for a marketing authorization for a medicinal product for human use and the evaluation of the environmental impact should be made also if there is an increase in the environmental exposure. Nevertheless, this guideline does not apply to medicinal products consisting of genetically modified organisms (GMOs).

The evaluation of risk assessment to the environment is a step-wise process, consisting of two phases. The first phase (Phase I) includes checking the exposure of the environment to the drug substance against the action limit assessment. If the result is lower than the limit assessment the ERA is finished. Alternatively, second-phase information about the fate and effect of the drug substance should be carried out. This Phase II is divided into two parts (Tier A and B). In Table 1, the phase approach of environmental risk assessment according to the guidelines of EMEA is shown [14]. Phase I is considered a pre-screening and it is independent of route administration, pharmaceutical characteristics, metabolism, and excretion. The calculation of PEC is restricted to the aquatic environment and some restrictions are considered:

- A market penetration factor (Fpen) is defined, the value can be a default value or refined according to specific data (eg. Epidemiological data).
- The amount is distributed along the year and the considered geographic area.
- The sewage system is the main route of entry for the substances.
- No biodegradation of the substance is taken into account during the treatment in the sewage treatment plant (STP).
- Metabolism in the patient is not considered.

For calculation of the PEC the following equation is applied [14]:

$$\mathrm{PEC}_{\mathrm{surfacewater}} = \frac{\mathrm{Dose}_{\mathrm{ai}} \times \mathrm{F}_{\mathrm{pen}}}{\mathrm{Wastewater}_{\mathrm{inh}} \times \mathrm{Dilution}} \tag{1}$$

**Table 1** The phase approach in environmental risk assessment according to the Committee for Medicinal Products for Human Use [14]

| Stage in regulatory evaluation | Stage in risk assessment | Objective | Method | Test/data requirement |
|---|---|---|---|---|
| Phase I | Pre-screening | Estimation of exposure | Action limit | Consumption data, log $K_{\mathrm{ow}}$ |
| Phase II Tier A | Screening | Initial prediction of risk | Risk assessment | Base set aquatic toxicology and fate |
| Phase II Tier B | Extended | Substance and compartment – specific refinement and risk assessment | Risk assessment | Extended data set on emission, fate and effects |

where $Dose_{ai}$ ($mg\,inh^{-1}\,d^{-1}$) is the maximum daily dose consumed per inhabitant; $F_{pen}$ is the percentage of market penetration and represents the proportion of the population being treated daily with a specific substance; $Wastewater_{inh}$ ($L\,inh^{-1}\,d^{-1}$) corresponds to the amount of wastewater per inhabitant and per day and Dilution is the dilution factor.

When the $PEC_{surfacewater}$ value is below $0.01\,\mu g\,L^{-1}$ and there are no other environmental concerns it is assumed that the pharmaceutical is not a risk. In the case where the $PEC_{surfacewater}$ is above this value, a Phase II environmental fate and effect analysis should be carried out. In drugs that have a $PEC_{surfacewater}$ lower than $0.01\,\mu g\,L^{-1}$ but may affect reproduction a strategy including Phase II evaluation should be carried out.

In the Phase II assessment, the evaluation of the PEC/PNEC ratio is based on aquatic toxicology data and predicted environmental concentration (Tier A). For drugs where a potential impact can be weighted a refinement of the values should be realized in Tier B. The guidelines for experimental bioassays of the Organization for Economic Cooperation and Development (OECD) or the International Organization for Standardization (ISO) should be followed and all relevant data about physical-chemical properties, metabolism, excretion, biodegradability, persistence, and pharmacodynamic processes must be taken into account.

For the aquatic effect analysis standard long-term toxicity tests in fish, *daphnia*, and algae are proposed (OECD 201, 210, and 211) [16] and to determine the $PNEC_{water}$ an assessment factor (AF) is applied to the no-observed effect concentration (NOEC).The AF applied is a default value of 10 and it represents the uncertainty associated to intra-species variability and inter-species sensitivities and extrapolation from lab to field studies.

The refinement of the risk when it has been identified in Tier A involves refining PEC and PNEC values for the compounds using data on transformation of the substance in the environment. The equation that should be applied is:

$$PEC_{surfacewater} = \frac{Elocal_{water} \times F_{stpwater}}{Waste_{inh} \times Capacity_{stp} \times Factor \times Dilution} \quad (2)$$

$$Elocal_{water} = Dose_{ai} \times F_{excreta} \times F_{pen} \times Capacity_{stp} \quad (3)$$

$Waste_{inh}$ = amount of wastewater per inhabitant per day

$Capacity_{stp}$ = capacity of local sewage treatment plant

$F_{stpwater}$ = fraction of emission directed to surface water

Factor = factor to take into account the adsorption to suspended matter

Dilution = dilution factor

$Elocal_{water}$ = local emission to wastewater of the relevant residue.

If the pharmaceuticals can be adsorbed on soil or sediment, an effect analysis on sediment-dwelling organisms should be carried out and compared

**Table 2** Terrestrial fate and effects studies recommended in Phase II Tier B, according to the Committee for Medicinal Products for Human Use [14]

| Study type | Recommended protocol |
|---|---|
| Aerobic and anaerobic transformation in soil | OECD 307 |
| Soil microorganisms: Nitrogen transformation test | OECD 216 |
| Terrestrial plants, Growth test | OECD 208 |
| Earthworm, Acute toxicity tests | OECD 207 |
| *Collembola*, Reproduction test | ISO 11267 |

to $PEC_{sediment}$ (OCDE 308) [16]. For compounds with $K_{OC} > 10\,000\,L\,kg^{-1}$, unless they are readily biodegradable, methodologies such as TGD [17] are recommended for risk assessment including $PEC_{soil}$ calculation. The bioassays recommended for Phase II Tier B in soils are shown in Table 2.

Recently, the European Risk Management Strategy (ERMS) work programme for 2008 and 2009 has been adopted, which will focus on improvement of the EU Pharmacovigilance system and the science and methodologies which give support to the safety monitoring of medicines for human use [17].

## 2.2 Regulations in USA

The National Environmental Policy Act of 1969 requires the Food and Drug Administration (FDA) to take into account the environmental impact of approving drug and biologic applications as an integral part of its regulatory process. A guidance was prepared by the direction of the Chemistry Manufacturing Controls Coordinating Committee, Center for Drug Evaluation and Research (CDER) and the Center for Biologics Evaluation and Research (CBER) and it represents the current thinking on environmental assessment. This guidance [18] involves several topics, among them: the content and format of environmental assessment (EAs), test methods and specific guidance for the environmental issues that are associated with human drugs.

According to this guidance, the EA is required when the estimated concentration of the compound is: (a) equal or higher than $1\,\mu g\,L^{-1}$; (b) when the substance occurs naturally but its application alters significantly its concentration or distribution or its metabolites and (c) when the expected exposure levels can potentially generate harm to the environment. A tiered approach is employed to assess the environmental fate and effects of pharmaceuticals (Fig. 2).

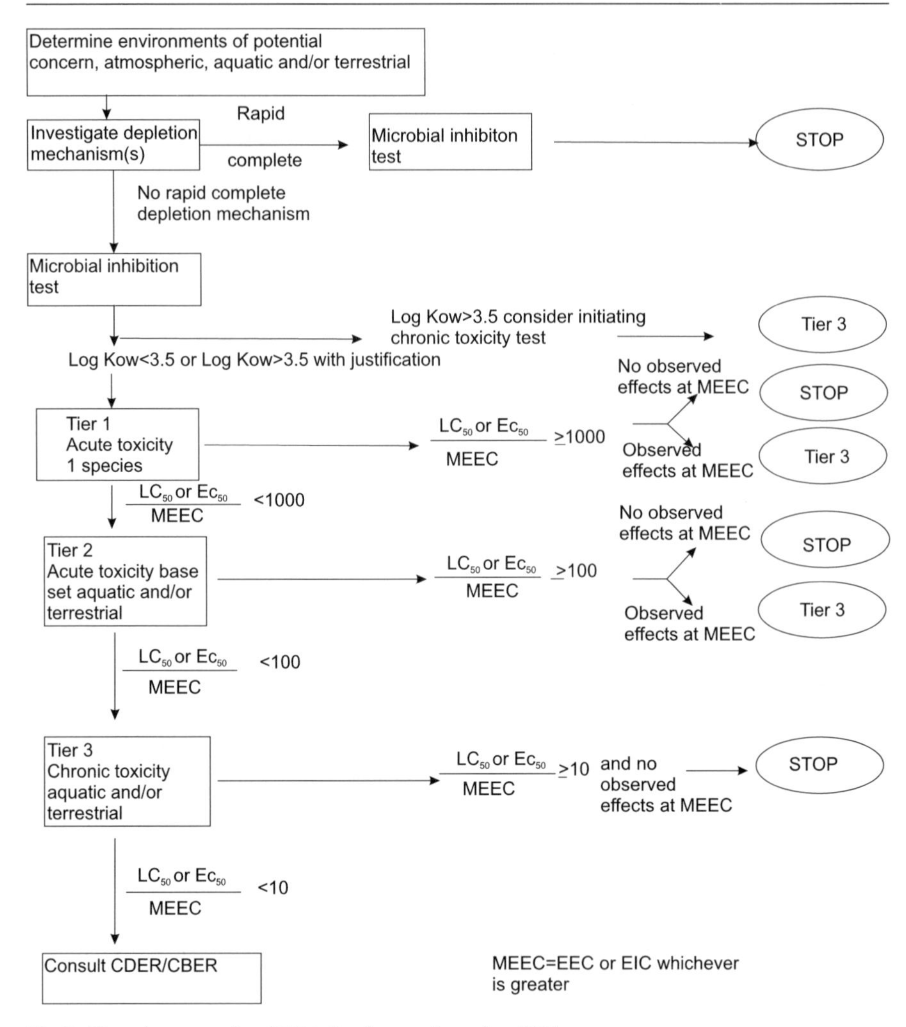


**Fig. 2** Tiered approach of FDA for fate and testing [18]

The expected introduction concentration (EIC) should be estimated and the method for calculating this value in aquatic media is:

$$\text{EIC} - \text{aquatic(ppb)} = A \times B \times C \times D$$

$A = \text{kg}\,\text{y}^{-1}$ produced for direct use
$B = 1/\text{L}$ per day entering in STP
$C = \text{year}/365$ days
$D = 10^9\ \mu\text{g}\,\text{kg}^{-1}$

Some kinds of drug may enter the terrestrial environment when biosolids from waste water treatment plant facilities with adsorbed material are applied to soil. The calculation of this concentration is carried out considering the typical treatment, disposal, and application processes. A metabolizing process (biodegradation) occurs during the waste treatment process and it should be considered for calculating EIC.

The PEC is calculated using EIC and taken into account are the processes which affect the compound (spatial or temporal variations, dilution, degradation, sorption, etc.). Normally, EPA applies a dilution factor of 10 to the EIC-aquatic to estimate the PEC.

In summary, the fate of the substance should be provided for the environmental compartment and the transport between compartments should be taken into account if it is of interest to the environmental behavior of the compound.

The evaluation of the effect of pharmaceuticals is oriented to the aquatic compartment because their effect will be on aquatic organisms. Nevertheless, for compounds with high adsorption capacity or high degradation rate, its effects in the aquatic environment could not be considered. For the terrestrial environment, fate and effects testing should be considered when the substance has a $K_{OC} > 10^3$.

Testing of the environmental effects of the pharmaceuticals should be carried out according to the tiered approach as was indicated in Fig. 2. If the compound is not removed from the environment quickly, its persistence and the associated toxic effects should be taken into account. A tiered approach should be used (as was proposed in the guidance), thus the ratio between $LC_{50}$ or $EC_{50}$ and the EIC or EEC is employed as the assessment factor (10, 100, and 1000) to carry out toxicity tests at different levels. The toxicity tests should be performed according to the protocols defined by FDA, OECD, and other peer-reviewed literature if they are appropriate for environmental studies.

## 3 Pharmaceutical Environmental Concentrations

### 3.1 Predicted Environmental Concentration

The ERA requires one to know the occurrence and concentration of compounds in the environmental compartments. The exposure assessment should take into account the fate of the substance released to the environment and predict the environmental concentration [19]. The lack of information about measured levels of pharmaceuticals in environmental compartments mean that to carry out the ERA for pharmaceuticals the $PECs_{surfacewater}$

have been estimated, in many cases, according to the recommendations of EMEA or FDA [14, 18]. A review of 111 substances, corresponding to the highest-selling human drugs that have annual sales in Germany of more than 5000 kg, has been carried out. For all compounds the values were higher than 0.01 $\mu g\,L^{-1}$ [20]. According to the scheme developed by EMEA a Phase II process should be carried out for evaluating the exposure. The $PEC_{surfacewater}$ for pharmaceuticals according to data for its use in Germany, Sweden, France and UK [19–23] are presented in Fig. 3. The differences among $PEC_{surfacewater}$ should be related to drug prescription patterns in the countries. These data correspond to the worst case because degradability is not considered. Thus, for paracetamol the PEC is 367.3 $\mu g\,L^{-1}$ [19], although a high degree of elimination, around 98%, has been observed during activated-sludge wastewater treatment [7]. On the other hand, for other compounds such as oxytetracycline (OTC), human metabolism is limited [24], and the compound will be excreted without transformation. It has been observed that biodegradation

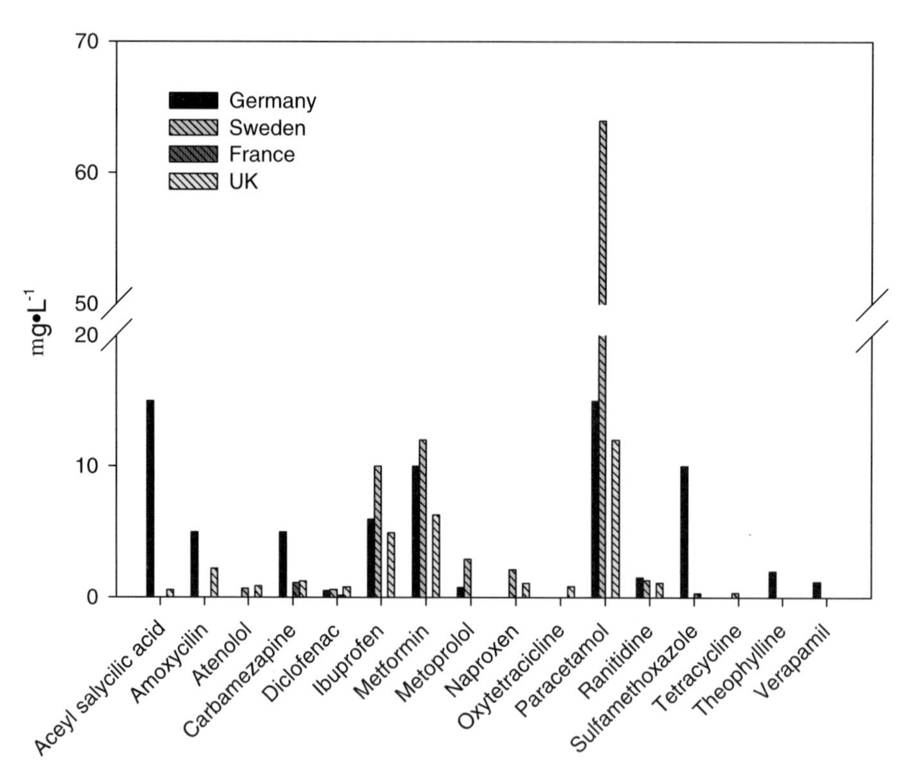


**Fig. 3** Predicted environmental concentration (PEC) for pharmaceuticals in surface water of several countries (Germany, Sweden, France, and UK). Data were extracted from [20, 21, 23, 57]

for OTC is limited [25]; and its PEC will be equal to 0.62 $\mu g\,L^{-1}$ after applying a dilution factor of 10.

## 3.2 Measured Environmental Concentration

### 3.2.1 Effluent Sewage Treatment Plant

The first work on the presence of drug residues in STP effluents was carried out in the USA and it was focused on clofibric acid, the metabolite of three lipid regulators: clofibrate, etofyllin clofibrate and etofibrate at $\mu g\,L^{-1}$ concentration levels in treated sewage [26]. Later, significant advances in analytical techniques have allowed one to measure pharmaceuticals in environmental compartments [27]. The main drawback of the conventional analytical approach is target-compound monitoring which is insufficient to assess the environmental relevance of emerging contaminants, and the lack of knowledge about the transformation products. Other problems relating to conjugated metabolites (e.g. glucuronides and sulfate conjugates) which can be deconjugated by microbial actions in STP have been pointed out [28].

The pharmaceutical levels in the effluents of STP in many countries are high. Table 3 presents information on the levels for individual compounds in the effluents of STP in Germany, Greece, Spain, and Switzerland. The highest concentrations were recorded in the effluent of STP in Seville (Spain) for two anti-inflammatory drugs, ibuprofen and naproxen, with concentrations of 48.2 and 4.3 $\mu g\,L^{-1}$, respectively [29]. The differences between influent and effluent showed the degradability of these compounds. The values recorded for ibuprofen in the Seville STPs are very high, because the concentrations are below 1 $\mu g\,L^{-1}$, normally. Acetylsalicylic can be degraded into its metabolites, although they are eliminated in the STP process; thus only the metabolite salicylic acid has been detected in sewage effluents [30, 31]. The ubiquity of target compounds can be related to the metabolism, sales, and practices carried out in each country. Therefore, analgesics and antibiotics are detected frequently because they are excreted as the unchanged parent compound; in addition the high loads of analgesic and anti-inflammatories, in comparison with other therapeutic groups is attributed to the higher consumption. The removal efficiency is related to the treatment applied in each plan and the compound physicochemical characteristics and hydraulic retention time [32].

### 3.2.2 Environmental Levels

In developed countries, production and use of pharmaceuticals are increasing annually [33]. The measurement of these compounds in environmental

**Table 3** Concentration range and mean concentration in $\mu g\,L^{-1}$ of pharmaceuticals and metabolites in effluents of municipal STPs of several countries

| Drug | Germany | Greece | Spain | Switzerland | Canada |
|---|---|---|---|---|---|
| Acetyl salycilic acid | 0.32–0.92 | na | na | na | na |
| Diclofenac | 0.21–1.11 | 0.20–0.34 | blq–0.38 | 0.1–0.7 | 0.015–0.039 |
| Ibuprofen | 0.32–0.58 | na | 0.78–48.24 | 0.005–1.5 | 2.2–3.5 |
| Naproxen | 0.12–0.53 | nd | 0.22–4.28 | 0.1–3.5 | 1.0–1.7 |
| Indometazine | 0.07–0.11 | na | na | na | 0.048–0.075 |
| Benzafibrate | 0.72–1.2 | nd–0.15 | na | na | 0.13–0.28 |
| Gemfribozil | 0.12–0.35 | na | na | na | 0.37–0.60 |
| Fenofibric acid | 0.32–0.44 | nd | na | na | na |
| Clofibric acid | 0.42–0.69 | na | na | nd–0.06 | na |
| Carmabezapine | 1.31–2.2 | na | blq–1.29 | 0.1–0.8 | na |
| Phenazone | 0.12–0.20 | na | na | na | na |
| Porpanolol | 0.34–0.48 | na | na | na | na |
| Metoprolol | 1.72–2.44 | na | na | na | na |
| Bisoprolol | 0.12–0.16 | na | na | na | na |
| Betaxolol | 0.14–0.20 | na | na | na | na |
| Terbutalin | 0.10–0.12 | na | na | na | na |
| Carazolol | 0.05–0.09 | na | na | na | na |
| Dihydrocodeine | 1.47 | na | na | na | na |
| Hydrocodone | 0.72 | na | na | na | na |
| Ketoprofen | nd | 0.27–0.82 | blq–3.48 | nd–0.20 | 0.015 |
| Mefenamic acid | nd | 0.08–0.22 | na | na | na |
| Primidone | nd–0.88 | nd | na | na | na |
| Propyphenazone | nd–0.74 | nd | na | na | na |
| Salycilic acid | nd–0.65 | 0.64–2.0 | 0.57 | na | 0.054–0.46 |
| Caffeine | na | na | 0.15–3.20 | na | na |

* Data were extracted from [6, 29, 36]
*na* not analysed, *nd* not detected, *blq* below limit quantification

compartments can improve knowledge about the occurrence and persistence of the compounds in the environment. The advances in analytical techniques have allowed one to measure extremely low concentrations of pharmaceuticals in surface water, rivers, streams, etc. [34]. The occurrence of organic wastewater contaminants is high in the environment, 80% of 139 streams sampled in the USA [35] showed at least one organic wastewater contaminant, although the authors pointed out that the results were influenced by the design of the study and it can not be considered as representative of the global situation in USA streams. The concentrations were, in general, less than 1 $\mu g\,L^{-1}$ but their presence in many streams indicated that compounds survived biodegradation.

Pharmaceuticals in effluents of wastewater treatment plants are diluted when entering river waters being detected in the $ng\,L^{-1}$ range. However, the same spectrum of compounds that are found in the STP are found in the Ebro river basin where analgesics (diclofenac, naproxen, ibuprofen), lipid regulators (gemfibrozil, bezaibrate), antibiotics (azithomycin, trimethoprim, and sulfamethoxazole), antipiletic (carbamezapine), antihistamic (ratidine), and $\beta$-blockers (atenolol and sotanol) are the recorded compounds, which are consumed at high levels in Spain [32]. Drugs in a large body of receiving water are in many cases below detection limits although in small receiving streams were around 15–30% effluent median concentration [36]. The availability of occurrence data for pharmaceuticals in estuarine or marine waters is less common than stream and river waters. In the North Sea, for clofibric acid concentrations of $1\ \mathrm{ng\,L^{-1}}$ have been reported, whilst in seawater samples ibuprofen has not been measured above $0.2\ \mathrm{ng\,L^{-1}}$ [37, 38]. Pharmaceutical residues are present as contaminants in UK estuaries [39], but the authors only detected above the detection limits the following targeted compounds/metabolites: clofibric acid, clotrimazolem dextropropoxyphene, dicofenac, ibuprofen, mefenamic acid propanolol, tamoxifen, and trimethoprim, with ibuprofen showing the highest detected concentration ($928\ \mathrm{ng\,L^{-1}}$). In the Victoria Harbor of Hong Kong, antibiotics (belonging to the class quinolones, macrolides, sulfonamides, $\beta$-lactam, and chloramphenicol) were mainly below the limit quantification (LOQ). However, they were found in the Pearl River during the high and low water seasons in the range $10–100\ \mathrm{ng\,L^{1}}$. The level of antibiotics in the high water season is controlled by daily sewage discharge patterns and in the low season may be controlled by water column dynamics [40].

There is less knowledge about pharmaceutical concentrations in soil and sediment than for the aquatic environment. This was due to the lack of suitable sensitive analytical methods for the detection of compounds [41]. The persistence of a drug in a sediment or soil mostly depends on its photostability, its binding and adsorption capability, its degradation rate, and leaching in water [42]. The main route of entry for antibiotics for human use is related to the use of sewage sludge for fertilizing the soil. The occurrence of fluoroquinolones, ciprofloxacin, and norfloxacin in sewage sludge has been detected at concentrations ranging between 1.4 to $2.4\ \mathrm{mg\,kg^{-1}}$ [43], which is in the same range as can be measured in digested sludge, indicating a high affinity to the solid phase. Most of the literature on pharmaceuticals in solid environmental samples is related to veterinary drugs, especially those employed in fish farming, which are principally antibiotics.

Pharmaceuticals, as other chemical compounds, can be accumulated by aquatic or benthic organisms. Oxytetracycline (OTC, tetracycline) and oxolonic acid (OA, quinolone) are accumulated by the blue mussel, preferentially being accumulated in the viscera for OTC and in the gills for OA. Bioaccumulation factors (BAF) were low ($< 0.5$) regardless of the analyzed bivalve part. The application of $K_{\mathrm{ow}}$ for antibiotic bioaccumulation can predict a weak accumulation in mussel for antibiotics with $K_{\mathrm{ow}} < 2$, whereas antibiotics such as macrolides with $K_{ow} > 2$ accumulate at a higher level [44]. Fluoxetine and sertraline are prescribed as antidepressants and their occurrence has been detected in surface water or effluent discharges [35, 45]. The analysis of these compounds in streams from a reference site and an effluent-dominated stream showed that these compounds were not detected in the reference site whereas they were detected in all tissues analyzed from fish from the effluent-dominated stream, including *P. nigromaculatus*, L. *macrochirus*, and *I. punctatus*, with a preferential accumulation in the brain, although they also accumulate in muscle at concentrations higher than the limits of quantitation, and subsequently an exposure route to humans in this way should be considered [46]. The influence of pH on the bioconcentration factor of fluoxetine in the fish *Oryzia latipes* has been analyzed [47], showing that BCF values were lower at pH 7 and higher at pH 9 because of an increase of hydrophobicity at pH values closer to $pK_a$.

## 4 Ecotoxicology of Human Pharmaceuticals

### 4.1 Acute Toxicity

Aquatic organisms are targets to analyze the effect of human pharmaceuticals because they are exposed via wastewater over their whole life. Drugs are designed to have a specific mode-of-action along the target pathway. Hypotheses about the mode-of-action in lower animals in many cases are not well supported, because many of the organisms lack the required receptors. Although a mode-of-action for a pharmaceutical should be taken into account when an experiment is designed, this approach may not be appropriate because the mode-of-action could be different or not well known [48].

The ecotoxicological effects of human pharmaceuticals are focused on acute and standard tests. More than three-hundred-and-six endpoints for pharmaceutical ecotoxicity data have been collected for macroinvertebrates, fish, and algae, and over one-hundred for human pharmaceuticals [12]. The selection of three trophic levels (algae, *Daphnia*, and fish) showed that sensitivity followed the order algae > *Daphnia magna* > fish. However, the range of acute toxicity endpoints varied from > 15 000 mg L$^{-1}$ (for atropine sulfate-anthicolorgenic/mydriatic) [49] to < 0.003 mg L$^{-1}$ for fluvoxamine (antidepressant) [50]. The ecotoxicity effects for therapeutic classes showed the following order: antidepressants, antibacterials, and antipsychotics [12]. A recent review [48] summarized the ecotoxicity data, taking into account the ecological relevance and the different classes of human pharmaceuticals: analgesic and non-steroidal anti-inflammatory drugs, beta-blockers, blood lipid-lowering agents, neuroactive compounds, and cytostatic compounds and cancer therapeutics. Seventeen percent showed acute toxicity below $100\,mg\,L^{-1}$ and 38% above $100\,mg\,L^{-1}$, which is classified as not harmful for aquatic organisms according to EU Directive 93/67/EEC. The rest of the compounds (45%) showed high variability in acute toxicity tests. The difference between the acute toxicity data and the environmental levels for human pharmaceuticals demonstrate that only in the case of spills will the toxicity be relevant.

## 4.2 Chronic Toxicity

The standard acute toxicity tests have as endpoints the lethality and they do not seem appropriate for risk assessment of pharmaceuticals, because of the nature of these compounds. The use of chronic tests over the life-cycle of organisms for different trophic levels could be more appropriate [51]. Nevertheless, the database for this kind of bioassay is very limited.

Most chronic aquatic toxicity data for human pharmaceuticals are available for algae because they are the quickest to perform and therefore less expensive. The sensitivity to antimicrobial substances is higher in Cyanobacteria such as *Microcystis aureginosa* than standard algal toxicity tests (*Pseudokirchneriella subcapitata*) although there are no differences for non-antimicrobial substances [52].

Only in the case of the synthetic steroid EE2, which is present in contraceptive pills, has an effect been observed at environmentally relevant concentrations. In a recent study [53], vitellogenin induction in fathead minnows was reported at an $EC_{50}$ value of $1\,ng\,L^{-1}$. The life-cycle exposure of zebrafish to $3\,ng\,L^{-1}$ EE2 provoked an increase of vitellogenin and caused gonadal feminization [54]. The exposure of some invertebrate taxa (snails) to EE2 also caused effects at very low concentrations $\sim 1\,ng\,L^{-1}$ [55]. Fish are also sensitive to other sex hormones such as methyltestosterone and beta-adrenergic receptor blockers [56].

Analgesic and non-steroidal anti-inflammatory drugs are the most-consumed drugs, and a chronic study with diclofenac has been reported in invertebrates [22, 57]. A chronic study with rainbow trout showed renal lesions at $5\,\mu g\,L^{-1}$ [58]. Regarding beta-blockers, propanolol showed chronic toxicity not only on the cardiovascular system in fish but also in the reproductive system [48]. The number of eggs released by fish was reduced at $0.5\,\mu g\,L^{-1}$ after four weeks of exposure but not at 50 and $100\,\mu g\,L^{-1}$ [59]. The blood lipid-lowering agents have been evaluated by traditional toxicity tests and NOEC in the range of $246\,\mu g\,L^{-1}$ to $70\,mg\,L^{-1}$ have been recorded for *B. caliciflorus* (2 days) and early life stages of zebrafish (10 days), respectively [57].

Chronic toxicity tests have been carried out with carbamezapine (an antiepileptic) and *C. dubia* showed a NOEC (7 days) = $25\,\mu g\,L^{-1}$ [57]. Lethal concentration in zebrafish was reported at 43 $\mu g L^{-1}$ [60]. Chronic studies have been carried out on selective serotonin re-uptake inhibitors (SSRI). Serotonin is a neurotransmitter found in vertebrates and invertebrates. SSRI may affect the function of the nervous and associated hormonal systems. The role of serotonin varies between phyla and in consequence also the effects of SSRI; in medaka (*O. latipes*) serotonin induced oocyte maturation [61] but the opposite action was reported in mummichog (*F. heteroclitus*) [62]. The chronic effects of SSRI on reproduction in fish and invertebrates are not yet clear, interference in the reproduction occurred at concentrations not ecologically relevant [48].

To date, chronic toxicity data using marine or estuarine species have been very scarce. The results with different classes of compounds (carbamezapine, acetaminophen, and ibuprofen) and the endpoint inhibition growth at 72 h for the marine microalgae *Phaeodactylum tricornutum* did not show toxicity below 2.0 $mg L^{-1}$.

Studies concerning the effects of mixtures of pharmaceuticals are very limited in the scientific literature [63, 64]. The mixture of diclofenac, ibuprofen, naproxen, and acetylsalicylic acid has been evaluated using *Daphnia* and algae, the toxicity of the mixture followed the concept concentration addition. Nevertheless, the effects of mixtures of compounds with different modes-of-action depends on the species and they do not all act in the same way. Few studies concerning the toxicity of mixtures of pharmaceuticals in realistic ecological systems (microcosms and mesocosms) have been carried out. The effect of a combination of eight pharmaceuticals at three levels on *Lemna gibba* and *Myriophyllium sibiricum* has been tested [65]. In a similar microcosm (periphyton, phytoplankton, zooplankton, algae, and benthic communities), three pharmaceuticals with different modes-of-action were analyzed at three levels [66]. At low concentrations (6–10 $\mu g L^{-1}$) only trends were appreciable and no significant effects could be recorded. The comparison of assayed treatment with current concentrations in the environment did not allowed to establish a risk situation for this mixture. Nevertheless, many pharmaceuticals are present in the environment and the effect of this "cocktail" could affect to aquatic communities.

## 5 Environmental Risk Assessment

The objective of environmental risk assessment is to determine the nature and likelihood of the effects of human actions (in this case the use of pharmaceuticals) on animals, plants, and the environment [67]. According to this principle, operational monitoring in support of this concept should be adequate for characterizing exposure and effects [68]. The two-tiered approach (EU and USA) is employed normally for risk assessment of pharmaceuticals (see Sects. 2.1 and 2). In both risk strategies trigger values are selected for further research via tiered assessment 0.01 $\mu g\,L^{-1}$ and 0.1 $\mu g\,L^{-1}$, respectively. The use of this value permits a reduction in the need to carry out many assessments which facilitates the release of new drugs to the market. However, for some compounds this trigger value is insufficient; this is the case for endocrine disruptors which at 1 $ng\,L^{-1}$ showed environmental effects, below the stricter trigger value.

The potential effect of pharmaceuticals is calculated according to the ratio between PEC and PNEC. The PEC is calculated in many cases using figures such as sales, density of population, etc., representing the worst case. In order to get a refinement of this value more precise environmental risk assessment should be carried out; data for biodegradation adsorption, and abiotic factors (pH, temperature) of the environment should be taken into account. The use of measured concentrations allows one to establish more realistic ERA. The other data which should be available is the PNEC, but the lack of chronic toxicity data has made it difficult to perform this assessment. The use of the assessment factor when only acute data are available involves the reduction of uncertainty associated with its use [22]. Though the use of a quantitative structure—activity relationship has been pointed out as a possibility for identifying hazard or prioritizing substances to be analyzed it is not sufficiently precise for risk assessment [48].

The risk of an acute toxic effect from pharmaceuticals in the environment is unlikely [21]. However, many drugs have been designed to affect specific biological systems in target organisms at relatively low dose and exposure concentrations. For this reason, the long-term sublethal effects of pharmaceuticals could be a greater potential concern than acute effects. With the exception of a limited number of drugs, available ecotoxicity data could be inadequate for risk assessment and an extensive suite of chronic sublethal tests may be necessary [69].

# 6 Concluding Remarks

Although human pharmaceuticals are found at low concentrations in the environment and acute toxicity is not frequent, a broad database with chronic and subtle toxicity tests is necessary to carry out the ERA of these compounds. A priori knowledge about the target biological pathway can identify compounds with higher priority for testing and the species, life stages, and endpoints suitable for testing. In this sense, the selection of estuarine and marine species should be considered.

On the other hand, biomarkers as responses to molecular or biochemical changes can be useful for ecological risk assessment. In vitro systems can be appropriate tools for screening the ecotoxicological effect of pharmaceuticals before fish toxicity testing is carried out. The lack of toxicity tests for pharmaceutical mixtures should be taken into account in order to improve the risk assessment because of the additive, antagonistic, or synergetic effects that can be present. Finally, new technologies such as proteomics and genomics, which are powerful tools for human diagnosis, are under development and they may be helpful to validate effects in the environment and should be included in the framework of ERA, although its use is limited by the current knowledge of the impacted biota.

## References


1. Field JA, Johnson CA, Rose JB (2006) Environment Sci Technol 40:1
2. Chapman PM (2006) Environ Toxicol Chem 25:2
3. Daughton CG, Termes TA (1999) Environ Health Perspect 107:907
4. CPMP (Committee for Propietary Medicinal Products) (2003) Note for Guidance on Environmental Risk Assessment of Medicinal Product for Human Use. CPMP/SWP/4447/00. EMEA, London
5. Halling-Sorensen B, Nielsen SN, Lanzky PF, Ingerslev F, Holten Lützhoft HC, Jorgensen SE (1998) Chemosphere 36:357
6. Díaz-Cruz S, Barceló D (2004) Ocurrence and Analysis of Selected Pharmaceuticals and Metabolites as Contaminants Present in Waste Water, Sludge and Sediments. In: Barceló D (ed) The Handbook of Environmental Chemistry, vol 5. Springer, Berlin Heidelberg New York, p 227
7. Ternes T, Hirsch R, Mueller JM, Haberer K (1998) Fresen J Anal Chem 362:329
8. Ternes T, Bonerz M, Schmidt T (2001) J Chromatogr A 938:175
9. Kolpin DW, Furlong ET, Meyer MT, Thurman EM, Zaugg SD, barber LB, Buxton HT (2002) Env Sci Technol 36:1202
10. Fent K, Weston AA, Caminada D (2006) Aquat Toxicol 76:122
11. Cunningham VL, Constable DJC, Hannah RE (2004) Environ Sci Technol 38:3351
12. Webb SF (2004) A data based perspective on the enviromental risk assessment of human pharmaceuticals I—Collation of available ecotoxicity data. In: Kümmerer K (ed) Pharmaceuticals in the Environment, Springer, Berlin, p 317
13. Crane M, Watts C, Boucard T (2006) Sci Total Environ 367:23
14. EMEA (European Medicines Angency) (2006) Guideline on the Environmental Risk Assessment of Medicinal Products for Human Use. Doc Ref EMEA/CHMP/SWP/4447/00. EMEA, London
15. US Environment Protection Agency (1998) Guidelines for Ecological Risk Assessment EPA/630/R-95/002F. USEPA, Washington DC
16. OECD (1996) Guidelines for testing chemicals. OCDE Publications Service, Paris, p 1040
17. EMEA (2007) European Risk Management Strategy: 2008–2009 work programme adopted. Doc Ref EMEA/564868/2007. EMEA, London
18. Guidance for Industry Environmental Assessment of Human Drug and Biologics Application (1998) http:// www.fda.gov/cder/guidance/index.htm, Accessed 25 March 2008
19. Webb SF (2004) A data based perspective on the enviromental risk assessment of human pharmaceuticals II—Aquatic Risk Characterisation. In: Kümmerer K (ed) Pharmaceuticals in the Environment, Springer, Berlin, p 345

20. Huschek G, Hansen PD, Maurer H, Krengel D, Kayser A (2004) Environ Toxicol 19:226
21. Carlsson C, Johansson AK, Alvan G, Bergman K, Kühler T (2006) Sci Total Environ 364:67
22. Ferrari B, Mons R, Vollar B, Frayse B, Paxeus N, Giudice RL, Garric J (2004) Environ Toxicol Chem 23:1344
23. Jones OAH, Voulvoulis N, Lester JN (2002) Water Res 36:5013
24. Dollery CT (1991) Therapeutic drugs, vol I and II. Churchill Livingstone, Edinburgh
25. Richardson ML, Bowon JM (1985) J Pharm Pharmacol 37:1
26. Ternes TA, Meisnhiemer M, McDowell D, Sacher F, Braunch HJ, Haist-Gulde B, Preuss G, Wilme U, Zulei-Seilbert N (2002) Environ Sci Technol 36:3855
27. Kot-Wasik A, Debska J, Namiesnik J (2007) Trends Anal Chem 26:557
28. Barcelo D. Trends Anal Chem 26:454
29. Santos JL, Aparicio I, Alonso E (2007) Env Int 33:596
30. Heberer T, Fhurmann B, Schmidt-Bäumler K, Tsipi D, Koutsouba V, Hiskia A (2001) Ocurrence of pharmaceuticals residues in sewage, river, ground and drinking water in Greece and Germany. In: Daughton CG, Jones-Leo T (eds) Pharmaceutical and Personal Care Products in the Environment: Scientific and Regulatory Issues. Symposium Series 791. American Chemical Society, Washington, DC, p 70
31. Farré M, Ferrer I, Ginebreda A, Figueras M, Olivella L, Tirapu M, Vilanova M, Barceló D (2001) J Chromatogr A 938:187
32. Gros M, Petrovic M, Barcelo D (2007) Environ Toxicol Chem 26:1553
33. Jones OAH, Voulvoulis N, Lester JN (2003) Bull WHO 81:768
34. Sedlak DL, Gray JL, Pinkston KE (2000) Environ Sci Technol 34:509A
35. Kolpin DW, Furlng ET, Meyer MT, Thurman EM, Zaug SD, Barber LB, Buxton HT (2002) Eviron Sci Technol 3:1202
36. Brun GL, Bernier M, Losier R, Doe K, Jackman P, Lee HB (2006) Environ Toxicol Chem 25:2163
37. Buser HR, Poiger T, Müller MD (1999) Environ Sci Technol 33:2259
38. Weigel S, Kulhmann J, Hühnrfuss H (2002) Sci Total Environ 295:131
39. Thomas KV, Hilton MJ (2004) Mar Pollut Bull 49:436
40. Xu W, Zhang G, Zou S, Li X, Liu (2007) Environ Pollut 145:672
41. Hamscher G, Pawelzik HT, Höper H, Nau H (2004). Antibiotics in soil: routes of entry, environmental concentration, fate and possible effects. In: Kümmerer K (ed) Pharmaceuticals in the Environment. Springer, Berlin, p 139
42. Díaz Cruz MS, López de Alda MJ, Barceló D (2003) Trends Anal Chem 22:340
43. Golet EM, Sthehler A, Alder AC, Giger W (2002) Anal Chem 74:5455
44. Bris HL, Puliquen H (2004) M Pollut Bull 48:434
45. Metcalfe CD, Miao XS, Koening BG, Struger J (2003) Environ Toxicol Chem 22:2281
46. Brooks BW, Chamblis CK, Stanley JK, Ramirez A, Banks KE Johnson RD, Lews RJ (2005) Environ Toxicol Chem 24:464
47. Nakamura Y Yamamoto H, Sekizawa J, Kond T, Hiai N, Tatarazako N (2008) Chemsphere 70:865
48. Fent K, Weston AA, Caminada D (2006) Aquat Toxicol 76:122
49. Calleja MC, Persoone G, Geladi P (1994) Arch Environ Contam Toxicol 26:69
50. Fong PP, Huminski PT, D'Urso ML (1998) J ExpZool 280:260
51. Halling-Sørensen B; Nielsen N, Lanzky PF, Ingerslev F, Holten-Lützhøft HC, Jørgensen SE (1998) Chemosphere 36:357
52. Boxakk ABA, Kolpin DW, Halling-Sørensen B, Tolls J (2003) Environ Sci Technol 37:286

53. Brian JV, Harris CA, Scholze M, Backhaus T, Booy P, Lamoree M, Pojana G, Jonkers N, Runnalls T, Bonfa A, Marcomini A, Sumpter JP (2005) Environ Health Perspect 113:721
54. Fenske M, Maack G, Schafers C, Segner H (2005) Environ Toxicol Chem 24:1088
55. Schulte-Oehlmann U, Oekten M, Bauchmann J, Oehlmann J (2004) Effects of ethinyloestradiol and methyltetosterone in prosobranch snails. In: Kümmerer K (ed) Pharmaceuticals in the Environment. Springer, Berlin, p 237
56. Zerulla M, Länge R, Steger-Hartmann T, Panter G, Hutchinson T, Dietrich DR (2002) Toxicol Lett 131:51
57. Ferrari B, Paxeus N, Lo Giudice R, Pollio A, Garric J (2003) Ecotoxicol Environ Safe 55:359
58. Schawaiger J, Ferling H, Mallow U, Wintermayr H, Negale RD (2004) Aquat Toxicol 68:141
59. Huggett DB, Brooks BW, Petreson B, Foran CM, Sclenk D (2002) Arch Environ Contam 43:229
60. Thaker PD (2005) Environ Sci Technol 39:193A
61. Iwamatsu T, Toya Y, Sakai N, Terada Y, Nagata R, Nahahama Y (1993) Dev Growth Differ 35:625
62. Cerda J, Subhedar N, Reich G, Wallace RA, Selman K (1998) Biol Reprod 59:53
63. Cleuvers M (2003) Toxicol Lett 142:185
64. Cleuvers M (2004) Ecotoxicol Environ Safe 59:309
65. Brain RA, Johnson DJ, Richards SM, Hanson ML, Sanderson H, Lam MW, Young C, Mabury SA, Sibley PK, Solomon KR (2004) Aquat Toxicol 70:23
66. Richards SM, Wilson CJ, Johnson DJ, Castle DM, Lam M, Mabury SA, Sibley PK, Solomon RK (2004) Environ Toxicol Chem 23:1035
67. Jones OA, Volvoulis N, Lester J (2004) Cri Rev Toxicol 47:71
68. Hansen PD (2007) Trends Anal Chem 26:1095
69. Ankley GT, Brooks BW, Huggett DB, Sumpter JP (2007) Environ Sci Technol

# Subject Index

Printing: Krips bv, Meppel, The Netherlands
Binding: Stürtz, Würzburg, Germany

CW01066851

# PIN-YP

gan Janice Jones

ISBN 1-904845-04-5



Dymuna'r cyhoeddwyr gydnabod cymorth
Adrannau Cyngor Llyfrau Cymru.

Diolch i Gyngor Celfyddydau Cymru am brynu'r hamdden i ddechrau ar y gwaith o ysgrifennu'r nofel hon, ac i fywyd go iawn yn ei holl ogoniant am beri iddi gymryd cyhyd i'w chwblhau.

*Janice Jones.*

Cyhoeddwyd ac argraffwyd yng Nghymru
gan Wasg y Bwthyn, Caernarfon

# CHWARAE CŴN CADNO

Neuadd y Sir
Ffordd yr Orsaf
Tresarn
Cedog
19 Mai 2003

Y Br Cledwyn Tomos
Cyngor Cymuned Nant-yr-Onnen
Y Llythyrdy
Nant-yr-Onnen
Cedog

Annwyl Syr,

Yn dilyn rhai wythnosau o drafod rhwng cynrychiolaeth o Gyngor Cymuned Nant-yr-Onnen a Chyngor Sir Cedog, hoffwn ddatgan safbwynt y Cyngor Sir i chi, gan ofyn i chi drosglwyddo'r wybodaeth i aelodau'r Cyngor Cymuned cyn gynted ag y bo modd.

Mae'r Cyngor Sir yn cadarnhau na allwn barhau i ysgwyddo baich costau cynnal a chadw neuadd bentref Nant-yr-Onnen pan ddaw'r flwyddyn ariannol gyfredol i ben. Fel y gwyddoch, mae'r neuadd mewn cyflwr diffygiol iawn, a byddai angen gwariant sylweddol cyn y byddai'r adeilad, a'r gwasanaethau oddi mewn iddo, yn cyrraedd safon foddhaol. Nid yw'r cyfalaf hwn ar gael ar hyn o bryd, ac nid yw'r Cyngor Sir yn ffyddiog y gellir ystyried buddsoddiad o'r fath yn y dyfodol, chwaith, wrth bwyso a mesur yr hinsawdd economaidd bresennol.

Serch hynny, mae'r Cyngor Sir yn fodlon i'r Cyngor Cymuned a phentrefwyr Nant-yr-Onnen barhau i fanteisio ar ddefnydd y safle am y rhent rhesymol sydd wedi'i bennu eisoes, hyd ddiwedd y flwyddyn ariannol bresennol. A phetai Cyngor Cymuned Nant-yr-Onnen yn medru cyflwyno cynlluniau pendant i ddatblygu'r safle er budd y gymuned leol erbyn diwedd y cyfnod penodedig yma, yna bydd y trefniad hwnnw'n parhau. Fodd bynnag, yn niffyg unrhyw strategaeth bositif, bydd yn rhaid i'r Cyngor Sir ailystyried y cynnig ac edrych yn fanwl ar y posibilrwydd o werthu'r safle.

Yr eiddoch yn ddiffuant,

Prys Harris

Dirprwy Uwch-Drysorydd
(ar ran y Cyngor Sir)

**Cofnodion Cyfarfod Brys**
**Cyngor Cymuned Nant-yr-Onnen, 24 Mai 2003**

| | |
|---|---|
| Yn bresennol: | Cledwyn Tomos (Cadeirydd) |
| | Albert Parry (Is-gadeirydd) |
| | Sara Watcyn (Ysgrifenyddes) |
| | Elsi Morgan (Trysorydd) |
| | Lewis Jones |
| | Wyndham Davis |
| | Ithel Roberts |
| Ymddiheuriadau: | Caryl Summers |
| | Dei Howells |
| | Sioned Ellis |

Darllenwyd a derbyniwyd cofnodion y cyfarfod diwethaf.

Darllenodd y Cadeirydd lythyr a dderbyniodd oddi wrth y Cyngor Sir yn datgan sefyllfa'r Cyngor parthed dyfodol neuadd y pentref.
Lleisiwyd siomedigaeth ac anfodlonrwydd y Cyngor Cymuned ynglŷn â'r penderfyniad.
Cafwyd nifer o awgrymiadau gan yr aelodau gyda golwg ar ddatrys y broblem.
Gwnaethpwyd cynnig hael gan yr Is-gadeirydd, sef derbyn cyfrifoldeb am y gwaith o ddymchwel y neuadd bresennol, clirio'r safle, a gosod caban symudol gyda gwasanaethau ar y safle at ddefnydd y pentref, tra bo'r Cyngor Cymuned yn gwneud ymholiadau ynglŷn â cheisiadau am grantiau i adeiladu neuadd newydd ar gyfer y gymuned leol.
Awgrymodd yr Is-gadeirydd y byddai swm o oddeutu £5000 yn ei ddigolledu am y gwaith.
Gadawodd yr Is-gadeirydd y cyfarfod er mwyn i'r aelodau fedru trafod y cynnig.
Wedi peth trafod penderfynwyd cyfarfod eto ymhen wythnos i geisio dod i gytundeb ynglŷn â'r ffordd orau i symud ymlaen wrth wynebu'r argyfwng yma.
Yn y cyfamser diolchwyd i'r Is-gadeirydd am ei gynnig a mynegwyd gwerthfawrogiad y Cyngor.

Dim materion eraill.

Dyddiad y cyfarfod nesaf 31 Mai 2003 am 7.30.

Prys Harris
Oddi wrth: albertparry@hotmail.com
At: prysharris@supanet.com
Pwnc: Neuadd bentref Nant-yr-Onnen
Gyrrwyd: 24 Mai 2003 21:48

Mae'r cynnig wedi ei osod gerbron. Mae ganddyn nhw gystal siawns o godi'r arian ag y byddai gan ddyn un goes o ennill Ras yr Wyddfa.
Ond yn araf deg mae dal iâr.
Byddaf mewn cysylltiad gydag unrhyw ddatblygiadau.

A.P.

# IONAWR

"Gad lonydd i'r bwced 'na, Glesni. Plîs."

Llusgodd Sioned y fechan o'r bwced am yr ugeinfed tro'r bore hwnnw. O leiaf.

Damia'r glaw 'ma. Doedd yr hen le ddim ffit i blant bach. Roedd yn oer, yn dal dŵr cystal â chyrtans net aflonydd Elsi Morgan, ac roedd y tai bach a'r gegin yn siŵr o fod wedi eu hetifeddu gan nain Methwsela. Doedd ryfedd fod y pethau bach yn hel annwyd byth a hefyd.

"Beth am gael panad fuan bora 'ma, Tina," awgrymodd Sioned, wedi iddi osod y bwced yn ôl yn ei phriod le er mwyn dal y diferion a dreuliai trwy'r nenfwd. "Os bydd raid i mi warchod y bwced 'ma fawr mwy, mi fydda i'n rhoi 'mhen ynddi."

Gwenodd Tina cyn diflannu i'r gegin hynafol i ddarparu ffics o gaffîn iddi ei hun a Sioned, a lluniaeth llesol i'r plant. Yna tra eisteddai hanner dwsin o drigolion ieuengaf Nant-yr-Onnen a'r cyffiniau yn eu seddi coch ger y bwrdd isel yn mwynhau byrbryd blasus o ffrwythau a diod o lefrith mewn llestri Tomos y Tanc, llymeitiodd Sioned a Tina eu coffi mewn cymharol heddwch.

Rhedodd Sioned ei bysedd trwy ei gwallt cwta browngoch.

"Tydw i ddim yn gweld sut y gallwn ni aros yma am aeaf arall," meddai, a'i llygaid glas yn llawn pryder. "Mae cyflwr y to'n mynd o ddrwg i waeth gyda phob sbel o dywydd mawr. Ac am y gegin a'r toileda . . ." Ysgydwodd Sioned ei phen.

"Beth am sgwennu at dîm *Changing Rooms*?" awgrymodd Tina.

"Fyddai Carol ddim hanner mor Smillie petai hi'n gweld stad y neuadd," atebodd Sioned. "A byddai angen cyllideb cyfres gyfan i roi mêc-ofyr i'r lle 'ma."

"Ond fedran ni ddim gwneud heb y neuadd," meddai Tina, gan grychu ei thrwyn smwt. "Gallwn ni gynnal cyfarfodydd Merched y Wawr yn y Llew Coch ar binsh, er wn i ddim sut y byddai'r mudiad yn ymateb i hynny, a gallai'r Cyngor Cymuned ffeindio rhywle arall i gyfarfod, mae'n siŵr. Ond byddai wedi ta-ta dominô ar glwb yr henoed, ac arnon ni fel cylch meithrin."

"Byddai, mae gen i ofn," cytunodd Sioned. "'Tydan ni wedi crafu'n penna ganwaith i drio meddwl am adeilad arall yn y pentre i gynnal cyfarfodydd aballu. Does 'na unlle. 'Dan ni wedi colli'r capel a'r ysgol – allwn ni ddim fforddio colli fan'ma hefyd."

Eisteddodd y ddwy gan syllu'n synfyfyriol i'w mygiau gweigion am eiliad neu ddau.

Nes i Glesni dywallt ei llefrith dros ben Aled.

Bu Sioned yn arweinyddes cylch meithrin Nant-yr-Onnen ers bron i bymtheng mlynedd, ers i Nesta'r arweinyddes flaenorol symud i'r canolbarth i fyw pan oedd gefeilliaid Sioned, Dylan a Ceri, yn dair oed ac yn mynychu'r cylch meithrin yn selog. Bellach roedd yr efeilliaid bron yn ddeunaw, ac yn sefyll ben ac ysgwyddau'n dalach na'u mam. Roedd y ddau'n tynnu ar ôl eu tad – yn dal a chyhyrog. Wrthi'n adolygu ar gyfer eu harholiadau lefel A oeddynt ac yna'n bwriadu gadael cartref i fynd i'r coleg. Roedd amser yn hedfan. Byddai Sioned yn gweld eu colli'n arw gan fod eu crio a'u cecru, eu cwmni a'u cariad wedi meddiannu pob cornel o'r aelwyd. Heb sôn am eu holl geriach bondigrybwyll. Byddai angen pantecnicon yr un arnynt i gario'u trugareddau o Nant-yr-Onnen i'w neuadd breswyl!

Gallai Sioned gofio'r diwrnod cyntaf yr aeth y ddau i'r "ysgol fawr". Fel ddoe. Roedd Ysgol Nant-yr-Onnen yn dal ar agor bryd hynny, ac i mewn â Dylan a Ceri trwy'r giatiau law yn llaw, gyda phrin amser i roi sws frysiog i'w mam. Cofiodd iddi ruthro adre i grio llond ei bol am eu bod mor annibynnol. Ond petaen nhw wedi gafael yn dynn ynddi, ac yna wedi cael eu llusgo i mewn i'r ysgol gan sgrechian, fel ambell un, mi fyddai hi wedi mynd adre a chrio 'run fath. Pwy faga blant, yntê? Diwrnod diderfyn oedd hwnnw. Aeth i'r cylch meithrin i roi trefn ar bethau ar gyfer y diwrnod canlynol; fe goginiodd deisennau bychain dros Gymru – ffefrynnau'r efeilliaid – ac aeth draw i weld ei mam. Ond bu bysedd y cloc yn greulon o araf deg; cymerodd o leiaf bythefnos iddynt gyrraedd tri o'r gloch. Ac yna'r breichiau bach mwythdew'n gafael yn dynn amdani, a'r lluniau o Mam a wnaethpwyd yn yr ysgol – un ohoni â gwallt gwyrdd a dwy lygad ddu, a'r llall efo bol mwy na Sir Fôn – yn cael eu cludo adre fel trysorau drudfawr a'u harddangos ar wal y gegin.

Gwenodd Sioned. Byddai'n falch iawn o'i gwaith yn y cylch meithrin mewn ychydig fisoedd, ac o gwmni'r genod yng nghangen Nant-yr-Onnen o Ferched y Wawr. Er nad genod mohonynt mwyach, ond merched yn anelu at ganol oed (toedd canol oed yn swnio'n uffernol o hen?), a rhai wedi pasio'r garreg filltir ddi-droi'n-ôl honno. Byddai pen-blwydd Dylan a Ceri yn ystod yr haf yn ei hatgoffa o hynny. Ond diolch am y genod, yn enwedig pan welai golli'r efeilliaid.

Ac am Dewi. Diolch amdano yntau. Un tawel oedd Dewi. Ond roedd wastad yno'n gefn iddi, trwy dda a drwg.

Oedd, cysidrodd Sioned, roedd hi wedi bod yn lwcus iawn. Doedd rhywun ddim yn ystyried y petha 'ma o ddydd i ddydd, ond rŵan, a'r plant ar drothwy bod yn "oedolion", roedd hi'n naturiol i edrych yn ôl ryw gymaint, a sylweddoli mor ffodus oedd hi a Dewi o fod gyda'i gilydd. Ac o gael dau o blant iach a dymunol, er mai dim ond cael eu benthyg oedden nhw.

A Mam hefyd. Roedd ei mam wedi bod yn gymorth mawr iddi pan oedd Dylan a Ceri'n fabis, er ei bod yn dechrau simsanu bryd hynny hyd yn oed. Roedd yn reit fusgrell bellach, ond wedi cael adfywiad ar ôl symud i un o'r fflatiau newydd yn yr hen gapel, ac wedi bod yn lwcus iawn cael gafael ar un ar y llawr gwaelod. Roedd pawb yn ei thrin fel ein nain ni oll, a rhywun yn picio i mewn am sgwrs ryw ben bob dydd. Roedd hi wrth ei bodd yno, ac wedi cael modd i fyw ar ôl ymuno â chlwb yr henoed o'r diwedd, wedi misoedd o ddatgan nad oedd hi'n ddigon oedrannus i fynd i'r fath le i gymysgu efo hen bobl!

Yr unig ddrwg yn y caws, meddyliodd Sioned, heblaw am Elsi Morgan, busnes pawb y pentre, a drigai yn ei byngalo *beige* ("hoff liw y Fam Frenhines, heddwch i'w llwch") nid nepell o waelod yr ardd, oedd cyflwr truenus neuadd y pentref. Roedd honno *yn* broblem. Byddai'n rhaid atgyweirio'r to neu byddai gaeaf arall yn ei droi'n ogr, ac am y gwasanaethau mewnol . . . Ond efo be?

Gwthiodd Sioned y benbleth honno i gefn ei meddwl. Am y tro roedd yn falch o gyfrif ei bendithion, ac yn eithaf bodlon ei byd.

"Na. Nefar in Iwrop. Dim ffiars o beryg."

Lleisiodd Sioned ei barn yn blwmp ac yn blaen, ac eisteddodd pawb am rai eiliadau gan syllu ar ei gilydd.

"Wel," meddai Sara Watcyn, a oedd yn y gadair ac yn wraig dal, unionsyth â thorch o wallt llwyd, "dyna un farn reit bendant. Beth am y gweddill ohonoch chi?"

"Tydw inna ddim yn bwriadu dangos fy mloneg i neb," datganodd Mererid yn bendant.

"Na finna," ategodd Gwyneth.

"Dwi'n synnu atoch chi, Sara," cwynodd Sioned, "am i chi hyd yn oed ystyried y fath beth."

"Dim ond taflu ambell syniad gerbron, Sioned fach," atebodd Sara'n bwyllog, "dyna i gyd."

Roedd cyfarfod misol Merched y Wawr Nant-yr-Onnen yn ei anterth. Yn ogystal â Sara Watcyn y cadeirydd, a Sioned yr ysgrifenyddes, roedd y selogion i gyd yn bresennol: Marian Tomos, trysorydd y gangen a chyd-berchennog siop a llythyrdy Nant-yr-Onnen; Gwyneth Taylor, perchennog y Llew Coch; Anwen Parry, gwraig Albert Parry y contractwr adeiladu lleol, a mam i ddau o blant ifanc; Tina Roberts, cynorthwywraig y cylch meithrin, barforwyn ran-amser y Llew Coch, a morwyn lawn-amser i'w gŵr di-waith a'i thri mab glaslancaidd; Caryl Harper, cydlynydd crand gwasanaeth pryd ar glud yr ardal, a mam Marc; Nia Morgan, gwraig i ffermwr, mam i dri o blant bach, a merch-yng-nghyfraith Elsi, baich gormesol os bu un erioed; Brenda Jones, cynorthwywraig ran-amser yn siop y pentref, gwraig tŷ, mam a nain a fendithiwyd â slafdod tragwyddol gan ei theulu di-hid; Phyllis Gruffudd, newydd ei gwneud yn ddi-waith, ac yn briod â gŵr a holltai flew â holl saint Ynys Enlli petai'r cyfle'n codi; Mererid Wynn, ysgrifenyddes i ŵr busnes yn Nhresarn a ddaeth i fyw i Nant-yr-Onnen yn lled ddiweddar wedi i'w phriodas fyrhoedlog dorri'n deilchion o'i hamgylch; ac Alison Rhys, nyrs gymunedol sengl a wasanaethai'r ardal.

Cyflwr neuadd y pentref oedd y pwnc dan sylw. Neu yn hytrach, cynllun chwyldroadol Sara i godi'r pum mil o bunnau a fyddai'n achub man cyfarfod y pentref dros dro.

"Mae'n rhaid bod yna ffyrdd eraill o godi'r arian," cysidrodd Marian, gan edrych o'i chwmpas yn obeithiol, "ac nid ni ydi'r unig gymdeithas sy'n defnyddio'r neuadd."

"Wel, go brin y gallwn ni enwebu aelodau clwb yr henoed ar gyfer Ras yr Wyddfa a cheisio codi'r arian trwy eu noddi," rhesymodd Sara, "a heb fwriadu unrhyw amarch tuag atat ti, Sioned, mae'r cylch meithrin yn ei chael hi'n anodd casglu digon o arian at ddibenion y plant, heb sôn am ddim arall. Ac os byddwn yn dibynnu ar yr hen drefn – bore coffi yma, te bach acw, ac ailgylchu rhyw anialwch o dombola i dombola – mi fyddwn ni wrthi hyd Sul y pys yn trio codi'r arian 'ma."

"Yr hyn sy'n fy synnu i," meddai Brenda gan benderfynu rhoi'i phig i mewn, digwyddiad oedd yn ddigon prin i ddenu sylw pawb, "ydi mai chi'r genod ifanc sy mor gadarn yn erbyn y syniad."

Cymerodd Sioned arni edrych y tu ôl i'w chadair. "Pa genod ifanc? Does 'na neb yma dan ddeg ar hugain ar ddiwrnod da. A rhai ohonon ni wedi heneiddio cryn dipyn yn ystod yr hanner awr ddiwetha."

"Twt lol," wfftiodd Tina, a'i llygaid gwyrddlas yn pefrio, "mi fydda i'n ddeugain 'leni, a'r awgrym yma ydi'r unig gyfle dwi wedi'i gael, neu'n debygol o'i gael, o fod yn ferch pêj-thrî cyn i 'mronna i ddod i gysylltiad rhy agos â mhenaglinia. Ac os ydi Sara'n barod i'w mentro hi, mi wna inna hefyd. Mae cynhyrchu calendar bronnoeth yn swnio'n lot mwy o sbort na chynnal hanner cant o forea coffi sychlyd. Ac mi rown ni halen ym mhotes y Cyngor Cymuned llugoer 'na – sori, mi wn i bod rhai ohonoch chi, neu'ch gwŷr, yn aeloda – a dangos iddyn nhw fod Merched y Wawr Nant-yr-Onnen yn barod i wneud unrhyw beth i warchod y gymuned."

Tawelodd Tina. Roedd honno wedi bod yn araith hir iddi a chilwenodd mewn embaras ar yr aelodau eraill.

Dechreuodd Sara, a oedd wedi bod yn chwerthin dan ei dannedd wrth wrando ar farn Tina am y Cyngor Cymuned, glapio. Ymunodd y merched eraill fesul un a dwy, ac aeth gweddill y cyfarfod heibio mewn dwndwr o drafod cynhyrfus a pharatoadau cychwynnol.

Cilagorodd Sioned ei llygaid. Gallai weld rhimyn o awyr las rhwng y llenni; roedd am fod yn ddiwrnod braf. Clywai anadlu dwfn Dewi wrth ei hymyl. Cymerodd gip ar y cloc larwm – hanner awr fach eto, diolch byth – cyn troi'n ddioglyd a chlosio at ei gŵr.

Agorodd ei llygaid eilwaith. Led y pen, fel clicied ar gamera wrth dynnu llun.

Doedd hi 'rioed wedi cytuno â'r ffwlbri a awgrymwyd yng nghyfarfod Merched y Wawr neithiwr. Suddodd ei chalon drwy'r fatres a glanio ar y llawr o dan y gwely.

Do, fe basiwyd y cynnig lloerig. Ar ôl i Tina ddweud ei dweud roedd y bleidlais yn unfrydol.

Ond byddai pawb wedi dod at eu coed erbyn bore 'ma, siŵr. Wedi sadio a gweld rheswm. Fyddent?

Roedd hi'n cochi dim ond wrth feddwl am y peth. Rhaid bod Sara wedi colli arni'i hun. Bendith y Tad iddi, roedd hi'n gyn-brifathrawes! Sara oedd pennaeth ysgol gynradd y pentre nes i'r ysgol honno gau ei drysau jyst ar ôl i Dylan a Ceri symud i'r ysgol uwchradd yn Nhresarn. Ac roedd hi'n ferch i weinidog capel Bethesda gynt. Nes i'r capel hefyd golli'r frwydr am ei gynhaliaeth. Roedd Sara'n credu hyd heddiw mai marw o dorcalon wnaeth ei thad, cwta flwyddyn wedi cloi drysau'r capel.

Ac, wrth gwrs, dyna'r eglurhad dros ei safiad anghymodlon ynglŷn â'r neuadd. A'i hymroddiad llwyr i wneud popeth o fewn ei gallu i achub man cyfarfod cymdeithasau'r pentref. Oedd, roedd y dirywiad cyson a arweiniodd at chwalfa mewn cymunedau bychain gwledig wedi costio'n ddrud i Sara. Ond a oedd hi, Sioned, yn barod i sefyll ar flaen y gad? Byddai colli'r neuadd yn cael cryn effaith arni hithau. A allai hi estyn yr un ymrwymiad â Sara i ddiogelu'r hyn oedd yn weddill o rwydwaith cymdeithasol y pentref?

A beth fyddai ymateb Dewi? A'r plant? A sut yn y byd y byddai hi'n dweud wrth ei mam? Beth petai'r datguddiad bod ei merch yn ystyried gwneud sioe ohoni ei hun yn ddigon iddi?

Na, lol botes oedd y cyfan. Byddai pawb wedi sylweddoli hynny bellach.

Tarfodd caniad cras y cloc larwm ar feddyliau dryslyd Sioned.

Yng nghegin Fflat 2, Annedd Hedd, Nant-yr-Onnen, eisteddai Sioned a Dilys, ei mam, gyferbyn â'i gilydd wrth y bwrdd. Roedd wyneb Sioned yn fflamgoch, a syllai'n fanwl ar y dail te yng ngwaelod ei chwpan. Wedi ychydig eiliadau dechreuodd Dilys bwffian chwerthin. Cododd Sioned ei hwyneb i edrych ar ei mam, a chyn pen dim roedd y ddwy'n chwerthin o'i hochr hi.

"Tydach chi ddim yn flin?" holodd Sioned, pan allai siarad unwaith eto.

"Bobol bach, nac ydw," atebodd Dilys. "Taswn i ugain mlynedd yn iau mi fyddwn i'n ymuno â chi. Roeddwn i wedi clywed y si, wsti, a disgwyl i ti alw oeddwn i, i mi gael dy berswadio i'w mentro hi efo'r lleill. Rwyt ti dal yn ifanc ac yn siapus. Amdani, Sioned, ddaw 'na ddim cyfle fel hyn eto."

Trawyd Sioned yn fud. Ei mam hi ei hun, a oedd wedi pasio oed yr addewid ac a fu yn llygaid Sioned yn wraig barchus a sidêt gydol ei hoes, yn annog ei merch i gymryd rhan yn y fenter wallgo 'ma.

"Efallai 'mod i'n fusgrell bellach," parhaodd Dilys gan wenu, "ond tydw i ddim yn ddwl nac yn ddall. Does gennych chi ddim gobaith codi pum mil o bunnau ar frys trwy forea coffi a nosweithia caws a gwin. Mae'r oes wedi newid. Dwi wedi dysgu llawer ers symud i fan'ma, wsti, ble mae cymaint o bobol ifanc o'm cwmpas. Chwara teg iddyn nhw, maen nhw'n ffeind iawn efo fi. A diolch i Albert Parry am sicrhau fflat i mi ar y llawr isa, neu mi fyddwn i wedi gorfod symud i un o'r bocsys bach llwyd 'na yn Nhresarn, a byddai hynny wedi bod yn loes calon i mi."

"Rydw inna'n falch eich bod chi'n dal wrth ein hymyl yma'n y pentre," cytunodd Sioned, "a 'dach chi'n symud gyda'r oes yn well na fi, yn ôl pob golwg."

"Be 'di barn Dewi am y plania?" holodd Dilys. "A Dylan a Ceri?"

"Wel, 'dach chi'n gwybod am Dewi, tydi o ddim yn deud llawer, ond tydw i ddim yn meddwl fod ganddo wrthwynebiad. Roeddwn i'n disgwyl i Dylan a Ceri godi reiat, a deud y bydden nhw'n marw o gywilydd, ond na, mae'r ddau'n hynod gefnogol, a reit falch ohona i, er mawr syndod."

"Da iawn nhw," cymeradwyodd Dilys. "Mae'n hen bryd i rywun wneud safiad i warchod cymdeithas gefn gwlad wsti, ac os byddwn ni'n dibynnu ar y dynion sy'n eistedd fel cyda' gwynt yn yr holl siopa' siarad 'ma, mi fydd hi wedi canu arnon ni."

Crychodd Sioned ei gwefusau wrth ystyried cyflwr sigledig ei hargyhoeddiad hi ei hun.

"Trendi a thanbaid," meddai. "Dwi'n dechra meddwl y byddech chi wedi bod yn fwy diogel yn un o'r bocsys bach llwyd 'na yn Nhresarn, Mam. Ond rydach chi'n deud y gwir bob gair. Byddai'n chwith gen i weld y cylch meithrin yn cau, yn enwedig gan fod y pentre wedi colli'i hysgol. A'r teuluoedd sydd â mwya o angen y cylch fyddai'n ei chael hi'n anodd teithio i Dresarn i geisio gwasanaeth tebyg. A does 'na unlle arall yn y pentre, heblaw am lolfa'r Llew Coch, ble gallai Merched y Wawr a chlwb yr henoed gyfarfod."

"Byddai hynny'n fy siwtio i i'r dim, wsti," awgrymodd Dilys yn ffugddiniwed wrth ei merch, "ond tydw i ddim wedi llwyddo i lusgo'r aeloda eraill i gyd i'r unfed ganrif ar hugain. Ddim eto."

Ysgydwodd Sioned ei phen a chwerthin drachefn. "Wn i ddim be 'dach chi arno fo ers symud i fan'ma, Mam, ond cadwch beth i mi, plîs."

"Dwed wrtha i rŵan," meddai Dilys. "Pwy arall sy'n mynd i gael eu hanfarwoli yn y calendar 'ma? Dwi'n gwybod dy fod ti a Tina a Sara'n cymryd rhan. A Mererid ac Alison i fyny'r grisia. Ond pwy ydi'r lleill?"

Cyfrodd Sioned yr enwau ar ei bysedd. "Marian, Gwyneth, Brenda, Phyllis . . ."

"Brenda a Phyllis?" torrodd Dilys ar ei thraws. "Da iawn, y ddwy ohonyn nhw; maen nhw wedi bod wrthi'n slafio ers blynyddoedd, ac yn cael eu trin fel baw am eu trafferth."

". . . Anwen, Nia . . ."

"O, mi fydd y ddwy ohonyn nhw'n wynebu rycsiwns gartre, mi gei di weld," rhybuddiodd Dilys.

". . . a Caryl," gorffennodd Sioned.

"Caryl? Dwi'n synnu ei bod hi'n bwriadu'i hymostwng ei hun i lefel y gweddill ohonoch chi."

Cododd Dilys yn araf deg i lenwi'r tecell. Er bod ei symudiadau braidd yn ansad bellach, roedd hi'n werth ei gweld. Cadwai ei chartref fel pin mewn papur, a hithau ei hun yr un modd. Roedd bob amser yn drwsiadus, a galwai'r ferch trin gwallt arni'n selog bob pythefnos, i osod ei gwallt llwyd yn donnau deniadol.

"Oes gen ti amser am baned arall, un sydyn?" gofynnodd i'w merch.

"Oes, siŵr," atebodd Sioned. "Fydd Dylan a Ceri ddim adre am sbel, ac mae Dewi'n gweithio'n hwyr, felly mae 'nhraed i'n rhydd am ychydig eto. Gyda llaw, Mam, ydach chi'n cofio rhywfaint o hanes Albert Parry? Dwi'n gwybod ei fod o wedi bod yn dda iawn yn cadw'r fflat 'ma i chi, ac wedi gwneud yn siŵr mai pobol leol yw canran uchel o'r perchnogion. Ac mae o'n deud mai felly y bydd hi gyda'r tai newydd mae o'n eu hadeiladu ar gyrion ei dir. Ond fedrith Dewi wneud dim â fo, a rydach chi'n gwybod un mor ddigyffro ydi Dewi. Wedi mynd yn rhy fawr i'w sgidia, medda fo, ac yn ei lordio hi o amgylch y pentre 'ma fel pe tasa fo'n sgweiar. Ac mae o ar sodla Sara druan yn barhaus, yn ceisio'i pherswadio i werthu'r tŷ capel iddo."

"Wel, wn i ddim a fyddet ti'n ei gofio fo yn yr ysgol," cysidrodd Dilys, "mae o ychydig yn hŷn na chdi. Hogyn bach del oedd o hefyd, gwallt cyrliog coch a llygaid glas, digon o sioe. Cael ei fagu gan ei nain a'i daid wnaeth o, wsti. Mi fuodd 'na ryw sgandal ynglŷn â phwy oedd ei dad o, a chafodd neb wybod hyd heddiw, ond dwi'n cofio'i fam o, Gwen, yn iawn. Roedden ni'n mynd i'r Band of Hope efo'n gilydd. Cyn yr helbul. Mi farwodd hitha pan oedd Albert ond ychydig fisoedd oed. Ond cafodd bob gofal gan ei nain a'i daid, wsti, er bod arian yn brin. Fo oedd cannwyll eu llygaid." Syllodd Dilys trwy'r ffenest yn synfyfyriol. "Ac mi ofalodd Albert am ei nain a'i daid pan oedd angen hefyd. Gweithiodd o'n galed i gyflawni'r hyn y mae wedi'i wneud. Wedi dechra heb ddim, a chrafangu i fyny'r ysgol . . . Ond mae Dewi'n iawn; *mae* o'n troi'n dipyn o ben bach. Mae llwyddiant yn gwneud hynny i bobol, wsti."

"M-m-m." Ystyriodd Sioned atgofion ei mam. "Ydi, mae'n siŵr. Ond mae o wedi gweithio'n galed, felly, ac wedi gwneud tipyn dros y pentre, chwara teg iddo. Byddai unrhyw gontractwr arall yn codi llawer mwy arnon ni am wneud y gwaith ar safle'r neuadd."

"Well i chi'i siapio hi efo'r sgiâm 'ma, felly," oedd ateb parod Dilys, "i ni i gyd gael rhywle clyd a chynnes i gynnal ein cyfarfodydd dros y gaea nesa 'ma."

Ysgydwodd Sioned ei phen eto dan wenu wrth i'w mam droi i wneud paned arall. Ac roedd ganddi baced o Jaffa Cakes ar gyfer achlysur i'w ddathlu fel hwn, tasa hi ond yn gallu cofio ble roedd hi wedi ei gadw . . .

". . . ac un bob un i'r genod dewr 'ma – a chitha'ch dwy hefyd, wrth gwrs. Duwcs, mae gan y pentre 'ma le i ddiolch i chi."

Ar noson dyner ar gynffon yr haf roedd lolfa'r Llew Coch yn dechrau llenwi, a Cledwyn Tomos, Cadeirydd y Cyngor Cymuned a chyd-berchennog swyddfa bost a siop-bob-dim Nant-yr-Onnen, yn prynu rownd i'r ŵyn cyn y lladdfa. Er mai ŵyn digon di-sbonc oedden nhw. Hyd yn oed Gwyneth, landlord y Llew Coch, a Tina, a weithiai y tu ôl i'r bar gyda'r nos.

Roedd nifer sylweddol o'r merched wedi amau doethineb y cynllun dodwy arian ers y cychwyn, ac erbyn hyn, a'r awr dyngedfennol ar eu gwarthaf, roedd pob un ohonynt yn barod i adael y wlad am bellafoedd byd. Ar fyr rybudd a chyda thocyn unffordd.

"Diolch, Cledwyn," meddai Sara, wrth iddo ef a Tina gludo'r ddau hambwrdd i'r gornel ble'r eisteddai'r merched o amgylch dau fwrdd crwn oedd wedi'u gwthio at ei gilydd. Mwmiodd y gweddill eu diolch yn llugoer.

"Wyt ti wedi clywed gan bobol y Loteri eto?" holodd Sara.

"Do, mi gefais alwad ffôn wsnos yma'n holi ynglŷn â

rhyw fân betha," atebodd Cledwyn. "Beryg na chawn ni wybod am sbel a fu'n cais ni'n llwyddiannus. Ond beth bynnag, mae cymdeithasa'r pentre'n sicr o fan cyfarfod dros y gaea 'ma. Diolch i chi."

Ond wynebau gwangalon iawn a welai Cledwyn o'i flaen, er gwaetha'r ganmoliaeth.

"Reit, genod, lawr â fo!" Ceisiodd Sara godi tipyn ar yr awyrgylch prudd. "A fi pia'r rownd nesa."

"Ia wir, dowch," cytunodd Sioned, gan geisio cymryd arni nad oedd ei hasgwrn cefn yn teimlo fel darn o seleri dair wythnos oed. "Rydan ni wedi dod cyn belled â hyn, fedran ni ddim rhoi'r ffidil yn y to rŵan. Dos iawn o tseribincs a bydd popeth yn edrych yn well." Gwagiodd ei gwydr lawr ei chorn gwddw. "Un arall o'r rheina, plîs, Sara."

Ac yn wir, erbyn stop-tap roedd y criw wedi ailafael yn eu gwroldeb. Teimlent yn fwy fel sêr teledu na changen o Ferched y Wawr. A dweud y gwir, roedd noson yn rhydd o ddyletswyddau cartref a theulu yn drît ynddo'i hun i ambell un, heb sôn am gael esgus gwerth chweil dros lowcio llawer mwy o'r ddiod felltith nag arfer, gan fod pob un o bentrefwyr Nant-yr-Onnen a alwodd i mewn i'r Llew Coch y noson honno wedi mynnu prynu diod i bob copa walltog ohonynt. A'u canmol i'r cymylau.

Fesul llwnc a chegaid bu'r ddiod gadarn yn falm effeithiol i'w hamheuon a'u pryderon.

O'r diwedd gwagiodd y dafarn a daeth Gwyneth a Tina draw i eistedd at y gweddill am dipyn, cyn dechrau ar y gwaith glanhau a thwtio. Safai'r gwydrau llawn a hanner llawn yn rheng ar y byrddau bach o hyd, gan nad oedd yfed y merched yn gallu cadw'n wastad â haelioni eu hedmygwyr.

"O, mae'n braf cael ista," ochneidiodd Gwyneth.

"Ydi wir," cytunodd Tina. "Mae 'nhraed i wedi'u piclo."

"Tydw i ddim yn cwyno, cofiwch," ychwanegodd Gwyneth. "Mae heno'n siŵr o fod wedi dyblu f'elw i am yr wsnos."

"Mi greda i," oedd ateb mingam Sara. Roedd wedi gobeithio y byddai pawb yn ymlacio, ond wedi gweld y gwydrau llawn yn cyrraedd y byrddau'n un rhibidirês, roedd yn poeni bellach bod ei phraidd wedi gor-wneud y plwc potel. Ymdrechodd i siarad yn gall â nhw. "Reit, ydach chi i gyd yn cofio'r drefn fory? Ac yn gwybod faint o'r gloch i gwrdd â Harri?"

Edrychodd nifer o'r criw ar ei gilydd fel pe na baent erioed wedi clywed am Harri'r ffotograffydd. Nac am fory.

"Waeth i chi heb â thrio cael shyn . . . shwn . . . shensh ohonyn nhw rŵan, Shara," meddai Sioned, a oedd â thueddiad at dafod dew wedi dim ond chwa o ddiod feddwol, heb sôn am y rhan orau o lond bol.

Chwarddodd Gwyneth a Tina a oedd, wedi'r cyfan, wedi gweithio gydol y noson, ac felly mewn tipyn gwell cyflwr na rhai o'u ffrindiau.

"Peidiwch â phoeni, Sara," gwenodd Gwyneth, "mi fydd popeth yn iawn. A synnwn i ddim na fydd pen mawr ambell un yn help i fygu swildod."

Nodiodd amryw mewn cytundeb, gan geisio'u gorau glas i edrych yn ddifrifol a deallus, ond difethwyd eu hymdrechion gan eu bochau cochion a'u hamrannau, a oedd yn ymddwyn fel petaent ar lastig; edrychent fel cnwd o'r cŵn fflocs rheini a nodiai eu pennau yng nghefn ceir.

"Iawn 'ta, genod," meddai Gwyneth gan godi drachefn. "Amser gwely, dwi'n meddwl."

"Ia, wir," cytunodd Sara, "neu mi fydd golwg y fall arnon ni i gyd fory."

Simsanodd Sioned ar ei thraed a chwifio'i llaw dros y gwydrau hanner gwag oedd yn weddill.

"Un am lwc," cyhoeddodd, a chodi'i gwydr. "I ni – amddinyffwyr cefn gwlad, abuchwyr neuadd y pentref, a phishish Nant-yr-Onnen!"

"I ni!" cytunodd pawb, a'u gwydrau'n tincian.

* * *

Gorweddai Sioned yn ei gwely. Teimlai fel petai wedi bod yn gorwedd yno ers oriau lawer, a'r cloc â'i dic-toc, tic-toc didrugaredd yn llusgo'i draed trwy oriau maith y nos, ac anadlu rheolaidd Dewi'n cyfrannu at ei hanhunedd.

Yn gorfforol roedd hi wedi ymlâdd. Wedi'r noson fawr neithiwr, a thensiynau'r dydd, roedd ei chorff yn sgrechian am orffwys. Ond roedd ei meddwl yn morio o emosiwn i emosiwn, fel llong ar ddyfroedd geirwon yn ceisio diogelwch tir a erys yn barhaol y tu hwnt i'w chyrraedd.

Cychwynnodd taflunydd ei meddwl y ffilm unwaith eto.

Roedd wedi codi'n hwyr, a'i phen yn bowndian. Roedd Dewi wedi gadael am ei waith a'r plant wedi mynd ar y bws i Dresarn i brynu offer ar gyfer y coleg. Ond ddim cyn sgrifennu "Hei lwc, Mam" ar yr oergell â llythrennau magnetig lliwgar. Daeth hynny â gwên i'w hwyneb, er gwaetha blas ffiaidd yr Alka-Seltzer. Roedd yn dechrau adfywio dan gawod achubol pan ffoniodd ei mam i ddymuno'n dda iddi. Chwarae teg i Mam, roedd ei chefnogaeth wedi bod yn agoriad llygad yn ystod yr wythnosau diwethaf. Heb hynny, ac anogaeth Dylan a Ceri, byddai wedi tynnu'n ôl lawer gwaith.

Roedd ei choesau fel uwd wrth iddi gerdded i'r neuadd, ac erbyn iddi gyrraedd, yn llawer rhy sydyn yn ei thyb hi, roedd yn laddar o chwys oer, a'i chrys wedi glynu wrth ei chefn. Roedd Tina wedi addo dod i afael yn ei llaw, a diolch amdani, neu byddai Sioned wedi troi ar ei sawdl cyn rhoi bawd ei throed dros drothwy'r neuadd.

Gŵr ifanc hynod ddymunol ac amyneddgar oedd Harri'r ffotograffydd. Roedd yn ffrindiau â Gwyn, brawd Mererid, ac yn dechrau ennill ei blwyf fel ffotograffydd medrus a mentrus. Ond er tringarwch pwyllog Harri, teimlai'n anhygoel o hyfriw'n eistedd yno yng nghanol taclau'r cylch meithrin yn fronnoeth, heblaw am ddau flodyn haul a grewyd â ffyn lolipop, platiau papur a phapur lliw. Diolch mai lluniau du a gwyn oeddynt oherwydd roedd ei bochau ar dân.

A beth fyddai ymateb cyfarwyddwyr y Mudiad Ysgolion

Meithrin i'r fath sioe? Byddai'n siŵr o gael ei hesgymuno. Neu ei dyrchafu'n Brif Swyddog Marchnata!

Clic clic y ffordd hyn, clic clic y ffordd acw, ac roedd yr ordîl drosodd.

Chwipiodd ei hun yn ôl i'w dillad a derbyn gwasgiad buddugoliaethus gan Tina, yna camodd allan i'r stryd gan gredu bod llygaid chwilfrydig yn sbecian arni o'r tu draw i bob cyrtan les. Croesodd y ffordd i gyfeiriad Annedd Hedd gan deimlo fel petai'n dal yn noeth; roedd arni eisiau croesi'i breichiau o'i blaen i gelu ei chywilydd.

Erbyn iddi gyrraedd cartref ei mam roedd ei choesau sigledig yn gwegian, a suddodd i'r gadair freichiau orau fel pechadur ym mhurdan. Cymerodd ddwy baned o goffi, tair Jaffa Cake a dos o gyffro ei mam iddi ddod ati ei hun, ond wrth iddi gerdded am adre teimlai fel Buddug, yn barod i drechu unrhyw un a'i heriai. Roedd hi, Sioned Ellis, wedi mentro, wedi strancio â chadwynau confensiwn. Ac wedi llwyddo.

Ar ôl cinio roedd yn rhaid mynd i dŷ Tina i ad-dalu'r gymwynas. Roedd Tina wedi mynnu cael tynnu ei llun ger y sinc yn ei chegin oherwydd taerai mai yno y treuliai bron pob awr o'i hamser pan oedd yn ei chartref. Ond er ei gwrhydri, pan ddaeth yr awr hollbwysig nid edrychai'n rhy hapus yn sefyll yn ei chegin a dim ond lliain sychu llestri, a mŷg a'r slogan anfarwol "Ciwb Oxo – deis y dyn dwl" arno i arbed ei gwyleidd-dra.

Wedi cadw cwmni i Tina roedd hi'n amser troi am adre drachefn a phlicio tatws ar gyfer swper, ac fe gâi Sioned hi'n anodd credu bod pawb yn mynd o gwmpas eu pethau, a bod y byd yn dal i droi, yn union fel cynt.

Cyrhaeddodd y plant adre 'run mor gynhyrfus â'u nain, gan geisio gwasgu pob manylyn allan o'u mam, ond bu'n rhaid iddyn nhw fodloni ar grynhoad. Ychydig iawn a ddywedodd Dewi, a doedd dim byd diarth yn hynny, ond roedd yn osgoi ei llygaid ac aeth allan i'r ardd i weithio'n syth ar ôl swper. Daeth Llinos draw i gadw cwmni i Ceri dros ryw fideo ac aeth Dylan allan i gwrdd â'i ffrindiau.

Gan nad oedd ar neb ei hangen, aeth Sioned i gael bath hamddenol i olchi tyndra a thensiwn y dydd ymaith. Bu'n socian yno am sbel, tra llepiai mân donnau gorfoledd a chywilydd, pryder a beiddgarwch o'i chwmpas. Erbyn iddi gyrraedd ei gwely roedd Dewi yno, yn cysgu a'i gefn ati. Neu'n smalio cysgu beth bynnag.

". . . a chyn bo hir mi gawn ni wared ar hon unwaith ac am byth."

Aeth Sioned â'r bwced i'r gegin i wagio'r diferion o'i gwaelod. Roedd bore prysur arall yn y cylch meithrin wedi dod i ben, a Tina a hithau'n tacluso cyn mynd adre.

"Wyt ti'n edrych ymlaen at weld y calendar yn ei holl ogoniant heno?" holodd Tina wrth i Sioned ddychwelyd i'r stafell.

"Ydw," atebodd Sioned, wrth osod y bwced yn ei lle, "dwi'n meddwl. Ond dwi reit nerfus hefyd."

Gwthiodd ei hamheuon a'i hansicrwydd ynghylch dieithrwch diweddar Dewi i gefn ei meddwl a chau'r drws yn glep arnynt

"A finna," cyfaddefodd Tina. "Ond dwi'n falch ein bod ni'n cael cyfle i weld y calendar gorffenedig yng nghwmni'n gilydd cyn y lansiad swyddogol fory. Mi fydd hwnnw'n brofiad od," myfyriodd, "ysgwyd llaw a bod yn neis efo pobol sy wedi dod yno i sbio ar lunia ohonon ni'n hanner noeth. Be wyt ti'n mynd i wisgo?"

"Wn i ddim eto," atebodd Sioned dan gilwenu, "ond mi fydda i'n dewis rhywbeth tipyn mwy swmpus na'r ddau flodyn haul 'na, mae hynny'n saff i ti!"

# CHWEFROR

Brathodd Mererid ddarn o'r ffon fara â'i dannedd gwynion twt. Dyna ben yr ast, a hwnnw cyn waced â llyfr adnod y diafol. Brathodd eto. A'i hysgwyddau, yn gulach na beiro Bic. A thamaid eto. Dyna fol y jaden, a hwnnw'n gafnog fel powlen gawl. Gwthiodd Mererid y darn olaf o'r ffon fara i'w cheg a'i grensian yn fileinig. Dyna gluniau'r sguthan, oedd ddim hyd yn oed yn cyfarfod yn y canol fel cluniau pobl normal. Pwy arall oedd â chluniau y gellid gweld awyr las rhyngddynt? Crychodd Mererid ei llygaid ambr. Ac am Berwyn . . . Byddai wrth ei bodd yn gwasgu ei geilliau trwy felin garlleg a'u clywed yn clecian cyn dod allan yn gyrbibion llipa.

Gwthiodd Mererid ei llaw i'r bocs ffyn bara. Dim byd ond briwsion. Llawn cystal, a hithau'n trio colli pwysau. Ond roedd hi wedi cael llond bol (neu ddim, yn hytrach) ar fwyd cwningen. A fyddai hi byth yr un siâp â Gwawr mi-sefa-i-wysg-f'ochr-a-diflannu. A phe byddai, ni ddeuai hynny â Berwyn yn ôl. A hyd yn oed petai o'n cropian yn ei ôl ati fory nesa trwy fôr o biswail, ac yn crefu arni am faddeuant, byddai'n ei wrthod. Y bastard.

Roedd Mererid Wynn a Berwyn Williams wedi bod yn gariadon ers dyddiau ysgol uwchradd. Bu'r ddau'n canlyn yn selog tra oeddynt yn y coleg, ond wynebodd y berthynas hoe o reidrwydd pan dderbyniodd Berwyn swydd ddysgu yng Nghaerdydd ar ôl graddio, tra cychwynnodd Mererid ar ei gyrfa fel ysgrifenyddes, cynorthwywraig bersonol a Siân-bob-swydd i un o gyflogwyr mwyaf Tresarn. Roedd Mererid yn gyfarwydd â'r ardal oherwydd iddi hi a'i brawd Gwyn dreulio tipyn o'u hamser yno pan oeddynt yn blant, yn dod ar wyliau i dŷ Taid a Nain, ac yn cael aros yno ar eu pennau eu hunain, heb Mam a Dad, a chael eu difetha'n llwyr.

Ymhen cwta ddwy flynedd sicrhaodd Berwyn swydd yn Ysgol Uwchradd Tresarn, ac wedi'r cyfnod o garu o hirbell, ailafaelodd y cwpwl ifanc yn eu serch yn eiddgar. Prynwyd tŷ o faint a chost cymedrol a threuliwyd penwythnosau a nosweithiau prysur a phoitshlyd yng nghanol paent a phâst a phapur wal. Dodrefnwyd y tŷ a'i lunio'n gartref clyd a chyfforddus, ac yno y bu Mererid a Berwyn yn byw mewn dedwyddwch dibriod am bron i bedair blynedd. Nes i rygnu achlysurol eu teuluoedd ddatblygu'n gytgan fynych, a phenderfynodd Mererid a Berwyn briodi. Yn rhannol i blesio'u teuluoedd, yn rhannol i'w plesio'u hunain, yn rhannol a'r chwiw o fagu teulu'n hofran yng nghefn eu meddyliau, ac yn rhannol er mwyn cael esgus am glamp o barti.

Ond rywle ar hyd y llwybr rhithiol hwnnw rhwng yr allor a'r neithior, y neithior a'r mis mêl, y mis mêl a bywyd bob dydd, trowyd mantol y berthynas.

Er nad oedd Mererid yn ymwybodol o hynny. Ar y pryd, beth bynnag. Roedd hi'n hapus yn ei hanwybodaeth, a'i bywyd yn dilyn yr un patrwm ag o'r blaen; yr unig wahaniaeth oedd y fodrwy aur a sgleiniai ar ei bys, yn ei hatgoffa'n feunyddiol, pe bai angen gwneud, o'i hymrwymiad parod i Berwyn. Nes iddi gyrraedd adre un nos Wener wedi treulio deuddydd mewn cynhadledd gyda Colin, ei chyflogwr, a darganfod bod ei gŵr wedi ei gadael; wedi symud pob blewyn o'i eiddo o'r tŷ, heb adael dim yn weddill ond nodyn swta ar fwrdd y gegin – dyrnaid o eiriau dideimlad, brad braen ar bapur.

Sylweddolai Mererid ei bod wedi goroesi'r penwythnos aflan hwnnw. A'r wythnosau canlynol. Rhaid ei bod wedi mynd i'w gwaith a dychwelyd i'r tŷ; rhaid ei bod wedi golchi a smwddio a chadw trefn; rhaid ei bod wedi byw pob dydd o'i ddechrau i'w ddiwedd.

Ond doedd hi'n cofio dim, dim ond teimlo'i bod ar ddisberod mewn gwacter diderfyn, mewn twll du a'i furiau'n lasrew llithrig, llathrog. Tan y bore hwnnw yn y swyddfa pan ofynnodd Colin – fel y gwnâi'n feunyddiol – sut oedd hi. A nodi bod golwg welw arni wrth iddo daro mygiad o goffi a dwy fisgeden o'i blaen. A chyda'r weithred honno, arwydd diniwed o gonsýrn, agorwyd argae ei theimladau, a thywalltodd allan ei phoen a'i phylni, ei phrinder a'i phenyd. A chwarae teg i Colin, wnaeth o ddim gwingo wrth iddi grio llond ei grys, dim ond ei dal a'i hanwesu nes i'r llif ostegu.

Roedd yn rhaid gwerthu'r tŷ oherwydd roedd pob gronyn o'i sylwedd yn atgoffa Mererid o'r hyn a fu, a'r hyn a ddylai fod. Roedd canfod yr arwydd "Ar werth" wedi ei blannu yn yr ardd yn afiach derfynol. A hithau ei hun wedi'i rhwygo o'i gwreiddiau. Bu'n ffodus i glywed am y fflatiau yn Nant-yr-Onnen trwy un o'i chydweithwyr. Roedd yn falch o adael Tresarn, a oedd yn graddol droi'n garchar o'i hamgylch gan fod arni ofn mentro'n bellach na'i swyddfa a'r stryd fawr, lle byddai'n rasio i ddiwallu ei hanghenion sylfaenol rhag ofn iddi daro ar Berwyn. Clywsai Mererid fod ganddo gariad newydd, ac ni allai wynebu'r gwaradwydd o'u cwrdd ar hap. Ond wedi symud ei chelfi i'w fflat yn Nant-yr-Onnen, dechreuodd ymlacio tipyn yn niogelwch ei chartref newydd ble câi gau'r drws yn glep ar y byd a'i brifodd.

Ac yno yn ei lloches dechreuodd ei chysuro'i hun yn ystod y nosweithiau blin ag ambell dafell o deisen, rhyw far neu ddau o siocled, a sleisen dew o fara ffres wedi ei orchuddio â haen sylweddol o fenyn cartref. Wedi misoedd o esgus byw, o stryffaglio o un pen diwrnod i'r llall a briwsion brau'r briodas yn bwrw trwy'i bysedd, dechreuodd ei chynnal ei hun gyda chymorth ffyn baglau Mars a Mr Kipling. Newidiodd o fod yn frwynen welw i fod yn ferch luniaidd unwaith eto. Ac yna'n fwy na llond ei chroen, a'i dibyniaeth ar ei swcr yn peri iddi orfod ei harllwys ei hun i'w dillad yn y bore, a phob botwm a sip yn crefu am ryddhad.

Nes iddi weld Berwyn a'i gariad newydd – Gwawr dwi'n-rhy-fain-i-daflu-cysgod-ar-grac-yn-y-palmant – un gyda'r nos wrth iddi fentro i'r archfarchnad ar ôl gadael y swyddfa. Roedd Mererid ar fin dod allan o'i char pan welodd y ddau'n gwthio'u troli'n hapus gytûn yn y gwyll, tra heriai hithau strydoedd y dre'n wrol unig, yn unswydd i geisio cysur y KitKat a'r *croissant*. Sylwodd 'run o'r ddau arni; roeddynt wedi ymgolli yn ei gilydd. Ond suddodd Mererid i waelod ei sedd, yn eu casáu â chas perffaith, ac yn ei chasáu ei hun am ildio i demtasiwn boddhad parod nes teimlo fel blomonj blonegog.

A nawr roedd cyfundrefn y bwyd cwningen yn ei hanterth – diflastod ar blât. Ond roedd Mererid yn benderfynol pe byddai'n taro ar Berwyn – gyda'r ferch a wnaed o fatsys neu beidio – y byddai ar ei gorau, yn edrych fel petai ar ben ei digon, ac yn barod i sathru drewgi fel Berwyn i'r baw dan ei sawdl. Petai hi'n werth y drafferth.

Yn araf deg dechreuodd ymdopi â'i bywyd newydd, dechreuodd wneud ffrindiau ymysg perchnogion y fflatiau eraill, dechreuodd gymryd rhan ym mywyd cymdeithasol y pentref.

Ac wedi misoedd maith o fodoli, roedd hi bron yn barod i ddechrau byw.

"Paid â rwdlan, Mererid," wfftiodd Alison. "Siapus wyt ti, nid tew. Byddwn i wrth fy modd yn cael corff fel d'un di, a hwnnw'n mynd mewn ac allan yn y llefydd iawn, yn lle bod mor ddi-siâp â phostyn giât. Ac mae dy groen di fel dilia rhos."

Ffliciodd Mererid ei phleth hir o wallt brown dros ei hysgwydd a gafael yn ei mỳg rhwng ei dwylo.

"Ond Sara," meddai, gan ysgwyd ei phen mewn anghredinedd. "Sara'n awgrymu'r fath beth?"

"Mi gefais inna dipyn o sioc," cyfaddefodd Alison. "Mae

hi'n ymddangos mor sidêt. Ond mae hi yn llygad ei lle; beryg na fydd unrhyw ddatblygiada pendant ynglŷn â'r cais Loteri am fisoedd. Ac ella bydd hynny'n rhy hwyr i rai mudiada . . . Ac wedi clywed hanes cau'r capel 'ma, a'r effaith gafodd hynny ar dad Sara, ac wedyn cau ysgol y pentre tra oedd hi'n brifathrawes arni . . ."

"M-m-m," ystyriodd Mererid. "Byddai Gwyn a finna'n arfer dod ar ein gwylia i dŷ Taid a Nain, rhyw ddwy filltir tu allan i Nant-yr-Onnen, pan oedden ni'n blant. Dwi'n cofio tad Sara'n iawn, gweinidog hen ffasiwn yn rhoi'n hael o'i amser a'i egni, ac yn ŵr bonheddig heb ei ail. Does gen i ond brith gof o Sara; rhaid ei bod hi yn y coleg, neu newydd ddechra dysgu ac yn byw oddi cartre yn ystod y tymor. Ond roedd hi'r un mor drawiadol bryd hynny – ei gwallt yn dorch o amgylch ei phen fel heddiw, heblaw ei fod fel edeifion o sidan melynwyn, ei llygaid glas tywyll craff yn pefrio, a chwa o lafant yn ei hamgylchynu. Roedd yn f'atgoffa i o lun o Nia Ben Aur a welais mewn llyfr yn yr ysgol. A deud y gwir, roedd arna i ychydig bach o'i hofn ar y pryd."

Roedd Mererid ac Alison yn mwynhau paned yn ystafell fyw Fflat 4, Annedd Hedd, ar ôl mynychu cyfarfod Merched y Wawr. Ac er i'r ddwy, wedi araith ysbrydoledig Tina, godi eu dwylo'n frwdfrydig dros gynnig chwyldroadol Sara, tybient yn awr eu bod wedi llwyr ddrysu – dros dro, beth bynnag.

"Tasa'r fath beth yn digwydd, a'r cynllun gwallgo 'ma'n dwyn ffrwyth, wyt ti'n meddwl byddai dy frawd yn gallu cael gafael ar y ffotograffydd 'na mae o'n gyfeillgar â fo i ni? Beth oedd ei enw fo eto?" holodd Alison.

"Harri? Wn i ddim. Mae o'n reit brysur, dwi'n credu . . . ond go brin yr aiff petha cyn belled â hynny."

"Ia," cytunodd Alison. "Roedden ni i gyd ar dân heno ar ôl i Tina ddeud ei deud, ond stori arall fydd hi bore fory, dwi'n siŵr."

Yfodd y ddwy eu te'n feddylgar.

Ers i Mererid symud i'w chartref newydd yn Fflat 3,

Annedd Hedd, roedd cyfeillgarwch cyfforddus wedi datblygu rhwng y ddwy ferch. Er mai adnewyddu hen gysylltiad â'r ardal oedd Mererid, ac Alison wedi symud i bentref Nant-yr-Onnen o'r newydd yn dilyn ei phenodiad fel nyrs gymunedol, roedd y ddwy wedi efelychu arferiad eu mamau a'u neiniau trwy ymuno â changen leol Merched y Wawr, gyda'r bwriad o wneud ffrindiau ac ymafael yn rhwydwaith cymdeithasol eu cynefin. A changen lwyddiannus a bywiog oedd cangen Nant-yr-Onnen; i ryw raddau oherwydd diffyg darpariaeth gweithgareddau hamdden lleol, ond i raddau helaeth oherwydd brwdfrydedd ac ewyllys da'r aelodau.

Gan eu bod yn gymdogion yn Annedd Hedd, dechreuodd Mererid ac Alison yr arfer o ddychwelyd i fflat y naill neu'r llall ohonynt am baned a sgwrs ar ôl y cyfarfod misol. Ac yna, wrth i'w cyfeillgarwch flodeuo, dechreuodd y ddwy gyfarfod o leiaf unwaith yr wythnos i adrodd eu helyntion ac ymlacio yng nghwmni ei gilydd.

Felly y dysgodd Mererid am y boddhad a gâi Alison yn ei swydd newydd, ei bodlonrwydd â'i bywyd yn Annedd Hedd ac yn y pentref, a'i chais a'i gobaith llugoer am "yr un iawn" wedi stribed o siomedigaethau. Clywodd Alison hithau am garwriaeth hir a phriodas fer Mererid, ac er mor ddiffwdan yr adroddwyd y stori, gallai Alison ddirnad, i ryw raddau, frath clwyf ei ffrind.

"Sgwn i be fydda ymateb Dilys," cysidrodd Alison, "tasa Sioned yn ymddangos yn y calendar 'ma?"

Gwenodd Mererid. "Tydi Dilys ddim hanner mor ddiniwed ag y mae'n ymddangos, tybiwn i," meddai. "Fyswn i'n synnu dim petai hi'n mynnu ymuno yn y fenter ei hun!"

"Wel, mi gawn ni weld sut siâp fydd ar ein hargyhoeddiad ni i gyd fory, gefn gola dydd," meddai Alison.

Gorffennodd Mererid ei phaned, ac wedi rhyw fân sgwrsio, dychwelodd i'w fflat ei hun. Ychydig yn ddiweddarach, wrth olchi'r mymryn llestri yn y sinc cyn ei throi hi am ei gwely, tybiodd Alison y gallai'r calendar arfaethedig fod yn achubiaeth o fath i Mererid, tasa'r cynllun yn cael gwynt dan ei adain. A gobeithiodd yn daer y byddai Berwyn y bradwr yn canfod yr hyn a daflodd o'r neilltu yn ei holl ogoniant.

"Iw-hw! Harri! 'Dan ni'n fan'ma!"

Gwingodd Mererid wrth i un o esgidiau Doc Marten Eira lanio ar fawd ei throed tra oedd honno'n trio tynnu sylw Harri. Roedd hi, ei brawd Gwyn a'i gariad Eira, yn cyfarfod â Harri'r ffotograffydd. Cododd yntau ei law ar y ddwy ferch o ganol llawr y dafarn cyn mynd at y bar, ble roedd Gwyn yn codi rownd. Ochneidiodd Mererid yn ddistaw pan eisteddodd Eira drachefn a rhyddhau ei bawd. Roedd hi'n hoff iawn o gariad ei brawd, ond bobol bach, roedd ei breichiau a'i choesau fel llafnau melin wynt. Merch fechan bryd tywyll oedd Eira, a chanddi wallt du cwta a sbectol drom â ffrâm ddu a ymddangosai fel cysgod prifathro blin ar ei hwyneb picsïaidd. Ffafriai ddillad lliwgar a llachar, a chariai fag llaw oedd bron cymaint â sach Siôn Corn a hwnnw'n llawn trugareddau hanfodol a olygai fod tasg megis estyn ei goriadau yn orchwyl tebyg i chwilio am gontact lens mewn pwcedaid o stwnsh rwdan. Roedd ei natur sionc yn siwtio Gwyn i'r dim, ac yntau mor lêd-bac nes ei fod bron ar wastad ei gefn.

Daeth Gwyn a Harri at y bwrdd gyda'r diodydd. Gŵr ifanc, cwrtais oedd Harri, a chanddo wallt byr cyrliog du a llygaid glas tywyll a olygai ei fod yn cael digon o sylw gan ddarpar gariadon gobeithiol. Ond canolbwyntio ar ei yrfa a wnâi Harri. Aeth rownd y bwrdd bach i eistedd ar bwys Mererid tra eisteddodd Gwyn wrth ymyl Eira.

"Rydw i wedi egluro'ch syniada i Harri," meddai Gwyn, ar ôl i bawb gyfarch ei gilydd a setlo'n gyfforddus, "ac mae o'n barod iawn i ymgymryd â'r gwaith, os gallwch chi drefnu dyddiad cyfleus. Ydw i'n iawn, Harri?"

"Yn llygad dy le, Gwyn," cytunodd Harri. "Cofia di,

Mererid, tydw i ddim wedi gwneud gwaith glamor o'r blaen. Dwi'n gwneud digon o bortreada a gwaith priodasol aballu . . . ond dwi'n edrych ymlaen at eich gweld chi i gyd."

Cochodd Mererid at ei chlustiau a dechreuodd Eira bwffian chwerthin a gwingo nes i bawb gipio'u gwydrau oddi ar y bwrdd rhag iddynt gael eu sgubo i ebargofiant.

"Bihafia dy hun," dwrdiodd Gwyn.

"Sori, Mererid," ymddiheurodd Harri â gwên, "daeth hynny allan yn gwbl chwithig. Edrych ymlaen at eich cyfarfod chi i gyd, a chael cyfle i wneud gwaith ychydig yn wahanol, dyna roeddwn i'n 'i feddwl."

Agorodd Mererid ei cheg i ateb, ond cyn iddi gael cyfle i ddweud gair, rhoddodd Eira ei phig i mewn eto, a'i breichiau'n fflapio fel adenydd hwyaden oedd newydd golli'r frwydr dros y darn olaf o fara. "Paid â newid dy stori," meddai wrth Harri. "Mi *ddylet* ti fod yn edrych ymlaen at dynnu llunia merched teca Nant-yr-Onnen!"

"Yli, dos i nôl paced o gnau mwnci neu rywbeth o'r bar," ceryddodd Gwyn hi, "neu chaiff Mererid byth air i mewn. A thyrd â rownd arall i ni tra wyt ti yno."

Tynnodd Eira ei thafod ar ei chariad a gwgu arno wrth godi a chyrchu tuag at y bar a'i bag diwaelod yn colbio pawb o fewn cyrraedd.

"Rŵan, 'ta," meddai Harri dan chwerthin, "dwi am drio eto. Diolch am roi'r cynnig cynta i mi. Bydd yn bleser gweithio gyda chriw mor wrol."

"Wn i ddim am hynny," atebodd Mererid yn fingam, "ond gan ein bod wedi dod cyn belled â hyn . . . Diolch i ti am gytuno; dwi'n gwybod dy fod ti'n brysur."

"Gora po gynta i ni bennu dyddiad," meddai Harri. "Wedi'r cyfan, fe fydd yna dipyn go lew o waith i'w wneud, hyd yn oed ar ôl tynnu'r llunia. Mae Gwyn wedi egluro'r rheswm am y brys. Beth am . . . ?"

Sylweddololodd Harri'n sydyn ei fod yn siarad ag ef ei hun. Dilynodd lwybrau llygaid Mererid a Gwyn nes i'w lygaid ei hun lanio ar Berwyn, a oedd newydd ddod i mewn trwy'r drws a Gwawr yn hongian ar ei fraich. Clywodd Gwyn yn sisial "y bastard" dan ei wynt wrth i'r gwrid bylu o wyneb Mererid.

Yn ystod yr ychydig eiliadau a gymerodd i Berwyn sylwi arnynt llithrodd Harri ei fraich y tu ôl i gefn Mererid a rhoi gwasgiad ysgafn i'w hysgwydd. Lled-drodd hithau ei phen ato. Winciodd Harri arni a gwenu'n ddireidus cyn plannu clamp o gusan ar ei boch.

Wrth ymlwybro'n ôl o'r bar dan ei sang gyda'r sgrepan bondigrybwyll, yr hambwrdd o ddiodydd a'r cnau mwnci, stopiodd Eira'n stond pan welodd Harri'n cusanu Mererid yn frwd. Wedi cael ei gwynt ati, ymlaen â hi am y bwrdd, heb sylwi ar y cwpwl a fentrodd ar ei thraws ar eu ffordd o'r drws at y bar.

Cytunodd y pedwar mai damwain anffodus oedd hi.

Oedd yna rywun yn curo ar ddrws y fflat? Doedd Dilys heb glywed neb yn dod i mewn trwy'r drws mawr i'r lobi. Ond gwell iddi wneud yn siŵr, rhag ofn iddi golli cyfle am sgwrs.

Tu allan i ddrws ffrynt Dilys safai Mererid a bag brethyn â dolenni pren yn ei dwylo. Wrth i Dilys agor y drws dechreuodd Mererid feichio crio, ond roedd Dilys yn ddigon tebol i ddelio â'r sefyllfa. Gafaelodd ym mraich Mererid a'i harwain yn dyner i'w chadair freichiau orau, gan sefyll yno a'i braich o amgylch ei hysgwyddau nes iddi ymlonyddu.

"Yli, 'mechan i," dywedodd ymhen sbel, "eistedda di'n fan'na. Mi a' i i wneud panad i ni."

Roedd Dilys yn ei hôl o'r gegin mewn chwinciad.

"Dyna chdi – bocsaid o hancesi papur. Cria di hynny leci di. Does 'na neb ond chdi a fi yma i wybod, wsti."

Erbyn i Dilys ddychwelyd o'r gegin eilwaith, gyda hambwrdd ac arno debot, dwy gwpan a soser, a'i hateb greddfol i bob digwyddiad o bwys – boed da neu ddrwg – platiad o Jaffa Cakes, roedd Mererid wedi meistroli rhywfaint ar ei dagrau.

"Mae'n ddrwg gen i, Dilys," mwmiodd yn floesg. "Doedd gen i ddim . . ."

"Twt lol, 'mechan i," wfftiodd Dilys. "Tyrd i mi gael tollti panad i ti. Does 'na ddim byd fel yr hen banad i wneud i rywun deimlo'n well, yn fy marn i." Tywalltodd Dilys gwpanaid o de trioglyd i Mererid ac estyn y gwpan a'r soser iddi. "Yfa di hwnna, wedyn mi gei di ddeud wrtha i be sy wedi ypsetio cymaint arnat ti. Os wyt ti isio, hynny ydi. Jaffa Cake, 'mechan i?"

Allai Mererid ddim llai na gwenu ar bresgripsiwn Dilys, ond yn wir, wedi dos o'r feddyginiaeth amgen, dechreuodd ddod ati ei hun.

"Galw fi'n fusnes pawb os mynni di," meddai Dilys yn y man, "ond dwi'n cymryd mai rhywbeth yn y bag 'na sydd wedi d'ypsetio di."

Nodiodd Mererid a theimlo'r dagrau'n codi i'r wyneb drachefn. Rhoddodd ei chwpan a soser i lawr ac yna estyn ei llaw i'r bag yn fud a thynnu allan ddarn o gynfas brodwaith. Roedd "Mererid" a "Berwyn" wedi'u brodio arno, ynghyd â dyddiad eu priodas, yng nghanol gwaith pwytho addurniadol manwl.

"Cefais hyd i hwn," sibrydodd. "Roedd gen i ryw hanner dwsin o focsys heb fynd drwyddynt ar ôl symud o'r tŷ, a phenderfynais wneud hynny pnawn 'ma."

Rhoddodd Dilys ei llaw ar ben-glin Mererid yn dyner.

"Yli, mae dod ar draws petha fel hyn, neu glywed darn o fiwsig arbennig, neu gofio am rywle y buoch chi gyda'ch gilydd, yn bownd o d'ypsetio di, wsti. A bydd yn digwydd am sbel eto, cofia. Ond rhag pob clwyf eli amser, er nad wyt ti'n gweld hynny rŵan. Mi ddaw petha'n well."

Llwyddodd Mererid i wenu'n wan cyn estyn i'r bag unwaith eto a thynnu darn arall o gynfas ohono – darn glân y tro hwn, wedi'i rolio'n daclus.

Brathodd ei gwefus cyn datgan yn ddistaw, "Ar gyfer ein babi."

"O, 'mechan i," cysurodd Dilys. "Paid â digalonni. Mi

fydd yna rywun i ti, siŵr iawn. Rhywun fydd yn ŵr ac yn dad tipyn mwy cadarn na'r sglyfath arall 'na. Mi fyddi di'n deud gwynt teg ar ei ôl o un diwrnod, wsti. Coelia di fi."

Chwarddodd Mererid yn ysgafn trwy weddillion ei dagrau wrth glywed Dilys yn ei dweud hi am Berwyn yn y fath fodd.

"Dyna welliant," cyhoeddodd Dilys. "Yli, rho hwnna i mi; dwyt ti ddim angen testun hel meddylia fel'na. Ond cadw di'r cynfas glân a'r edeifion," rhybuddiodd, "nes daw'r amser pan fydd eu hangen nhw arnat ti."

"Rydach chi'n werth y byd, Dilys, ac mae'n wir ddrwg gen i am lanio ar eich stepan drws fel'na," ymddiheurodd Mererid eto, "ond pan gefais i hyd i'r sampler 'na . . ."

"Bobol bach, does dim isio, siŵr iawn," sicrhaodd Dilys hi. "'Dach chi i gyd y tu hwnt o ffeind efo fi yma, a dwi'n falch o gael talu'r gymwynas yn ôl. Yli, mi wna i banad ffres i ni, ac mi gei di ddeud wrtha i pwy ddysgodd di i frodio mor gywrain."

Erbyn i ddail te'r ail baned ddod i'r amlwg, roedd Dilys wedi trafod cynllun y calendar yn drylwyr gyda Mererid, ynghyd â hel achau ei theulu i gyd a sylweddoli iddi fod yn yr ysgol gyda chwaer nain Mererid, a'i bod yn adnabod y teulu'n iawn.

"A dy nain ddysgodd di i frodio?"

"Ia, er bod Mam yn brodio hefyd, ond doedd ganddi fawr o amser sbâr ar y pryd, druan ohoni. Roedd hi'n gweithio ac yn cadw tŷ, heb sôn am drio cadw trefn ar Gwyn a finna."

"Dwi'n cofio Beti, chwaer dy nain, a finna yn yr ysgol," synfyfyriodd Dilys, "yn trio dysgu pwytho ar gynfas bras – binca oedd o'n cael ei alw, os dwi'n cofio'n iawn – ac yn gwneud traed moch ohono fo. Ac yn cael slaes ar draws cefna'n bysedd am ein trafferth. Mi wnaethon ni wella," parhaodd, "ond doedden ni ddim patsh ar dy nain. Dwi'n cofio gweld hancesi a chlustoga roedd hi wedi'u haddurno; roedden nhw'n werth eu gweld."

Dychwelodd Dilys i'r presennol.

"Dwi'n dangos f'oed rŵan," meddai'n ysmala, "yn hel straeon am y gorffennol fel hyn. Ond mae golwg well arnat ti bellach."

"O, 'dach chi wedi achub fy mywyd i pnawn 'ma," sicrhaodd Mererid hi.

"Mae hi wedi bod yn bleser, 'mechan i," mynnodd Dilys. "Paid ti â cholli ffydd; mae 'na ddigon o bysgod yn y môr, wsti. Ond cadwa'n glir rhag slywod o hyn allan."

Ceisiodd Mererid ei hatal ei hun rhag gwenu.

"Rydach chi'n iawn," cytunodd yn ddifrifol. "Cyngor da iawn, Dilys."

"A dweud y gwir," datganodd Dilys, "mae meddwl am y gwersi brodio 'na wedi codi hiraeth arna i. Tasa'n llygaid i dipyn yn fwy craff mi fyswn i'n creu sampler i chi – genod y calendar – i ddathlu'ch dewrder."

Yn nes ymlaen, wedi iddi ddychwelyd i'w fflat ei hun ar ôl sicrhau Dilys bod dwy Jaffa Cake wedi bod yn foddion i'w llwyr iacháu, rhedodd Mererid ei bysedd trwy'r edeifion amryliw cyn rholio'r cynfas glân drachefn a'i gadw'n becyn twt yn y bag. Aeth i'r oergell, ond ymddangosai'r toreth o bethau gwyrdd a lechai yno'n ddiflas iawn o'u cymharu â Jaffa Cakes Dilys. Caeodd ddrws yr oergell yn glep ar y gybolfa iachusol. Byddai'n gohirio'r wledd am dipyn.

Sampler i ddathlu'r calendar, ystyriodd, wrth roi'r bag brethyn ar ben y wardrob. Dyna syniad.

Roedd Mererid wedi mynychu llawer o gyfarfodydd a chynadleddau gyda Colin, a doedd hi byth yn blino ar aros dros nos, pan fyddai angen, yn y gwestai moethus a fynnai Colin iddynt ill dau. Hyd yn oed cyn i Berwyn ei gadael, byddai wrth ei bodd yn archwilio pob twll a chornel o'i hystafell, fel merch fach â'i thraed yn rhydd mewn ffatri dda-da. Ac yn awr, heb ŵr gartref iddi hiraethu amdano, cymerai fwy o ddiléit mewn bodio a byseddu'r mân foethusion y deuai ar eu traws.

Yng nghanol prysurdeb deuddydd o gyfarfodydd pwysig, a chyda swper blasus (ond nid eithafol) yn ei bol, dywedodd Mererid wrth Colin, gan ddylyfu gên, ei bod am gael noson gynnar. Doedd dim at ei dant ar y teledu, felly penderfynodd gael bath ymlaciol a manteisio ar y pecynnau amrywiol a'r persawr a adawyd yn yr ystafell ymolchi.

Trodd y botymau ar y radio yn ei hystafell wely nes iddi ddod o hyd i gerddoriaeth oedd wrth ei bodd, yna suddodd i'r trochion cysurlon, a chyn hir roedd wedi ymlacio'n llwyr. Yn wir, roedd bron â hepian cysgu.

Ni chlywodd y tap-tap ysgafn ar ddrws ei hystafell wely, ble safai Colin a photel o siampên a dau wydr yn ei ddwylo.

Mentrodd Mererid agor un llygad. Iawn, hyd yma. Mentrodd agor y ddwy. Teimlai ei cheg fel petai rhywun wedi teilsio ei thu mewn hi â chrîm cracyrs. Byddai'n cymryd ymdrech oruwchnaturiol i godi ei phen oddi ar y gobennydd, ac roedd ei stumog yn troi a throelli fel corddwr menyn Nain erstalwm. Ond efallai mai nerfau oedd yn rhannol gyfrifol am hynny, ddim jyst y plwc potel. A dweud y gwir, ar ôl noson fel neithiwr roedd hi'n haeddu teimlo'n llawer gwaeth. Diolch ei bod wedi cadw at y gwin gwyn. Ddim fel rhai, a oedd wedi llowcio pob diferyn o fewn golwg. Byddai'n edifar ganddynt bore 'ma. Heb sôn am yr hyn oedd o'u blaenau.

Ati hi roedd Harri'n dod gyntaf, diolch byth. Wiw iddi dynnu'n ôl rŵan a gadael y criw yn y baw, ond ofnai na fyddai ei hargyhoeddiad tila'n para tan y pnawn. Byddai 'run fath wrth fynd at y deintydd neu'r meddyg – bob amser yn gwneud apwyntment buan, rhag iddi gael gormod o amser ymlaen llaw, a chyfle i newid ei meddwl. Gwenodd Mererid. Harri druan, yn cael ei gymharu â deintydd.

Cawod. Dyna'r cam cyntaf. Ac yna paned. Neu ddwy.

Ymhen ychydig funudau clywyd sgrech fuddugoliaethus

o ystafell molchi Fflat 3, Annedd Hedd. Er gwaetha'r gwin gwyn roedd Mererid wedi cyrraedd ei phwysau delfrydol, ac ar y diwrnod hwn, o bob un. Roedd y glorian wedi'i dyrchafu'n aelod teilwng o'r hil ddynol drachefn, a llifai hunanhyder newydd trwy ei chorff wrth iddi yfed ei choffi yn ei gŵn gwisgo tra gwrandawai ar Jonsi'n parablu. Roedd arni eisiau ffonio'r BBC a gofyn i Jonsi gyhoeddi i'r byd ei bod hi, Mererid Wynn, yn denau unwaith eto. Roedd hynny'n swyddogol. A'i bod hi'n mynd i gael tynnu ei llun ar gyfer calendar bronnoeth ar yr union bore hwnnw, a dim ond darn o gynfas brodio rhyngddi hi a'r byd.

Byddai'n dweud y newydd da wrth Eira heno. Roedd Gwyn ac Eira'n mynd â hi allan i ddathlu tynnu'r lluniau. Nid bod Eira'n gorfod poeni am ei phwysau ei hun, ond mi fyddai hi'n deall. Doedd pwrpas yn y byd sôn am bethau felly efo Gwyn; ysgwyd ei ben yn anobeithiol arni fyddai ei ymateb o.

Roedd Mererid wedi gofyn i Alison a hoffai ddod allan gyda nhw, ond dywedodd fod ganddi rywbeth ar y gweill. Roedd Mererid yn amau bod ganddi gariad newydd, gan fod yna sioncrwydd yn ei cham a sglein yn ei llygaid yn ddiweddar, ond byddai'n rhaid aros iddi hi rannu ei chyfrinach yn ei hamser ei hun.

Gwenodd Mererid wrth i Jonsi ffarwelio â hi, gan ei siarsio i'w chadw'i hun yn bur, ac yna sylweddolodd y byddai Harri yno cyn hir. Brysiodd i newid a phincio cyn estyn ei bag brethyn lle roedd y sampler a symbylwyd gan sylw Dilys ar ei hanner. Dewisodd un o'r edeifion lliwgar, ac roedd newydd ei roi yn y nodwydd pan ganodd cloch y drws.

Safai Mererid wrth y ffenest yn stiwdio Harri'n gwylio'r gwynt yn ceisio blingo'r coed. Roedd Harri wedi gofyn iddi alw heibio ar ôl gwaith i gymryd cipolwg ar y lluniau gorffenedig cyn iddo'u danfon at yr argraffwyr.

"Dyna ti," meddai Harri gan estyn mỳg iddi, "panad o

goffi. Yli, mae'n ddrwg gen i 'mod i ar ei hôl hi. Fedri di roi rhyw funud neu ddau i mi orffen ffeilio'r gwaith papur 'ma, neu fydd gen i ddim syniad be 'di be erbyn bora fory."

Setlodd Mererid mewn cadair esmwyth ger y ffenest ac agorodd Harri'r cwpwrdd mawr llwyd a safai yng nghornel yr ystafell. Clywodd bwff o chwerthin, a throdd i weld Mererid yn cymryd arni gael pwl o besychu.

"Be sy?" gofynnodd yn ffug ddiniwed.

Pwyntiodd Mererid at y cwpwrdd. "Ffeilio," meddai cyn i chwiw arall o chwerthin ei sigo.

"Dwi'n gobeithio nad wyt ti'n chwerthin am ben fy ffeilio i," rhybuddiodd Harri hi'n llym, ond â'i lygaid yn dawnsio'n ddireidus. "Mae wedi cymryd misoedd i mi lunio'r system yma," a chwifiodd ei law i ddangos perfedd y cwpwrdd oedd yn blith draphlith â phapurau a ffeiliau a bocsys.

Ysgydwodd Mererid ei phen. "Sut ar wyneb daear wyt ti'n cael hyd unrhyw beth?" holodd.

"Tipyn o chwilio a chwalu. A lot o lwc," cyfaddefodd Harri. "Dwi wedi prynu cyfrifiadur er mwyn trosglwyddo popeth iddo. Mae o gartre gen i. Yn ei focs. Fedra i wneud pen na chynffon ohono."

Ysgydwodd Mererid ei phen eto. "Yli, tyrd ti â fo i'r stiwdio ac mi dreuliwn ni ychydig nosweithia, neu benwsnos, os oes gen ti un rhydd, yn ei roi ar ei draed. Yna bydd gen ti system ffeilio heb ei hail."

"Diolch," atebodd Harri, "diolch yn fawr. Ond fedra i ddim hawlio dy amser hamdden di fel'na."

"Mi fydda i'n falch o gael helpu," sicrhaodd Mererid o. "Wedi'r cyfan, roeddet ti'n barod iawn i'm helpu i allan o dwll."

Cochodd Harri wrth feddwl am ei hyfrdra'r noson honno. Ac am y wefr arbennig a brofodd wrth dynnu llun Mererid a'i hosgo diymhongar tra eisteddai o'i flaen yn brodio, a'i llygaid ambr yn pefrio.

"A beth bynnag," parhaodd Mererid, "fydden i ddim yn lecio meddwl amdanat yn agor dy gwpwrdd ac yn cael dy gladdu dan y lluwch gwaith papur 'na."

Sgubodd Harri'r papurau oddi ar ei ddesg a'u stwffio i'r cwpwrdd at eu tebyg.

"Reit," meddai, gan gau'r drws ar y ffiasgo ffeilio, "os wyt ti wedi gorffen yfed fy nghoffi a gwawdio'm system ffeilio, tyrd i gael golwg ar y llunia."

Oedd, meddyliodd Mererid, roedd Harri'n llawn haeddu ei enw da. Roedd pob un o'r criw, er gwaetha'u swildod a'u pennau mawr, yn edrych yn odidog. A serch gweithgareddau amrywiol y merched, roedd y lluniau gorffenedig yn hynod chwaethus.

"Maen nhw'n werth chweil, Harri," cymeradwyodd Mererid. "Rwyt ti'n siŵr o fod wedi caboli tipyn arnyn nhw i wneud i ni edrych cystal."

"Naddo, wir," sicrhaodd Harri hi. "Dim o gwbl. Roeddech chi'n ddigon o sioe fel yr oeddech, pob un ohonoch."

"Wel, wel," meddai Mererid wrth iddi syllu ar y lluniau. "Pwy fasa'n meddwl? Criw bach dibrofiad Nant-yr-Onnen yn edrych fel petaent yn hen lawia ar y busnes tynnu llunia 'ma."

"Ti'n blês, felly," meddai Harri'n ansicr. "Ti'n meddwl bydd y lleill yn fodlon?"

"Bodlon? Mi fyddan nhw wrth eu bodda!"

Trodd Mererid at Harri a'i gofleidio.

"Diolch o galon i ti, Harri. Am bob dim."

Gwenodd Harri fel giât.

"Tamad o swper i ddathlu?" cynigiodd.

# MAWRTH

Go drapia! Roedd hi'n dechrau bwrw. Camodd Brenda ar draws y gegin yn frysiog, gan geisio peidio baglu dros deganau Lowri. Cipiodd y fasged olchi ac aeth i'r ardd.

Doedd hi ddim yn bwrw llawer wedi'r cyfan.

Bwrw haul, dyna fyddai hi'n galw tywydd fel hyn pan oedd hi'n blentyn. Credai bryd hynny mai'r tylwyth teg oedd yn rhyw hanner chwerthin, hanner crio ar ôl clywed stori chwerwfelys, a'u bod am gasglu blodau coed drops i wneud ambaréls iddyn nhw'u hunain.

Gwenodd Brenda iddi ei hun. Peth hyfryd oedd diniweidrwydd plentyn.

Dyna un peth roedd Brenda'n ei fwynhau ynglŷn â chael Lowri'n byw gyda nhw. Hynny a'r breichiau bach fyddai'n ei chofleidio, y llygaid glas llawn ymddiriedaeth a'i dilynai o gwmpas y tŷ, y gwallt cyrliog melyn angylaidd, a'r holl swsys a roddai Lowri i'w nain.

Roedd Wil, gŵr Brenda, a Robin a Sandra'r plant, yn ei sathru dan draed, ond roedd hi wedi gwirioni ar ei hwyres fach deirblwydd oed. Bu andros o ffrwgwd pan ddychwelodd Sandra adref yn feichiog, ond unwaith y gwelodd Brenda'r fechan anghofiodd am ddiflastod ei bywyd, a bu'n fam ac yn nain i Lowri.

Oedd, mi roedd hi am gawod wedi'r cyfan. Dechreuodd Brenda gasglu'r dillad, gan gymryd gofal arbennig o'r dillad gwely Tweenies. Sylwodd bod ei chynfasau ei hun wedi dechrau gwisgo'n denau, a chofiodd sut y byddai ei mam yn torri cynfasau oedd wedi treulio yn eu hanner, ac yn eu gwnïo'n ôl gyda'r ochrau at y canol, er mwyn ymestyn rhywfaint ar eu hoes. Roedd y dyddiau hynny wedi hen ddiflannu. Er, byddai Wil yn berffaith fodlon gwylio Brenda'n addasu aceri o gynfasau pe golygai hynny y câi ef sbario rhoi ei law yn ei boced.

Ysgydwodd Brenda ei phen, codi'r fasged ddillad, a dychwelyd i'r tŷ.

Pethau digon od oedd y Tweenies 'ma, ystyriodd Brenda, wrth iddi fwynhau paned tra oedd Lowri'n dilyn anturiaethau ei hoff gymeriadau ar y teledu, ond roedd y fechan wedi gwirioni'n lân efo nhw.

Bob bore danfonai Brenda ei hwyres i'r cylch meithrin, yna treuliai deirawr yn gweithio yn siop-bob-dim y pentref cyn dychwelyd i'w nôl hi. Anaml y byddai Sandra ar gael i ofalu am ei merch, felly byddai Brenda'n gwneud cinio iddi hi ei hun ac i Lowri ac yna'n ei diddanu drwy'r prynhawn, cyn paratoi swper i'r teulu, rhoi bath i Lowri a dweud stori wrthi tra oedd hi'n setlo am y noson. Yna byddai'n cwblhau ei gwaith tŷ tra eisteddai Wil, Robin a Sandra o flaen y teledu – os nad oeddent wedi hel eu traed i rywle.

Hyd yma, doedd heddiw'n ddim gwahanol i unrhyw ddiwrnod arall. A go brin y deuai unrhyw ryfeddodau i ran Brenda weddill y dydd.

Er ei bod hi'n ddydd pen-blwydd ei phriodas.

Ochneidiodd Brenda dros ei phaned a thaflodd Lowri gipolwg pryderus arni. Gwenodd Brenda.

"Gwylia di be mae'r Tweenies yn ei wneud tra bo Nain yn gorffen ei phanad, Lowri fach. Wedyn mi wnawn ni deisen i de."

Gwenodd Lowri'n ôl ar ei nain a bodloni ar wylio diwedd y rhaglen.

Doedd gan Brenda fawr o awydd hel atgofion am yr holl amser a aethai heibio er dydd ei phriodas, ond ar ddiwrnod fel heddiw roedd hi'n anodd peidio.

Safai llun ohoni hi a Wil ar y silff ben tân, er na chymerai neb unrhyw sylw ohono bellach. Roedd y llun, fel Brenda ei hun, mor gyffredin a disylw â'r brwsh llawr. Patiodd Brenda ei gwallt cwta brown golau, estyn ei sbectol a'i gosod ar ei thrwyn. Edrychai'n llawer iau yn y llun, wrth reswm. Yr adeg hynny roedd ei gwallt wedi'i dorri'n bòb ffasiynol at ei hysgwyddau. Pefriai ei llygaid gwinau'n ôl arni, yn disgleirio gan lawenydd. Ond roedd golwg un o'r doliau hynny a ddefnyddir i guddio papur lle chwech arni, yng nghanol yr holl ffrils yna! Doedd Wil ddim yn edrych yn ddrwg. Yn wir, yn ei siwt newydd (ac ni phrynasai siwt ers hynny), a'i law'n gafael yn ei braich yn berchnogol, a'i wên yn ymestyn o glust i glust, ymddangosai'n eithaf golygus.

Beth ddigwyddodd rhwng y diwrnod hwnnw a heddiw? Ble'r aeth yr holl flynyddoedd? Ble'r aeth yr holl obeithion, yr holl addewidion, yr holl ddyheadau? Roedd ochr ramantus, ystyriol Wil wedi hen ddiflannu cyn iddynt ddychwelyd o'u penwythnos o fis mêl yn Blackpool, a disodlwyd disgwyliadau uchelgeisiol Brenda gan fisoedd, ac yna blynyddoedd, o rygnu byw.

Bu'r plant yn gysur ac yn gwmni iddi tra oedden nhw'n fychan. Bendithiodd hi nhw â'r holl gariad oedd ganddi i'w roi. Gwneud gormod drostyn nhw wnaeth hi, efallai. Hwyrach mai dyna pam roedden nhw'n ei thrin hi fel morwyn erbyn hyn. Neu dichon mai jyst dilyn esiampl eu tad oedden nhw.

Sut bynnag, ochneidiodd Brenda, wrth edrych ar Lowri a oedd wedi ei swyno gan y bodau bach lliwgar ar y teledu, roedd ganddi rywun arall i'w charu, ac i'w charu hi'n ôl, erbyn hyn.

"Diar mi, mae golwg benisel iawn arnat ti bora 'ma."

Roedd Brenda wedi danfon Lowri i'r cylch meithrin ac roedd hi'n barod i ddechrau ar ei shifft yn y siop. Wrth iddi gerdded i mewn gwelai Marian Tomos, cyd-berchennog y siop a'r llythyrdy, a chyflogwraig Brenda, yn sefyll y tu ôl i'r cownter yn edrych fel petai newydd glywed fod siopau'n mynd allan o ffasiwn.

"Hisht," sibrydodd Marian, ac amneidiodd ar Brenda i ymuno â hi yn y gornel y tu ôl i'r cownter, fel y gallent sgwrsio heb i Cledwyn, gŵr Marian, eu clywed. Roedd o'n brysur yn y rhan o'r siop a wasanaethai fel swyddfa bost.

"O, Brenda," meddai Marian yn ofidus, "wnaethon ni 'rioed gytuno â ffwlbri Sara neithiwr, naddo? Dwed wrtha i mai hunlle gwallgo oedd y cyfan."

Gwenodd Brenda. "Dwi'n gobeithio ddim, wir. Dwi wedi cymryd at y syniad. Mae'n swnio fel hwyl."

Ochneidiodd Marian. "Ond be am Wil? A'r plant? Be maen nhw'n feddwl?"

Cododd Brenda'i hysgwyddau. "Tydw i ddim wedi deud wrthyn nhw. Ac a deud y gwir, taswn i'n tynnu 'nillad pob cerpyn ac yn ista'n noethlymun gorn mewn dysglaid o bwdin reis ar y bwrdd o flaen Wil, dim ond deud wrtha i am symud cyn iddo oeri y bydda fo. Be 'di barn Cledwyn?"

Ochneidiodd Marian drachefn. "Tydw inna heb ddeud dim chwaith. Dyna 'di'r broblem. Mae rhywun yn siŵr o ddod i mewn a chrybwyll y peth yn ystod y bora."

"Wel, gwell i ti ddeud wrtho fo'n reit handi, felly," rhybuddiodd Brenda.

Crychodd Marian ei gwefusau. "Byw mewn gobaith oeddwn i, y byddai pawb wedi callio erbyn bora 'ma. Yna fyddai dim rhaid i mi sôn am y . . ."

"Ydach chi'ch dwy'n gweld eich ffordd yn glir i wneud rhyw fymryn o waith bora 'ma?" holodd Cledwyn o'r tu ôl i gownter y swyddfa bost. "Duwcs, 'dach chi'n gweld eich gilydd bob dydd. Be sy gennych chi i drafod eto, deudwch? Rydach chi'n union fel dwy hogan fach ddrygionus yn sibrwd yn y gornel 'na."

Gwridodd y ddwy, a brysiodd Brenda i lenwi'r bylchau ar y silffoedd a sicrhau fod popeth yn yr oergell yn drefnus a'r nwyddau ffres i gyd o fewn eu dyddiad gwerthu. Gwenodd wrth sylwi olwg mor euog oedd ar Marian, a bod ei chyflogwraig yn rhoi naid fach bob tro y deuai rhywun trwy'r drws. Ond wedi awr neu fwy reit ddistaw, doedd neb a allai ollwng y gath o'r cwd wedi galw i mewn, a dechreuodd Marian ymlacio.

Daeth Elsi Morgan i mewn, fel brenhines yn cylchdeithio ac yn ymostwng i amlygu ei haelioni i'w deiliaid.

"Dydych chi byth wedi dechrau stocio hadog melyn, Marian." Ysgydwodd Elsi'i phen yn siomedig. " Dyna oedd hoff frecwast y Fam Frenhines, heddwch i'w llwch, wyddoch chi."

"Does 'na fawr o ofyn amdano fo a bod yn onest, Mrs Morgan," atebodd Marian.

"Wel!" ffromodd Elsi. "Rydw *i*'n gofyn amdano fo, Marian. Ac rydw i'n f'ystyried fy hun yn gwsmer selog yn y siop 'ma. Nid un o'r rheini sy'n picio mewn am bapur a hen sigaréts ffiaidd ydw i, Marian, naci, wir, ond cwsmer sy'n gwario'n sylweddol yma ar ddanteithion o'r ansawdd gorau. Ac yn un o hoelion wyth y pentre 'ma, hefyd: Trysorydd y Cyngor Cymuned fel y gwyddoch yn iawn, dinaswraig gydwybodol yn cadw llygad ar y trueiniaid hynny sy'n byw ar eu pennau'u hunain, ac aelod blaenllaw o'r cynllun Gwarchod y Gymdogaeth. A fedrwch chi ddim hyd yn oed mynd i'r drafferth o ordro ychydig bach o bysgod i mi!"

Trodd Elsi ar ei sawdl a gadael y siop, wedi pwdu'n llwyr.

"Cynllun Gwarchod y Gymdogaeth, wir!" ebychodd Marian. "Yr unig gynllun sy gan honna yw cynllun i roi'i phig ym mhotas pawb."

"Mae'n amlwg nad ydi *hi* wedi clywed am syniad Sara eto," sylwodd Brenda.

"Ddim eto. Ond os bydd y cynllun yn dwyn ffrwyth, bydd bywyd Nia'n uffern ar y ddaear unwaith y daw Elsi i wybod am y calendar," cyhoeddodd Marian.

"Pa galendar?" holodd Cledwyn, a oedd wedi sleifio o'r swyddfa bost draw at y ddwy, gan feddwl yn siŵr ei bod hi'n amser paned.

Cyn i Marian gael cyfle i ateb ei gŵr, daeth Sara i mewn i'r siop.

Diflannodd Brenda i'r cefn i wneud y te.

Clywodd Brenda su rhythmig y fan yn dod i lawr y stryd. Rhedodd i fyny'r grisiau i roi crib trwy'i gwallt a rhoi llyfiad o finlliw – un newydd sbon, lliw Turkish Delight – ar ei gwefusau. Cymerodd gipolwg arni'i hun yn y drych, yna rhedodd ei dwylo dros ei siwmper binc ysgafn a'i sgert las â phatrwm blodau mân. Cipiodd ei llyfrau a'i sbectol oddi ar y cwpwrdd bach wrth erchwyn y gwely a dychwelyd lawr y grisiau. Roedd Lowri wedi mynd adref gydag un o'i ffrindiau ar ôl bod yn y cylch meithrin, felly am unwaith roedd Brenda â'i thraed yn rhydd.

Oedodd yn y gegin am ychydig, gan edrych ar y cloc. Doedd hi ddim am gyrraedd pan fyddai'r fan yn llawn dop. Chwarddodd yn ysgafn ac ysgwyd ei phen.

Diar mi, roedd hi'n ymddwyn fel merch ysgol yn ei harddegau, wedi mopio'i phen ar rywun am y tro cyntaf. Be fyddai Wil yn ei ddweud pe gwyddai am yr hyn oedd ar ei meddwl? Pe gwyddai am y teimladau roedd hi'n eu celu. Ni fyddai'n malio dim, mae'n siŵr, cyn belled â bod ei de ar y bwrdd bob nos, a'i ddrôr dillad isa'n llawn o sanau a thronsiau glân.

Ond roedd Alun yn wahanol.

Roedd o'n ofalus o'r hen bobl a ddeuai i'w lyfrgell deithiol; rhoddai gymorth i'r rhai musgrell gyda'r stepiau serth, ac roedd yn cofio pwy oedd hoff awduron ei gwsmeriaid ac yn gwneud ei orau glas i'w plesio. Byddai Alun wrth ei fodd yn ymweld â phentrefi'r ardal yn ei fan, yn eistedd yn sedd y gyrrwr yn wên i gyd ac yn gwisgo un o'i siwmperi patrwm Fair Isle – rhai llewys hir ar gyfer y gaeaf, a rhai heb lewys ar gyfer yr haf.

Roedd Brenda wedi arfer mynychu llyfrgell Tresarn yn rheolaidd nes i Sandra ddod â Lowri adre o'r ysbyty. Wedi hynny roedd hi'n ormod o halibalŵ iddi fentro i lyfrgell y dref ar y bws gyda'r babi a'r goets, heb sôn am ei llyfrau a'r negesau roedd arni eu hangen. Felly ymunodd â'r llyfrgell deithiol, a daeth i adnabod Alun. Byddai hi'n ymweld â'r fan bob tro y deuai heibio. Gan amlaf byddai'n rhaid iddi fynd â Lowri gyda hi, a byddai Alun wrth ei fodd yn dweud straeon wrth y fechan tra porai ei nain trwy'r silffoedd. Ond sawl tro'n ddiweddar, llwyddodd Brenda i gydgordio ymweliadau achlysurol Lowri â'i ffrind gorau ag ymweliadau Alun a'i lyfrgell deithiol.

A dweud y gwir, roedd Brenda ar bigau'r drain heddiw. Byddai Alun yn siŵr o fod wedi clywed am y calendar, gan Dilys neu rywun felly. Beth fyddai ei ymateb? A fyddai'n meddwl ei bod hi wedi drysu'n lân drwy ystyried gwneud y fath sioe ohoni'i hun? Roedd Brenda'n amau a oedd Wil wedi clywed am y calendar, heb sôn am sylweddoli bod y cynllun yn dechrau cael ei gefn ato. Ond sêl bendith Alun. Roedd hynny'n hanfodol.

Edrychodd Brenda ar y cloc drachefn.

Dyna ni, mi fyddai'r rhan fwyaf o'r selogion wedi gorffen bellach.

Roedd Alun wrthi ar ei benliniau'n cadw'r llyfrau a ddychwelwyd pan glywodd sŵn traed yn dringo'r tair stepen i'r fan. Pan welodd Brenda crychodd ei lygaid tywyll a lledaenodd gwên ar draws ei wyneb crwn rhadlon.

"Roeddwn i'n dechrau meddwl nad oeddet ti am ddod heddiw."

Teimlodd Brenda'r gwrid yn codi i'w hwyneb.

"Disgwyl iddi dawelu oeddwn i."

"Wel tyrd i mewn. Dwi wedi clywed petha mawr amdanat ti a gweddill y criw 'na sy'n perthyn i Ferched y Wawr."

Cochodd Brenda'n fwy byth.

"Rydach chi wedi bod yn goleuo'ch canhwylla ac yna'n eu cuddio dan lestri, yn ôl y sôn," pryfociodd Alun.

Teimlodd Brenda ddagrau annisgwyl yn dod i'w llygaid. Lled-drodd ei phen oddi wrth Alun.

Roedd yntau ar ei draed mewn chwinciad.

"Be sy, Brenda? Paid â chrio. Dim ond tynnu coes oeddwn i."

Trodd Brenda'n ôl ato, a'i llygaid yn llawn pryder.

"Wyt ti'n meddwl 'mod i'n ffôl, Alun, yn cytuno i gymryd rhan yn y cynllun 'ma? Wyt ti'n meddwl mai gwraig ganol oed wirion ydw i?"

Edrychodd Alun yn synfyfyriol arni am eiliad neu ddau. Wedyn camodd yn bwrpasol at ddrws y fan, ei gau a'i folltio. Yna gafaelodd yn llyfrau Brenda a'u gosod ar y bwrdd bach a ddefnyddiai fel cownter cyn lapio'i freichiau o'i hamgylch nes bod ei hwyneb yng nghanol esmwythdra ei siwmper Fair Isle, a'i wefusau'n gogleisio'i thalcen.

"Dwi'n gwybod nad oedd o ddim yn wir, Bren, ond roedd y diwedd yn lyfli, toedd? Ac mi roedd o'n bishyn."

"Tyrd, Phyl. Rydan ni bron yno."

Roedd Brenda wedi mynnu bod Sandra'n gofalu am Lowri am un noson er mwyn iddi hi gael mynd i'r pictiwrs yn Nhresarn efo Phyllis. Roedd Phyllis wrth ei bodd gyda ffilmiau a llyfrau rhamantus ble byddai'r arwr golygus yn achub y ferch brydfeth o enau trychineb yn ddi-ffael, a'r ddau'n byw'n hapus byth wedyn.

Dihangfa, tybiai Brenda.

Ond roedd angen dihangfa ar Phyllis druan, meddyliodd, gyda gŵr oedd yn lladd ar bopeth a wnâi. Roedd yn well ganddo ef ei hun ogleisiadau'r *Hustler* na'r *Herald*, a'r *Cutie Girls* na'r *Cymro*, er na fyddai Phyllis yn cyfaddef hynny wrth unrhyw un, ddim hyd yn oed Brenda. Ond mewn pentref bach fel Nant-yr-Onnen, buan y deuai pawb i wybod busnes ei gilydd, ac roedd tuedd Tom at lenyddiaeth liwgar yn chwedlonol, heb sôn am . . .

Sgrialodd y bws i stop swnllyd yng nghanol y pentref.

"Awn ni am Babycham cyn mynd adre, Bren?" awgrymodd Phyllis wedi i'r ddwy ddisgyn o'r bws.

"Ia," cytunodd Brenda. "Pam lai?"

Roedd y Llew Coch yn reit brysur. Yn y bar roedd gêmau

pŵl a dartiau ar eu canol, ac yn y lolfa roedd nifer o grwpiau'n sgwrsio o amgylch y byrddau crynion.

Aeth Phyllis at y bar tra eisteddodd Brenda wrth fwrdd gwag. Tu draw i'r bar, ble roedd Tina'n gweini ar Phyllis, gallai Brenda weld Wil, a Tom, gŵr Phyllis, yn gwylio rhyw gêm bêl-droed ar y teledu lloeren roedd Gwyneth wedi ei osod yno'n ddiweddar.

Daeth Phyllis at y bwrdd gyda'r diodydd, yn dal i ganmol y ffilm.

"Mi fyddwn ninna'n sêr o fath cyn bo hir," meddai Brenda, gan geisio newid trywydd y sgwrs.

"Wel byddwn, erbyn meddwl," cytunodd Phyllis. "Er dwi'n ama'n gry weithia a ydw i'n gwneud y peth iawn."

"A finna. Ond does wiw i ni dynnu'n ôl rŵan," pwysleisiodd Brenda. "Dim ond dwsin ohonon ni sy 'na."

"Wyt ti wedi deud wrth Wil?" holodd Phyllis.

"Naddo. Tydi o heb sôn 'run gair am y peth, chwaith, felly dwi'n cymryd nad ydi o'n gwybod 'mod i'n un o'r criw sy'n mynd i gael eu hanfarwoli."

"'Run fath efo Tom," meddai Phyllis, "a go brin y byddai o'n cadw'n ddistaw tasa ganddo fo achos i godi twrw."

"Bydd y ddau ohonyn nhw'n cael tipyn o syrpreis, felly," meddai Brenda. "Wyt ti am un arall o'r rheina cyn i ni fynd?"

Roedd hi'n braf cael dod allan am newid bach, meddyliodd Brenda ar ôl dychwelyd o'r bar. Roedd o'n gwneud lles i rywun.

"Wyt ti wedi cael unrhyw lwc wrth chwilio am swydd arall?" gofynnodd Brenda i Phyllis yn y man.

Ysgydwodd Phyllis ei phen. "Dim byd," atebodd. "Ac roeddwn i mor hapus yn gweithio yn y Siop Pethau Bychain. Roeddwn i wrth fy modd yn tendio ar y mama a'r babanod a ddeuai i'r siop, ac roedd y dillad bychain yn ddigon o sioe."

"Mae'n biti gweld cymaint o siopa bach Tresarn yn cau," meddai Brenda. "Ar yr archfarchnad newydd 'na mae'r bai, dwi'n siŵr."

"Ti'n iawn," cytunodd Phyllis. "Dwi am ddal i chwilio, cofia. Ella y ca i fwy o lwc yn yr hydre. Bydd siopa isio rhagor o weithwyr at y Dolig erbyn hynny."

O gornel ei llygad roedd Brenda wedi sylwi ar Wil a Tom yn gadael y bar ac yna'n brasgamu ar draws y lolfa a golwg gandryll arnynt.

"Be 'di'r blydi lol 'ma rydan ni wedi'i glywed?" dechreuodd Tom.

"Amdanoch chi'ch dwy a rhyw galendar," parhaodd Wil.

"Tydi o 'rioed yn wir?" Chwarddodd Tom. "Pwy fydda'n talu pres da i'ch gweld chi'ch dwy'n fronnoeth?"

"Mi fydda'n rhaid i chi gael eich smwddio'n gynta," ychwanegodd Wil yn sbeitlyd.

Edrychodd Brenda a Phyllis ar ei gilydd.

Mewn un symudiad cododd y ddwy, tywallt eu diodydd dros bennau'u gwŷr, a cherdded allan o'r dafarn.

Roedd Brenda'n chwys domen dail.

Roedd bwrdd y gegin dan ei sang â brechdanau (tri math – ham, jam a chaws), sosejys bach, creision, ciwbiau o gaws wedi'u gosod ar ffyn gyda darnau o ffrwythau lliwgar deniadol, tafelli o bitsa, a llu o fisgedi siocled o wahanol siapiau. Gorweddai llieiniau gwyn glân dros y cyfan.

Go drapia! Ble roedd Sandra wedi mynd? Roedd hi wedi addo y byddai hi yno i helpu gyda'r parti pen-blwydd. A hithau Brenda bellach yn writgoch, yn siwgwr eisin o'i chorun i'w sawdl, yn coginio dros Gymru, gyda Lowri yn y stafell fyw yn gwylio'i hoff fideo, a dim golwg o'i merch.

Byddai ffrindiau Lowri yma ymhen yr awr.

Ac roedd Brenda'n cael trafferth gyda'r gacen.

Roedd wedi ei gorchuddio ag eisin gwyn plaen. Ond ni fu ei hymdrechion i greu Tweenie ag eisin glas tywyll yn llwyddiant ysgubol. Ymddangosai'r creadur fel petai wedi dioddef sioc drydanol go egr! Byddai'n rhaid gadael i'r gampwaith sadio am ychydig, yna trio twtio rhywfaint arno cyn gosod y canhwyllau, a gobeithio'r gorau. Diolch nad oedd Lowri a'i ffrindiau'n debygol o fod yn agos mor feirniadol ag y buasai gweddill y teulu.

Paned sydyn. Yna byddai'r rholiau sosej yn barod, byddai'r eisin yn haws ei drin, a gallai estyn y balŵns a gweddill trugareddau'r parti o'r twll dan grisiau, cyn mynd i molchi a newid.

Dros ei phaned diolchodd Brenda fod Lowri'n ferch fach mor fodlon, neu byddai wedi bod yn amhosib iddi gael trefn ar bethau ar ei phen ei hun.

Clywodd gnoc ysgafn ar y drws cefn. Wrth fynd i'w agor, gobeithiodd nad oedd neb wedi cyrraedd yn gynnar.

Ar y rhiniog safai Alun, a pharsel sgwar wedi ei lapio mewn papur amryliw yn ei ddwylo.

Roedd Brenda'n fud, ei meddwl yn lobsgows o deimladau. Roedd wedi gwirioni o'i ganfod yno. Ond beth petai rhywun yn ei weld? Ac roedd golwg drychinebus arni hi – ei hwyneb yn goch a siwgwr eisin dros ei dillad.

"Wel, ydw i am gael dod i mewn?" holodd Alun.

"Wyt, wyt, siŵr. Sori."

Setlodd Alun ar gadair yn y gegin tra gwnaeth Brenda baned ffres iddo, cyn prysuro i achub y rholiau sosej o'r popty a gosod y canhwyllau ar y gacen.

"Dwyt ti ddim i fod i alw yn Nant-yr-Onnen heddiw, wyt ti?" gofynnodd Brenda.

Cymerodd Alun arno edrych yn drist.

"Dwyt ti ddim yn swnio'n falch iawn o'm gweld i."

"O, ydw," sicrhaodd Brenda ef. "Ydw, dwi yn falch o dy weld ti. Wedi synnu, dyna i gyd."

"A deud y gwir wrthat ti," cyfaddefodd Alun, "dwi i fod yn y swyddfa'n dal i fyny â gwaith papur, ond roedd gen i rywbeth bach i Lowri . . . ac roeddwn i isio dy weld ti."

Gwridodd Brenda. "A golwg y fall arna i," chwarddodd. "Yli, bydd yn rhaid i mi fynd i molchi a newid cyn i ffrindia Lowri gyrraedd."

"Aros funud."

Estynnodd Alun ei fys a chodi sbotyn o eisin oddi ar drwyn Brenda. Yna llyfodd ei fys yn frwdfrydig.

Pan ddychwelodd Sandra i'r tŷ, a'r parti ar fin dod i ben, roedd hi'n llawn disgwyl pryd o dafod, ac fe'i synnwyd pan welodd ei mam a rhyw ddyn dieithr – Yncl Alun, yn ôl Lowri – yn mwynhau eu hunain yng nghanol y plant.

Isio'i smwddio, ia? Wel, sgwn i beth fyddai gan Wil i'w ddweud pan welai hi'n smwddio ar y calendar? Dim bod fawr o ots gan Brenda bellach. Gall hyd yn oed y gwannaf wingo. A gall hetar crasboeth fod yn arf go effeithiol os oes rhaid.

Roedd hi wedi cael bore'n rhydd o'r gwaith ar gyfer tynnu'r lluniau. Roedd Lowri'n mynychu Ysgol Gynradd Tresarn bellach ac yn cael pàs yno gan fam ei ffrind pennaf. Roedd wrth ei bodd yng nghanol y plant eraill ac, yn ôl ei hathrawes, roedd yn siarp ac yn sionc, felly roedd Brenda'n rhydd i osod ei stondin ar gyfer y llun, ac i bincio, er gwaetha cysgod y pen mawr oedd ganddi. Roedd Phyllis wedi cynnig dod draw i gadw cwmni iddi, ond roedd yn well gan Brenda wynebu Harri ar ei phen ei hun. Byddai Phyllis wedi'i mwydro.

Ond roedd y tŷ'n rhyfedd o ddistaw heb Lowri.

Lowri ac Alun. Yr unig ddau oedd â thipyn o feddwl ohoni.

Rhoddodd Brenda'i llaw ym mhoced ei gŵn gwisgo a thynnu allan y bocs bach coch a roddodd Alun iddi ddoe. Ynddo, ar glustog o wlân cotwm gwyn, gorweddai pâr o glustdlysau ar ffurf sêr.

"I'm seren ddisglair i," meddai Alun yn ddistaw, wrth eu rhoi iddi.

Doedd traed Brenda heb gyffwrdd y llawr ers hynny.

Roedd wedi deall mewn amrantiad beth a welai Phyllis yn yr holl ffilmiau rhamantus yna. Sylweddolodd fod ganddi hithau deimladau; teimladau a gladdwyd dan haenau llethol slafdod a diflastod. Ond rŵan, rŵan roedd hi'n bwysig i rywun, yn rhan o rywbeth mwy na hi ei hun. Clymwyd cadwyn o amgylch ei chalon, cadwyn frau i'w chyffwrdd â blaenau ei bysedd ac â pharch a rhyfeddod a thynerwch. Nid oedd am weld y gadwyn hon yn cael ei thrawsnewid i fod yn debyg i'r hual a'i caethiwodd ers blynyddoedd, a hithau'n cael ei rheibio a'i rhwygo'n ddiamynedd a di-hid.

Gwisgodd Brenda ei chlustdlysau newydd. Am lwc.

Chwarae teg i Harri. Roedd o wedi gofalu ei bod yn gyfforddus (neu mor gyfforddus â phosib dan yr amgylchiadau) ac yna wedi gwneud ei waith yn effeithlon a di-lol. Derbyniodd y cynnig o baned o goffi ar unwaith. Bu ar ei draed ers ben bore, a gwyddai fod ganddo ddiwrnod hir o'i flaen.

Gwisgodd Brenda'n frysiog yn y llofft tra oedd y tecell yn berwi.

"Wyt ti'n difaru cael dy berswadio i wneud y gwaith 'ma?" holodd Brenda wrth estyn mỳg i Harri. "Mae'n ymddangos fel tipyn o strach i mi."

"Tipyn bach," cytunodd Harri, "ond yn llai helbulus na llawer i briodas. Ac am dynnu llunia babis . . ." Ysgydwodd Harri ei ben. "Pob siwgwr candi mami ohonynt yn sgrechian dros y stiwdio ar ôl y fflach gynta."

Chwarddodd Brenda.

"Mae hi'n fraint ac yn bleser cael bod yn rhan o'r cynllun," sicrhaodd Harri hi, "a gweithio gyda merched beiddgar fel chi. A deniadol," ychwanegodd.

Gwenodd Brenda. "A 'run ohonon ni wedi cael ein smwddio," meddai'n llon.

Edrychodd Harri'n hurt arni.

"Nos da, Lowri."

"Nos da, Nain."

Swatiodd Lowri dan y dillad gwely gyda Tedi. Cadwodd Brenda'r llyfr stori ar y silff cyn diffodd y golau mawr yng nghanol y stafell a gadael lamp fechan ynghynn ar y cwpwrdd bach ger y gwely. Aeth allan gan adael y drws yn gilagored.

Yn ei llofft ei hun newidiodd Brenda'n ddioed. Roedd wedi gweld ffrog goch dywyll yn ei chatalog rai wythnosau ynghynt, ac wedi'i phrynu fel trît yn dilyn achlysur tynnu'r lluniau.

Roedd y ffrog yn gweddu i'w gwallt brown a'i llygaid gwinau i'r dim. Disgleiriai sêr Alun ar ei chlustiau fel petaent yn goleuo'i llwybr tuag at y dyfodol. Dyfodol ansicr, dyfodol llawn ofnau, dyfodol addawol. Gwisgodd Brenda ychydig o golur, gan ddal ei sbectol ar ben ei thrwyn er mwyn coluro'i llygaid yn effeithiol. Dyna ni! Twtsh o finlliw a chwa o bersawr ac roedd hi'n barod.

Teimlai'n reit nerfus ynglŷn â gweld y calendar y noson honno, ond roedd hynny'n well na'i weld am y tro cyntaf y noson ganlynol yn y lansiad swyddogol. Roedd Mererid wedi'u sicrhau bod y lluniau'n werth eu gweld, ond . . .

Yn y stafell fyw roedd Wil, Robin a Sandra'n sownd wrth y teledu. Hawliai'r bwda swat eu holl sylw. Safodd Brenda yn y drws am funud yn eu gwylio. Ddywedodd 'run ohonynt air nes iddi droi am y cyntedd i estyn ei hesgidiau.

"Dwyt ti ddim yn bwriadu mynd i'r Llew Coch i weld yr hen galendar coman 'na, gobeithio. Mi fyddwch chi wedi dychryn y cwsmeriaid i gyd."

"Dwi isio 'nghrys glas erbyn fory, Mam."

"Ac mae Lowri angen bocs bwyd fory. Mae'r dosbarth yn mynd ar drip i rywle."

Sylwodd 'run o'r tri ar Brenda'n cau'r drws yn ddistaw.

Wrth wisgo'i hesgidiau gwelodd Brenda ben bach melyn yn sbecian trwy'r canllaw ar ben y grisiau.

"'Nôl i dy wely, Lowri, yn hogan dda i Nain," sibrydodd, gan chwythu sws i'r fechan.

Chwythodd Lowri sws yn ôl i Brenda cyn diflannu i'w gwely drachefn.

Gafaelodd Brenda yn ei bag ac allan â hi trwy'r drws ffrynt i gyfarfod Phyllis.

# EBRILL

Wedi golchi'r ddwy gwpan goffi, gafaelodd Nia yn y cadach llestri a'i droelli'n ffyrnig rhwng ei dwylo. Petai Elsi'n iâr mi fyddai wedi gafael ynddi a rhoi tro i'w chorn gwddw. Jyst fel'na. A phetai hi'n dod i'r tŷ unwaith eto a rhedeg ei bys ar hyd y silff ben tân i chwilio am olion llwch, byddai'n ei chrogi efo'r dystar beth bynnag.

Ceisiodd Nia ymbwyllo.

Cwta ddwyawr oedd ganddi bellach, wedi ymweliad Elsi, cyn y byddai'n amser nôl Glesni o'r cylch meithrin, ac roedd yn rhaid iddi wneud y deisen heddiw neu fyddai byth wedi sadio digon i'w haddurno. A doedd wiw iddi adael cwsmer lawr, neu byddai si'n lledu nad oedd yn ddibynadwy. Ac roedd yr hen gwpwl o'r tu draw i Dresarn a oedd ar fin dathlu eu priodas aur mor annwyl – a'u merch, oedd yn trefnu'r dathliad. A byddai Ifan yn cyrraedd am ei baned ganol bore cyn bo hir. A dyna ugain munud arall wedi hedfan o'i gafael.

Doedd hi ddim eisiau dechrau cwyno wrtho fo am ei fam eto, er bod yr hen gnawes yn sbio lawr ei thrwyn ar holl ymdrechion Nia i gyfuno magu plant a bod yn wraig fferm a cheisio dod ag ychydig o arian i'r tŷ trwy wneud teisennau i bobl ar gyfer achlysuron arbennig. Ac erioed wedi cynnig codi bys na bawd i helpu. Chwarae teg i Ifan, roedd o wastad yn ei chefnogi, er bod y fferm yn mynd â chymaint o'i amser a'i sylw. Gwyddai y byddai'n dweud wrthi am anwybyddu sylwadau snobyddlyd ei fam fusneslyd fel y gwnâi bob tro y byddai Elsi'n tynnu blewyn o'i thrwyn. Roedd Nia'n rhyw amau mai mater o ddweud unrhyw beth er mwyn heddwch oedd hynny. Gwyddai fod

Ifan yn poeni am ddyfodol y fferm, a'u dyfodol hwythau. Roedd pethau wedi newid cymaint yn ystod y deng mlynedd ers iddynt briodi ac ymgymryd â gwaith y fferm. A chyda thri o blant i'w magu . . . Ond roedd anwybyddu Elsi'n dipyn o orchwyl pan oedd yr hen ast yn ei gormesu o bared i bost yn feunyddiol.

Mewn Eisteddfod Ffermwyr Ifanc y cyfarfu Nia ag Ifan Morgan. Roedd hi wedi ennill y wobr gyntaf am wneud teisen ac roedd ef, a dyrnaid o'i gyfeillion, ar ôl ymweliad answyddogol â'r dafarn leol, wedi dychwelyd i'r babell ble arddangosid y cynnyrch buddugol, i geisio lluniaeth rhad ac am ddim i flotio rhywfaint ar y cwrw. Collodd Nia ei limpin yn llwyr pan welodd y llafnau'n llowcio'i champwaith. A chollodd Ifan ei ben drosti pan welodd ei llygaid tywyll yn fflachio dan ei gwallt du cyrliog cwta.

Buan y sylweddolodd fod ganddo lawer o dir i'w adennill wedi trychineb y gacen. Holodd ym mha gangen o'r mudiad roedd Nia'n aelod. Yna llwyddodd i berswadio arweinwyr ei gangen ei hun y byddai twmpath dawns yn weithgaredd poblogaidd, ac mai da o beth fyddai iddynt wahodd aelodau o ganghennau cyfagos i ymuno â nhw. Cynigiodd sgubor ei dad fel lleoliad.

Bu wrthi am ddyddiau'n glanhau'r sièd wair ac yn trefnu ac aildrefnu'r byrnau gwair a'r addurniadau a'r goleuadau a fenthycodd gan ei ffrindiau. Cwyno'n ddi-baid am ei ddifaterwch ynghylch ei waith ar y fferm wnaeth ei fam, Elsi, ond pan oedd fwy neu lai'n fodlon ar ei baratoadau daeth ei dad i'r sgubor i sbecian ar ei ymdrechion. Rhoddodd winc i Ifan cyn rhoi ei law ar ei ysgwydd a dweud, "Gobeithio'i bod hi'n werth yr holl drafferth 'ma, 'machgan i."

Mi roedd hi.

Wedi gosod y lluniaeth ysgafn a'r diodydd ar eu stondin, cyfarch pawb yn barchus ac annog y grŵp gwadd i chwarae eu gorau glas, aeth Ifan ati i wahanu Nia oddi wrth y criw o ferched oedd yn piffian chwerthin o'i hamgylch. Ac wedi ei dosio ag ychydig o hylif ymlaciol, a'i chwipio o amgylch y llawr i guriadau'r gerddoriaeth werin, derbyniodd faddeuant am ei bechod.

Byr fu eu carwriaeth. Doedd Ifan ddim yn un i oedi unwaith y rhoddai ei fryd ar rywbeth. Roedd yn awyddus i setlo lawr, ac roedd am weld modrwy ar fys Nia cyn i neb arall geisio bachu ei anwylyd. Roedd ei dad wedi dechrau potsian ag un o'r siediau na fu mewn defnydd ers tro, gyda'r bwriad o greu byngalo iddo ef ei hun ac Elsi ar gyfer eu hymddeoliad, a bwriwyd iddi i gwblhau'r gwaith ar fyrder er mwyn i'r cwpwl ifanc ei gael yn gartref clyd a chyfleus iddyn nhw eu hunain.

Nid oedd ar Nia ofn gwaith caled, corfforol. Bob penwythnos byddai yno yn ei chanol hi'n llafurio gydag Ifan a'i dad a'r crefftwyr oedd yn gyfrifol am y gwaith a oedd angen sylw arbenigol. Cyfareddwyd tad Ifan gan ei brwdfrydedd a'i hymroddiad, ac roedd ei hasbri a'i sirioldeb fel chwa o awyr iach o'u cymharu â chwyno a beirniadu syrffedus ei wraig.

Syfrdanwyd Elsi pan welodd ei darpar ferch-yng-nghyfraith mewn welingtons a hen ddillad blêr yn gweithio ochr yn ochr â'r dynion. Fe'i siomwyd, oherwydd roedd hi wedi disgwyl pethau mawr gan Ifan, ac wedi dyheu iddo briodi merch syber goeth a fyddai'n ysgrifenyddes i reolwr banc ac wrth ei bodd yn trafod dillad a charpedi a llenni gyda hi. A dweud y gwir, dyfalai Elsi'n aml beth a ddaeth dros ei phen pan gytunodd i briodi ffermwr, oherwydd gwelai ei hun yn gymwys fel gwraig rheolwr banc dylanwadol. Fu ganddi erioed ddiddordeb na thuedd at fod yn wraig fferm. Doedd y dillad ddim yn ei siwtio i ddechrau arni, a'r unig adeg y closiai at y stoc oedd pan oeddynt wedi eu diesgyrnu a'u datgymalu ar gyfer cinio dydd Sul. Gwyddai fod Nia tua hanner ffordd trwy gwrs coginio yn y coleg; ddim cystal â chwrs ysgrifenyddol bach neis, ond dyna ni, fel'na roedd y genod ifanc yma y dyddiau hyn, a'u pennau'n llawn syniadau am yrfaoedd a bod yn annibynnol. Ond pan welodd Nia'n torchi ei llewys ac yn gweithio fel labrwr. Wel!

Trefnwyd y briodas ar gyfer yr haf, pan fyddai Nia'n rhydd o'r coleg a gwaith y fferm ryw ychydig yn ysgafnach. Dim ond y gwaith peintio oedd ar ôl i'w orffen yn y byngalo, ac yna byddai'n barod i'w ddodrefnu ar gyfer y cwpwl ifanc pan ddychwelent o'u mis mêl.

Roedd Ifan a Nia am gynnal eu neithior yn y sgubor ond parodd y syniad ffit binc i Elsi, felly cytunwyd ar gyfaddawd – fe fyddai Elsi'n cael ei gwledd mewn gwesty crand, ond gyda'r nos byddai twmpath dawns i'r bobl ifanc, a'r rhai ifanc eu hysbryd, yn y sièd wair.

Ond yna, cwta ddeufis cyn y briodas, bu farw tad Ifan. Yn ddistaw, yn ddisymwth, yn ddi-gŵyn. Bu Nia'n gefn i Ifan ac Elsi yn eu trallod, gan fod yr un mor barod i eistedd gydag Elsi ag yr oedd i helpu Ifan â gwaith y fferm. Roedd Elsi a Nia am ohirio'r briodas, ond mynnodd Ifan barhau â'r trefniadau, yn wyneb hunandosturi truenus ei fam. Gwyddai y byddai Elsi'n ceisio ei glymu wrth linyn ei ffedog am byth pe na byddai'n priodi rŵan, gan chwarae ar gydwybod ac ewyllys da Nia. Ac nid oedd Ifan am ei cholli.

Peintiwyd a dodrefnwyd y byngalo at ddant Elsi, felly, a symudodd y pâr ifanc i'r tŷ fferm. Gohiriwyd y mis mêl am gyfnod amhenodol, a gohiriodd Nia flwyddyn olaf ei chwrs coginio er mwyn dysgu bod yn wraig fferm.

A rŵan dyma hi, a chanddi dri o blant dan ddeg oed, fferm a ddeuai â llai a llai o incwm iddynt bob blwyddyn, cyfyngiadau amser ac adnoddau ar ei hymdrechion ei hun i ennill tamaid, mam-yng-nghyfraith a drethai Iesu Grist ei hun, a llond ei chroen o rwystredigaeth.

"Mam, Mam, faint eto nes bydd Dafydd a Gruffudd adra?"

Roedd hi newydd basio pedwar o'r gloch. Byddai Dafydd a Gruffudd yn cyrraedd ar y bws ysgol o Dresarn unrhyw funud, ar eu cythlwng fel arfer; roedd Ifan a'r contractwyr cneifio wedi bod yn yfed te ac yn bwyta'n ddi-baid ers ben bore; roedd y gacen ar gyfer dathliad pen-blwydd priodas yr hen gwpwl hoffus yn dal yn ei noeth ogoniant yn y gegin. Ac roedd Nia ar fin tynnu gwallt ei phen.

Erbyn pump roedd hi wedi gwneud pentwr o frechdanau, rhag ofn i'r bechgyn drengi cyn amser swper; wedi golchi llwyth o ddillad pêl-droed mwdlyd; wedi smwddio tomen o ddillad glân; wedi darparu pecynnau bwyd i Dafydd a Gruffudd ar gyfer y diwrnod canlynol a'u storio yn yr oergell; ac wedi dechrau plicio tatws dros Gymru.

Erbyn chwech roedd y contractwyr wedi gadael am ryw gymanfa gneifio, ac roedd pawb yn eistedd wrth y bwrdd yn bwyta'u swper.

"Mam, mae Gruffudd yn 'y nghicio i o dan y bwrdd."

"Nac'dw i. Wir yr. A beth bynnag, hi wnaeth gynta."

"Mae gen i lond tudalen o sỳms i'w gwneud erbyn fory, Mam."

"Rydw i angen hel gwybodaeth am y Rhufeiniaid."

"Ac mae Anti Sioned wedi gofyn i ni hel canol toilet rôls."

"Glesni, defnyddia dy gyllell a fforc, plîs. A hogia, peidiwch â llowcio'ch bwyd fel'na – does 'na neb yn mynd i'w ddwyn o."

Eisteddai Ifan gan gnoi'n synfyfyriol, heb sylwi ar y sgwrs a sbonciai ar draws y bwrdd bwyd. Gwyddai Nia ei fod yn ceisio amcangyfrif costau'r contractwyr a'u gosod yn erbyn y pris a gâi am y gwlân.

"Mam, ga i gacen rŵan? Dwi wedi bwyta pob dim." Roedd Glesni wedi gosod rhywfaint o'i llysiau gwyrdd yn llechwraidd o dan ei chyllell a fforc.

Doedd gan Nia mo'r amser na'r awydd i ddadlau dros lwyaid o bys, maethlon neu beidio, ond roedd yr hogiau wedi digio.

"Sut mae hi'n cael darn o gacen heb orffen ei bwyd?"

"Tydan ni ddim yn cael gwneud hynna."

Edrychodd Nia ar Ifan am gefnogaeth, ond llithrodd y cecru heibio iddo fel slipars ar rew.

Erbyn saith roedd Nia wedi golchi'r llestri a thwtio'r gegin, ac wedi cynorthwyo gyda thri darn o waith cartref, gan gynnwys lapio nifer o roliau tŷ bach am ei gilydd i wneud caseg o rolyn er mwyn arbed sterics gan Glesni. Yna aeth i chwilio am Ifan i'w atgoffa ei bod hi'n noson cyfarfod Merched y Wawr.

Daeth o hyd i'w gŵr yn y sgubor, yn edrych yn betrusgar ar y twmpathau trefnus o wlân. Edrychodd Nia arno am sbel fach cyn mynd draw a rhoi gwasgiad iddo.

"Mi fydd petha'n iawn, mi gei di weld," cysurodd ef.

Rhoddodd Ifan ei fraich o amgylch ei hysgwyddau.

"Byddan . . . byddan, gobeithio," atebodd, heb unrhyw argyhoeddiad yn ei lais.

Wedi slemp sydyn, a siarsio'i gŵr i beidio â mynd i grwydro'r caeau a gadael y plant yn y tŷ ar eu pennau'u hunain, na mynd â nhw efo fo a dod â nhw'n ôl yn fwd o'u corun i'w sawdl, cerddodd Nia lawr at y giât lôn i ddisgwyl am Anwen, gwraig Albert Parry, er mwyn i'r ddwy ohonynt gydgerdded i'r cyfarfod, yn ôl eu harfer. Wrth iddi ddisgwyl, gwelodd gyrtans les Elsi'n aflonyddu, ond er ei bod wedi ei bendithio â hen wrach o fam-yng-nghyfraith, meddyliodd, fyddai byth yn ffeirio'i gŵr a'i phlant gyda neb. Ond bobol bach, roedd hi'n braf cael dihangfa am ychydig.

"Wn i ddim, Nia. Mi fyddai Albert yn lloerig petawn i'n gwneud y fath beth."

"O tyrd, Anwen. Mi fydd hi'n hwyl. Ac roeddet ti'n ddigon cefnogol neithiwr."

Rhoddodd Anwen ei dwylo dros ei llygaid ac ysgwyd ei phen yn anobeithiol.

"Paid â'm hatgoffa i."

"Yli," anogodd Nia hi, "anaml iawn y byddi di na fi'n cael cyfle i wneud rhywbeth gwahanol. Tendiad ar ein teuluoedd o fora gwyn tan nos mae'r ddwy ohonom fel rheol."

Gwnaeth Anwen stumiau ar ei ffrind. Roedd hi wedi dychwelyd i'r fferm gyda Nia ar ôl i'r ddwy ohonynt ddanfon eu plant ieuengaf i'r cylch meithrin y bore wedi'r cyfarfod. Eisteddai Anwen yn y gegin yn llymeitian ei choffi tra dechreuodd Nia ar y gwaith o addurno'r gacen.

"Beth am Ifan? Be mae o'n feddwl?"

"Wel, mi wnes i drio deud wrtho fo bora 'ma, ond roedd o ar gymaint o frys i fynd 'nôl at y cneifio, dwi'n ama a glywodd o'r un gair. Beth bynnag, mi gyflawnodd o gryn dipyn o gampia gwirion ei hun pan oedd o'n aelod o'r Ffermwyr Ifanc, felly hyd yn oed tasa fo'n gwrthwynebu, fyddai ganddo'r un goes i sefyll arni. A bydd y ffaith 'mod i'n cymryd rhan mewn cynllun mor goman yn bownd o godi sterics ar Elsi," ychwanegodd Nia'n orfoleddus.

"Mi fydd hi'n horlics pan gaiff hi wybod," cytunodd Anwen. "Ond ti'n gwybod am Albert," parhaodd, "mae o mor henffasiwn weithia." Edrychodd yn synfyfriol ar Nia'n taenu'r marsipán melyn melys dros y deisen. "A deud y gwir," ychwanegodd, "dwi'n methu dod dros Sara'n cynnig y fath beth."

"Wel, mae angen rhywun ag ychydig o weledigaeth i warchod cefn gwlad," meddai Nia, "rhywun nad oes arnyn nhw ofn mentro. A chofia di, fedran ni'n dwy ddim fforddio colli'r cylch meithrin."

"Na fedran, ti'n iawn," cytunodd Anwen. "Oes gen ti rywbeth i'w yfed yma?" gofynnodd wedyn yn ddirybudd.

"Oes, dwi'n meddwl." Crychodd Nia ei haeliau wrth drio cofio. "Oes, mae gen i ryw ychydig o frandi fydda i'n ei ddefnyddio i gadw teisenna ffrwytha'n feddal ac i roi blas arnyn nhw. Yn y cwpwrdd acw. Ac ella bod yna rywfaint o sieri hefyd."

"Gymerwn ni sieri, 'ta?" holodd Anwen.

"Deg o'r gloch ydi hi! Tydi hi ddim braidd yn fuan?"

"O, tyrd 'laen. Wneith un bach ddim drwg. A bydd angen ffeindio hyder yn rhywle os ydan ni am gymryd rhan yn y busnas calendar 'ma."

Gosododd Nia'r gacen o'r neilltu'n ofalus tra estynnodd

Anwen y sieri a'r gwydrau. Sipiodd Nia o'i gwydr yn amheus wrth wylio'i ffrind yn llowcio â chryn frwdfrydedd. Yn y man cychwynnodd Anwen am adre, yn llawn ffydd yn y cynllun bellach, ac yn ei gallu ei hun i wrthsefyll protestiadau ei gŵr. Roedd Nia wrthi'n clirio'r gwydrau ac yn gofidio am ei ffrind pan drawodd Ifan ei ben heibio'r drws a chyhoeddi ei fod ef a'r contractwyr ar fin marw o newyn a syched yn y sgubor.

". . . a dyma fi'n deud wrthi, wir, fyddwn i ddim yn caniatáu i'm merch-yng-nghyfraith wneud sioe ohoni ei hun yn y calendar coman 'ma."

Gwenodd Nia'n serchog. "Does dim angen eich caniatâd arna i, Elsi," meddai. "Mae gen i hawl i wneud fel y mynnaf. Ac mi'r ydw i *yn* bwriadu bod yn un o griw'r calendar."

Meddyliodd Nia'n siŵr y byddai Elsi'n ffrwydro fel ŵy yn cael ei ferwi mewn meicrodon.

"Ond beth amdana i?" sbladdriodd Elsi. "Be fydd pobl y pentre 'ma'n meddwl ohona i, a titha'n llusgo enw da'r teulu trwy'r baw?"

"O, tydw i ddim yn meddwl bydd eu barn ohonoch chi'n newid yr un iot," sicrhaodd Nia hi. "Maen nhw'n eich adnabod chi'n iawn, Elsi."

Roedd Elsi wedi rhuthro am y ffermdy wedi clywed y newydd rhyfeddol yn y siop tra oedd hi'n plagio Marian am yr hadog melyn unwaith eto. Fe'i syfrdanwyd pan ddeallodd fod Nia'n un o'r criw dethol y bwriadwyd eu hanfarwoli, a phenderfynodd ei bod yn bryd iddi roi ei throed i lawr gyda'i merch-yng-nghyfraith danllyd nad oedd, yn ei thyb hi, yn dangos hanner digon o barch ac ymostyngiad tuag ati hi a'i statws fel matriarch y teulu. Yn union fel y Fam Frenhines, heddwch i'w llwch, twt-twtiodd. Chawsai honno, druan ohoni, ddim byd ond trafferth efo'i thylwyth anufudd, anniolchgar hithau.

Digwyddodd Elsi ddal Nia'n rhoi dillad ar y lein. Ond er

i Elsi gynnwys pob amrywiad ar flacmel emosiynol yn ei pherfformiad, gwelodd ei hymdrechion i roi ei throed i lawr yn bownsio fel peli ping-pong oddi ar unplygrwydd Nia, a oedd yn parhau yn ddiwyro.

Dychwelodd Nia i'r tŷ gyda'r fasged olchi, a sglein buddugoliaeth yn ei llygaid. Er gwaetha strancio Elsi, a gymerodd arni fod y newydd wedi rhoi sioc farwol iddi (ond byddai hynny'n ormod i obeithio amdano), am unwaith roedd Nia wedi cael y gorau ar yr hen ast.

"O, mae hi'n werth ei gweld! Diolch o galon i chi, Mrs Morgan."

"Nia, plîs. Dwi'n falch eich bod chi'n hapus efo'r gacen."

"Ydw, wir. Mi fydd Mam a Dad wrth eu bodd."

"Ydyn nhw'n gwybod bod 'na barti ar y gweill?"

"Wel, maen nhw'n rhyw ama, dwi'n meddwl, ond yn cymryd arnynt wybod dim i'm plesio i."

"Chwara teg iddyn nhw."

"Oes gynnoch chi amsar am banad? Ynteu 'dach chi ar frys?"

Cymerodd Nia gip ar ei horiawr.

"Wel, mae angen un neu ddau o betha arna i, ac mae'n rhaid i mi fod 'nôl mewn pryd i godi Glesni o'r cylch meithrin am hanner dydd. Ond byddwn yn ddiolchgar am banad."

Dros y baned fe glywodd Nia am holl drefniadau Siân ar gyfer parti ei rhieni, ac yna holodd Siân am ei hynt a helynt hithau. Clywodd am ei phryderon ynglŷn â'r fferm; am anturiaethau'r plant, a adroddwyd a thinc o falchder yn ei llais; clywodd am Elsi a'i stumiau snobyddlyd, ac am ddyhead Nia i ddychwelyd i'r coleg, ac i ddechrau busnes coginio o ddifrif.

Nid oedd Nia wedi arfer cael clust oedd mor barod i wrando, a theimlai fel petai'n cael sesiwn rhad ac am ddim gyda pherson trin pennau. O'r diwedd edrychodd ar ei horiawr drachefn a neidiodd ar ei thraed.

"Sori, Siân, mae'n rhaid i mi fynd neu mi fydda i'n hwyr yn nôl Glesni. Ond mi wnes i fwynhau'r sgwrs."

"A finna hefyd," sicrhaodd Siân hi. "Ylwch, Nia, rhowch y bil i mi rŵan, i mi gael ei setlo."

Wrth hebrwng Nia at y drws, cofiodd Siân yn sydyn am y si a glywsai.

"Gyda llaw, ydi hi'n wir bod rhai o ferched Nant-yr-Onnen yn mynd i gynhyrchu calendar bronnoeth i achub neuadd y pentre? Clywodd Mam ryw sibrydion am y peth pan oedd hi'n ymweld â'r llyfrgell deithiol y diwrnod o'r blaen."

Cochodd Nia at ei chlustiau.

"A-ha. Mae'r stori'n wir, felly. A 'dach chi'n cymryd rhan?"

Nodiodd Nia.

"Da iawn chi." Roedd Siân wedi cynhyrfu'n lân a rhoddodd wasgiad i Nia. "Mwy o ferched mentrus fel chi sy isio arnon ni; merched sy'n barod i sefyll dros eu cymuneda a gwarchod bywyd cefn gwlad."

Ochneidiodd Nia wrth wasgu'r sbardun ar y ffordd 'nôl i Nant-yr-Onnen. Ai gwarchod bywyd cefn gwlad oedd ei bwriad wrth gymryd rhan yn y fenter 'ma? Ynteu codi gwrychyn Elsi oedd ei phrif a'i hunig gymhelliad?

"Tyrd ti efo fi rŵan, Glesni. Mi gei di ddod acw i chwara efo Aled bora 'ma, tra bo Mam yn cael tynnu'i llun."

"Na, na, dwi isio aros yn fan'ma. Dwi isio tynnu llun hefyd."

"Yli, mi wna i dynnu llun ohonoch chi'ch dau yn tŷ ni, ac yna pan ddaw'r llun allan o'r camera mi gei di o i ddod adra."

Llusgwyd Glesni'n anfoddog i dŷ Anwen, gan ddal i brotestio.

Gan fod y cylch meithrin ar gau ar y diwrnod tyngedfennol, roedd Nia ac Anwen wedi trefnu trwy'i gilydd i warchod Glesni ac Aled tra oedd y gofwy'n digwydd. Ni wyddai Nia, pan geisiodd agor ei llygaid y bore hwnnw, p'run fyddai waethaf – dilyn y drefn a gytunwyd ganddi hi ac Anwen a chael tynnu'i llun yn ystod y bore, a golwg fel drychiolaeth arni wedi'r noson fawr yn y Llew Coch, ynteu ffeirio gydag Anwen a chael ychydig oriau i'r lliw ddychwelyd i'w gruddiau yng nghwmni didrugaredd y plant, cyn wynebu Harri ar ôl cinio. Er, ni fyddai'n gallu sbio ar ginio heddiw. Na brecwast na swper, petai hi'n dod i hynny. Ac o ran cael tynnu ei llun yn cymysgu teisen – ych a fi! Roedd peryg mawr iddi chwydu i'r bowlen.

Sut oedd golwg cystal ar Anwen? Roedd hi wedi yfed mwy na llond cratsh neithiwr, ac eto roedd golwg lon a lluniaidd arni tra oedd Nia'n cael trafferth enbyd i gadw'i llygaid yn ei phen a'i stumog yn ei bol. Roedd Ifan wedi rhoi brecwast i'r plant, chwarae teg iddo, ac wedi gwneud yn siŵr bod y bechgyn yn dal y bws i'r ysgol cyn cynnig yn gellweirus goginio bacwn ac ŵy a bara saim iddi hi. Ond doedd dim hwyl cael ei herian ar Nia, a diolchodd pan ddiflannodd Ifan at ei waith ar ôl ei chofleidio'n gefnogol.

Mentrodd Nia gael cawod ar ôl i Anwen a'r plant adael, ond bu'n rhaid iddi orwedd ar ei gwely wedi'i lapio yn ei thywel wedyn cyn meiddio gwisgo; roedd ei phen yn troelli fel cymysgydd sment, a'i stumog yn chwyrlïo fel peiriant golchi.

Yn y man llwyddodd i lithro i'w dillad, ac yna llusgodd ei charcas bregus yn ôl i'r stafell molchi i nôl moddion gwrthwenwynol.

Byddai sbel fach ddistaw a'i thraed i fyny ar y soffa'n siŵr o roi hwb i'r feddyginiaeth.

Y peth nesaf a glywodd Nia oedd Harri'n curo ar y drws ac yn gweiddi, "Oes 'na bobol?"

Ond fel y dywedodd Nia wrth Anwen yn nes ymlaen, bendith gudd oedd ei phen mawr y bore hwnnw. Ni chafodd gyfle i ystyried ei hofnau na'i hamheuon cyn ei bedydd tân yn y byd modelu bronnoeth.

Doedd dim rhaid rhoi'r cynhwysion yn y bowlen,

sicrhaodd Harri hi dan wenu, dim ond eu gosod ar y bwrdd a dal y ddesgl a'r llwy bren yn strategol. A chyn iddi gael cyfle i bwyso a mesur goblygiadau ei phenderfyniad diwyro i gythruddo'i mam-yng-nghyfraith, roedd y cwbl drosodd yn ddiffwdan ac yn ddi-oed. Erbyn iddi glywed y plant yn sgwrsio ac yn chwerthin ar eu ffordd yn ôl tua'r tŷ, roedd Nia'n eithaf hyderus y byddai'r ymarfer heriol yn talu ar ei ganfed pan welai Elsi'r calendar ar werth yn y man. A phan agorodd y drws i groesawu Anwen a'r plant, a gweld ei mam-yng-nghyfraith yn cyrchu ar hyd y llwybr y tu ôl iddynt a golwg y fall arni, roedd yn berffaith sicr nad ofer fu ei hymdrech.

Dim ond tair gwaith roedd un o'r plant wedi rhuthro i mewn i'r stafell molchi i achwyn ar gownt rhyw fisdimanars tra oedd Nia'n ceisio ymlacio yn y bath, a doedd hynny ddim yn record o bell ffordd. Roedd hi wedi gofyn i Ifan osod bollt ar ddrws y stafell molchi ers . . . ers blynyddoedd bellach. Ond dyna ni, roedd ganddo amgenach pethau ar ei feddwl. Caeodd ei llygaid a gadael i'r swigod sebon sent donni trosti.

Oedd hi'n edrych ymlaen at weld y calendar heno? Oedd. A nac oedd.

Wrth sbio'n ôl methai Nia â chredu ei bod wedi cymryd rhan yn y fath fenter. Ac wfft i'w hegwyddorion arwynebol ei hun, heb sôn am ymroddiad i warchod y cylch meithrin a bywyd cefn gwlad a holl fwriadau da eraill gweddill y genod; gwyddai yn y bôn mai ymuno â'r criw'n unswydd i gorddi Elsi a wnaeth. Diar mi, rhaid bod cysylltiad parhaus â'r hen widdan wedi troi ei hymennydd yn botes maip.

Eisteddodd i fyny er mwyn estyn y llyfryn roedd wedi ei guddio ar y stôl o dan ei gŵn gwisgo. Llawlyfr y coleg yn Nhresarn oedd o. Agorodd y llyfryn ar ei liwt ei hun ar y dudalen lle bu Nia'n pori'n ddyddiol yn ystod yr wythnos neu ddwy ddiwethaf – "Cwrs Coginio ac Arlwyo (Blwyddyn 3), yn cynnwys arweiniad, a sesiynau ymarferol i'r rhai sydd am redeg eu busnes eu hunain".

Ymhen llai na blwyddyn fe fyddai Glesni'n mynd i'r ysgol yn Nhresarn a gallai Nia ei danfon hi a Dafydd a Gruffudd yno ar ei ffordd i'r coleg. Byddai'n rhaid i Ifan eu cwrdd oddi ar y bws yn y prynhawn a'u gwarchod am ryw awr nes y deuai hi adref, ond ar ôl blwyddyn gallai geisio rhoi ei busnes ar ei draed go iawn. Byddai pob ceiniog ychwanegol yn dderbyniol, ac fe godai rywfaint o bwysau oddi ar ysgwyddau Ifan druan. Roedd yn gas gan Nia ei weld mor ofidus, ac mor gyndyn o rannu baich eu problemau â hi. A beth bynnag, meddyliodd Nia wrth newid yn ei llofft yn y man, byddai'n rhaid iddi wneud rhywbeth unwaith roedd Glesni yn yr ysgol, oherwydd byddai bod gartref ar drugaredd Elsi yn feunyddiol yn ei gwneud yn honco bost.

Wedi siarsio'r plant i fihafio i'w tad, a gorfodi Ifan i ailadrodd y catecism cadw'r-plant-yn-lân wrthi, fe gafodd Nia gofleidiad o gymeradwyaeth ganddo. Yna dili-daliodd i lawr at y giât lôn yn y gwyll i gyfarfod Anwen, gan geisio cofio pryd y cafodd hi ac Ifan gyfle i wisgo i fyny a mwynhau cwmni ei gilydd ddiwethaf, ond allai hi ddim. Ta waeth, byddai pethau ar i fyny cyn hir, roedd Nia'n siŵr o hynny, ac ni allai hyd yn oed y cip a gafodd o gornel ei llygad ar gyrtans Elsi'n ysgwyd yn fusneslyd effeithio ar ei hwyliau da. Wrth i Anwen ddod i'r golwg, sylweddolodd Nia fod y broblem flynyddol o beth i'w brynu'n anrheg Nadolig i Elsi wedi ei datrys am eleni.

# MAI

Un gwasgiad, un goflaid, un gusan.

Oedd hynny'n ormod i'w ofyn, yn ormod i obeithio amdano?

Syllodd Alison arni ei hun yn y drych.

Syllai pâr o lygaid llwyd yn ôl arni'n ddolefus.

Roedd hi'n drysu, wedi colli arni ei hun – dyna'r unig eglurhad.

Teimladau a berthynai i ieuenctid ffôl, nwydwyllt oedd y rhain, nid i aeddfedrwydd cymedrol a phwyllog. Nid oedd wedi teimlo fel hyn erioed, ddim hyd yn oed pan oedd hi gyda Jo, er iddynt fod gyda'i gilydd am dros bedair blynedd, a hithau'n credu bryd hynny mai Jo oedd "yr un", a bod cyfaredd cariad tragwyddol wedi ei chyffwrdd. Ac ar ôl colli Jo dilynodd dyddiau, wythnosau, misoedd o bydru byw, o gogio gwenu a goroesi, o drochi mewn triog, o gredu na fyddai byth yn gallu teimlo'r un fath am neb eto.

Ysgydwodd Alison ei phen arni ei hun nes bod ei gwallt gwinau'n chwifio o amgylch ei hysgwyddau. Beth am yr addewid cyfrinachol i beidio â mynd yn rhy agos at neb eto? Beth am y penderfyniad oeraidd i gymryd yr hyn a'i plesiai o berthynas o hyn allan, a gadael i rywun arall gasglu'r gweddillion? Ni allai gredu ei bod dros ei phen a'i chlustiau mewn cariad eto. Yn hollol ddirybudd roedd y byd yn rheitiach lle; roedd lliwiau'n fwy llachar, sawrau'n fwy ffein, blas pethau'n fwy melys, cerddoriaeth yn fwy pêr, a'i henaid cyn ysgafned â swigod hud plentyn. Roedd gweld gwrthrych ei serch yn ddigon i yrru ias trwyddi; ni allai ymdopi â manylion bywyd bob dydd heb ildio i bensynnu gwallgof.

Ac ni wyddai Alison beth i'w wneud nesaf.

Roedd Alison Rhys wrth ei bodd yn ei chartref newydd yn Fflat 4, Annedd Hedd, Nant-yr-Onnen. Wedi gweld angen am newid cynefin, ymgeisiodd am nifer o swyddi cyn ei phenodi'n nyrs gymunedol yn y fro, yn un o dîm a wasanaethai Dresarn a'r cyffiniau. Roedd yn mwynhau ei gwaith ac wedi ymgartrefu yn yr ardal. Dechreuodd fynychu cangen Nant-yr-Onnen o Ferched y Wawr gyda Mererid, ei chymydog, a chael yno griw croesawgar, clên, a'r to hŷn yn falch o weld wynebau ifanc newydd yn eu plith. Buan y daeth Alison i werthfawrogi eu cwmni yn y cyfarfod misol ac mewn gweithgareddau cymdeithasol achlysurol. A chan fod ei gwaith yn mynd â hi hwnt ac yma drwy'r ardal, daeth i adnabod y gymdogaeth, y cleifion a'u gofalwyr yn dda.

Cymerodd ddiléit yn y dasg o beintio'i fflat a'i addurno, a gosod ei thrugareddau ynddo. Roedd wedi tybio ar ôl colli Jo mai bywyd sengl fyddai ei ffawd, ac felly trefnodd ei chartref, ei gwaith a'i bywyd cymdeithasol i'w phlesio'i hun. Dechreuodd gymryd diddordeb mewn cadw'n heini hefyd. Byddai'n ymweld â'r ganolfan hamdden yn Nhresarn yn aml i nofio, i fynychu dosbarth aerobig, neu dreulio rhyw awr yn y stafell ffitrwydd, a phan fyddai eu hamser rhydd yn cyd-daro byddai Mererid weithiau'n ymuno â hi, er bod cael paned yn y caffi yn bwysicach i Mererid na'r ymarfer corff.

O bryd i'w gilydd fe âi Alison allan i Dresarn gyda'r nos gyda'i chydweithwyr, gan fwynhau a chael sbort gyda'r gorau. Roedd y tîm o nyrsys a wasanaethai'r ardal yn grŵp hawdd gwneud â nhw, a byddai Alison wrth ei bodd yn eu cwmni yn ystod eu nosweithiau hwyliog. Ac o dro i dro deuai Linda, chwaer Alison, draw gyda Bethan, ei merch fach. Roedd Alison yn gwirioni ar ei nith ac yn mynd â hi am dro o amgylch y pentre yn ei choets gan ganu ei chlodydd wrth bawb a welai. Byddai'n canu hwiangerddi ac yn chwarae'n braf gyda Bethan ar y mat o flaen y tân, ac yna'n magu'r fechan yn fodlon nes iddi syrthio i gysgu yn ei breichiau.

Ond pan fyddai Alison ar ddi-hun trwy'r nos wrth ei gwaith, byddai trymedd unigrwydd yn ei llethu. Ers i Jo ei gadael fu dim rhamant na dim agosatrwydd corfforol ym mywyd Alison. Doedd yno neb i ddweud wrthi ei bod hi'n werth y byd, i'w chofleidio a'i chysuro pan fyddai'r byd yn troi'n gas, i rannu manylion ei dydd dros swper a gwydraid o win. Ond, rhesymodd Alison, wedi claddu pob atgof cyffyrddadwy am Jo mewn cas dan y gwely yn y llofft sbâr, roedd hi'n lwcus o'i chymharu â llawer. Roedd ganddi ffrindiau a chydweithwyr gwerth chweil; roedd ganddi nith fach annwyl a welai'n rheolaidd; roedd ganddi gartref clyd a gyrfa oedd wrth ei bodd.

Ond yn nüwch didrugaredd y nos, a hithau'n dyheu am freichiau i afael yn dynn ynddi, a rhywun i sibrwd yn gariadus yn ei chlust, doedd rheswm yn fawr o gysur nac o gwmni.

"Ydi hwnna'n gyfforddus, Mrs Jones?"

"Ydi diolch, del, mae'n llawer mwy esmwyth rŵan."

"Dyna ni, 'ta. Mi gliria i 'mhetha. Ac os byswn i'n cael golchi 'nwylo?"

"Siŵr iawn. Ylwch, mae Emrys wrthi'n gwneud panad. Cymerwch un efo ni cyn mynd."

Erbyn i Alison lwyddo i'w rhyddhau ei hun o garedigrwydd diffuant Mr a Mrs Jones, roedd wedi cael eu holl hanes gan wraig y tŷ. Clywodd am ddamwain anffodus Mrs Jones gyda'r radell boeth tra oedd wrthi'n gwneud teisennau cri, a'i rhwystredigaeth tra oedd yn aros i'r clwyf fendio. Clywodd am y parti priodas aur oedd yn cael ei drefnu gan Siân y ferch, a sut roedd Mr a Mrs Jones yn cymryd arnynt wybod dim am y peth er mwyn ei phlesio. Cafodd Alison adroddiad cynhwysfawr o'u hanner can mlynedd o fywyd priodasol dedwydd, a chlywodd farn Mrs Jones bod "pobl ifanc y dyddia 'ma'n rhy barod o'r hanner i roi'r ffidil yn y to os ydynt yn profi rhyw fymryn o dalcen caled".

O'r diwedd eisteddai Alison yn ei char ar ei ffordd adref a phoeriadau'r glaw yn brychu'r ffenest. Gwenodd wrthi ei hun ac ysgwyd ei phen mewn anghredinedd. Hanner can mlynedd! A'r ddau'n heini a hoffus, ac yn amlwg yn dotio ar ei gilydd. Gwyn eu byd.

Teimlodd Alison ddeigryn yn cronni yng nghornel ei llygad. Twt lol, doedd hi ddim am ddechrau teimlo'n sentimental. Trodd y radio ymlaen. Damia, rhyw gân serch, cân roedd hi'n ei hadnabod yn iawn, cân fu'n ffefryn ganddi hi a Jo. Stopiodd y car ar ochr y lôn ac ildio i'r pwl o ddagrau, gan obeithio na fyddai neb oedd yn pasio yn sylwi arni'n gwneud ffŵl ohoni ei hun.

Yn y man sychodd ei dagrau a chychwyn am adre drachefn. Byddai'n rhaid iddi styrio. Roedd hi'n noson cyfarfod Merched y Wawr a byddai Mererid yn disgwyl amdani ymhen rhyw awr. Rhoddodd ei throed lawr a'r dagrau dichellgar yn dal i bigo'i llygaid a'r glaw yn dal i ddafnio'r ffenest. Pam, yng nghanol prysurdeb bywyd bob dydd, meddyliodd, roedd yn rhaid iddi deimlo mor blydi unig?

"Wel, pob lwc i chi, ddeuda i."

Roedd Linda, chwaer Alison, wedi dod draw am y prynhawn gyda Bethan, ac roedd cynllun y calendar dan ei chwyddwydr.

"Dwi ddim yn meddwl y byddai gen i ddigon o blwc i gymryd rhan mewn rhywbeth fel'na. Heb sôn am be fyddai Meirion yn ei ddeud. Ond isio mwy o bobol fel chi sydd – pobol â digon o gyts i sefyll ar eu traed a gwarchod eu cymuneda."

Ers noson y cyfarfod roedd gyts Alison wedi bod yn

gwneud eu gorau glas i ymadael â hi, felly gwenodd yn chwerw ar ei chwaer. "M-m-m. Un o'r syniada 'na sy'n wych ar y pryd, ond ei apêl yn pallu ffwl sbîd gefn gola dydd ydi o, mae gen i ofn. Fel adduneda blwyddyn newydd i stopio bwyta siocled, neu fynd ar y wagan. A deud y gwir, mae argyhoeddiad y rhan fwya ohonon ni'n sigledig iawn erbyn hyn."

"Tyd 'laen. Dwi'n reit genfigennus ohonoch chi. A mi ga i frolio – fy chwaer Alison, wyddoch chi, mae hi'n fodel pêj-thrî."

Taflodd Alison glustog at ei chwaer gan ofalu peidio â tharo Bethan a oedd yn cysgu'n dawel yn ei choets.

"Rwyt ti wedi setlo'n iawn yma, felly," meddai Linda. "Mae'n rhaid dy fod ti, os wyt ti wedi dy gynnwys yn un o'r criw dethol."

"Do," cytunodd Alison. "Dwi'n hapus iawn yma. Mae'r fflat yn braf, ac mae'r gwaith wrth fy modd."

Gwenodd Linda. Roedd hi'n gwybod cyfnod mor anodd a brofasai ei chwaer wedi colli Jo, ac roedd yn falch o weld gwên ar ei hwyneb unwaith eto.

"Ac rwyt ti wedi dod i nabod pobl, a dechra gwneud ffrindia."

Nodiodd Alison gan wenu. Pan oeddynt yn iau, byddai'n gwylltio'n gacwn gyda holl holi a stilio Linda, gan ystyried ei chwaer yn rêl busnes pawb. Dim ond yn lled ddiweddar roedd wedi deall mai pryder amdani oedd y tu ôl i'r holl gwestiynu.

"Do," cytunodd eto. "Ac mae'r tîm sy yn y gwaith yn griw llawn hwyl."

Cymerodd Linda arni siarad yn ddidaro.

"Wyt ti wedi . . . Oes gen ti rywun sbesial?"

Teimlai Alison y gwrid yn codi i'w hwyneb.

"Nac oes . . . ddim eto."

"Dwyt ti ddim yn dal i hiraethu am Jo?"

Ymlaciodd Alison.

"Nac'dw, Linda. Wir i ti. Mae Jo'n hen hanes bellach."

Ond doedd hynny ddim yn bodloni'r arch chwilyswraig.

"Rhywun newydd, 'ta? Dwyt ti 'rioed wedi cyrraedd yr oed pan wyt ti'n dechra cael pylia o wres a chochi ar ddim."

Wedi i Linda ddal y clustog nesaf yn ddeheuig, cyfaddefodd Alison, "Oes, mae 'na rywun, rhywun annwyl iawn, rhywun arbennig."

"A?"

"Dyna fo. Mae gen i ofn nad ydw i'n pori yn y cae iawn. A phetawn i'n rhoi fy nhroed ynddi . . ."

"Pwyll pia hi felly, ia? Ond cofia di, Alison, weithia mae'n rhaid mentro tipyn bach, os ydi'r cyfle'n ei gynnig ei hun."

"Os bydd o, mi fydda i'n neidio amdano, paid â phoeni. Mae bywyd yn rhy fyr, tydi?"

Nodiodd Linda.

Teimlai Alison don o ryddhad yn llifo drosti. Roedd hi wedi bod yn braf cael rhannu ei chyfrinach â rhywun. A phwy a ŵyr, efallai bod Linda'n iawn, ac y byddai pethau'n gweithio allan rywsut.

"Wel," parhaodd Linda, a oedd yn gyndyn o roi'r gorau i'r ymchwiliad heb wasgu pob manylyn posib allan o'i chwaer, "wyt ti am ddeud wrtha i be 'di enw'r 'rhywun arbennig' 'ma?"

Ond yr eiliad honno fe ddeffrodd Bethan yn ei choets a dechrau crio.

Safodd Alison dan y gawod a'r dŵr poeth yn tonni drosti. Roedd newydd dreulio awr yn y stafell ffitrwydd ac roedd yr ymarfer wedi codi ei hysbryd.

Wedi sychu a newid aeth draw i'r caffi i gwrdd â Mererid.

Ceisiodd lacio'i chalon o'i chorn gwddw pan ddarganfu Mererid yn eistedd wrth fwrdd yn y caffi'n mwynhau paned gyda Gwyneth, perchennog y Llew Coch. Chwifiodd y ddwy eu dwylo arni, a gosododd Alison wên addas ar ei hwyneb wrth fynd at y cownter i nôl paned cyn ymuno â nhw.

"Sbia pwy welais i yn y pwll nofio," meddai Mererid wrthi.

Gwenodd Gwyneth yn swil. "Wedi dianc ydw i. Bydd Tina'n dechra gweithio yn y cylch meithrin eto cyn hir, felly mi achubais ar y cyfle i adael y lle yn ei gofal am ychydig pnawn 'ma a dod i'r ganolfan."

"Mi fydda i'n mwynhau dod o bryd i'w gilydd," meddai Mererid, "a bydd Alison yn dod yn reit reolaidd, byddi?"

"Byddaf," cytunodd Alison.

"Mi ddylwn i ddod yn amlach," meddai Gwyneth, "ond rhwng pob dim . . ."

"Mae'n anodd ffeindio'r amser," cytunodd Mererid, "ond mi fydda i'n teimlo'n well wedi gwneud yr ymdrech."

"Doeddet ti ddim yn nofio heddiw, Alison?" holodd Gwyneth.

"Na . . . na, ddim heddiw," atebodd Alison.

Roedd gwagle yn ei phen ble'r arferai ei hymennydd fod, a theimlai mor dafodrwym â merch ysgol yn cyfarfod ei hoff ganwr pop yn y cnawd.

"Wel, bydd angen i ni i gyd edrych ar ein gora cyn hir," aeth Gwyneth yn ei blaen, "er, dwi'n meddwl 'mod i wedi'i gadael hi braidd yn hwyr i ddechra cyfundrefn cadw'n heini, a dim ond ychydig wsnosa'n weddill cyn y sbloet."

"Os dwi'n cofio'n iawn, ti, fi a Sioned oedd y rhai ucha'u cloch yn erbyn y syniad," meddai Mererid, "ac a deud y gwir, dwi'n dal i deimlo'n reit sâl wrth feddwl am yr awr fawr."

"O, dwi wedi 'ngorfodi fy hun i wthio'r embaras enbyd i gefn fy meddwl a chau'r drws yn glep arno," meddai Gwyneth, "neu mi fyddwn i ar biga drain yn barhaus. Beth amdanat ti, Alison?"

"Llithro i'r cynllun yn eich sgil chi wnes i, dwi'n meddwl," atebodd Alison. "Wedi'r cyfan, rhyw gyw aelod o'r gangen ydw i."

"Naci, siŵr," protestiodd Gwyneth. "Rwyt ti'n un ohonon ni bellach. Rydan ni'n hynod falch o bob un o'n haeloda, hen a newydd."

Roedd Mererid yn dal i bryderu.

"Gobeithio na fydd neb mor ddi-asgwrn-cefn â thynnu'n ôl," gofidiodd, "neu mi fydd hi wedi canu arnon ni."

"O, dwi'n siŵr y byddai Dilys yn fodlon camu i'r bwlch," sicrhaodd Gwyneth hi yn ffug ddifrifol, "neu ella y byddwn ni'n gallu mynd ar ofyn Elsi!"

Bu ond y dim i Alison a Mererid dagu ar eu paneidiau wrth ddychmygu'r peth.

Yn y man ffarweliodd Gwyneth â'r ddwy, ac aeth Alison a Mererid i wneud ychydig o siopa cyn dychwelyd i Nant-yr-Onnen.

Ni allai Alison yn ei byw roi ei meddwl ar siopa; roedd wrthi'n casglu pob manylyn o'r hyn a ddigwyddodd a'r hyn a ddywedwyd ac yn eu gwasgu i'w mynwes i'w dadansoddi drwyddi draw ar ôl mynd adref. Ac yn y car ar y ffordd yn ôl i Nant-yr-Onnen syllodd Mererid mewn penbleth ar ei ffrind wedi iddi orfod ei hailadrodd ei hun unwaith eto. Roedd hi'n amlwg bod rhywbeth ar feddwl Alison ers iddynt adael y ganolfan, ond ni allai Mererid ddyfalu beth a ddigwyddodd yno a allai fod wedi cael y fath effaith ar ei ffrind.

Dyna hwnna drosodd.

Ni wyddai Alison yn iawn beth oedd ei theimladau ynglŷn â chael tynnu ei llun yn fronnoeth, yn ei throwsus ymarfer cwta, a thywel wedi ei daflu'n hynod ofalus o ddidaro dros ei hysgwyddau. Roedd yn fodlon cydnabod mai ei hawydd a'i hymdrech i gael ei derbyn, i fod yn un o'r criw, oedd wrth wraidd ei pharodrwydd i ymuno â chynllun y calendar yn y bôn, nid egwyddorion ac ewyllys da, ynghyd â gobaith y byddai'r weithred rywsut yn dod â hi'n nes at wrthrych ei dyheadau.

Roedd Mererid wedi ei gwahodd i ymuno â hi a Gwyn ac Eira am noson allan i ddathlu wedi'r deheubrawf, ond roedd wedi penderfynu mai ar ôl tynnu'r lluniau roedd yr amser i fentro, i afael mewn bywyd gerfydd ei sgrepan a'i sgytian nes ei fod yn gwichian ac yn gwingo. Ond nawr, wedi clywed Mererid yn mynd allan, a dim ond chwinclin pen mawr y bore'n gwmni, roedd ei phenderfyniad wedi mynd a'i gadael.

Ffliciodd trwy'r sianelau ar y teledu, ond doedd dim arno i'w phlesio. Dili-daliodd o amgylch y fflat a'r lle'n cau amdani hi, yn casáu ei hofnau a'r teimlad o ddiymadferthedd llethol. Roedd ar fin ildio i'r llond bol o ddagrau a oedd yn bygwth ei thagu pan glywodd gnoc ar y drws.

Ar y landin y tu allan safai Linda; tusw anferth o flodau yn un llaw, potel o win yn y llall, a chlamp o wên ar ei hwyneb. Taflodd ei breichiau o amgylch ei chwaer gan wasgu'r blodau'n rhacs a rhoi clec egr i'r botel win yn erbyn y drws.

"Wel? Sut aeth petha?" holodd, cyn camu'n ôl ac edrych i wyneb Alison. "Hei, be sy?" holodd drachefn, wrth weld yr olwg ddigalon oedd ar ei chwaer. "Be ddigwyddodd? Oedd o'n ofnadwy?"

Dechreuodd Alison chwerthin, er ei gwaethaf, dan y gawod o gwestiynau.

"Na, aeth popeth yn iawn," atebodd. "Ond be wyt ti'n neud yma?"

"Wedi dod i weld sut hwylia sydd ar fy chwaer fach wedi diwrnod caled dan lygad ddidrugaredd y camera, siŵr. Ac wedi llwyddo i ddwyn perswâd ar Meirion i warchod. Am unwaith."

"Dwi'n cymryd ein bod ni am agor hon," meddai Alison, wedi achub y botel win. "Er, ar ôl y noson gawson ni yn y Llew Coch neithiwr, bydd chwa ohoni'n ddigon i mi!"

Wedi atgyfodi'r blodau ac agor y botel, setlodd y ddwy chwaer yn stafell fyw Alison â gwydraid yr un i drin a thrafod digwyddiadau cyffrous y dydd. Yna trodd Alison drywydd y sgwrs i holi am helyntion ei nith. Ond ni ddaethai Linda'n bencampwraig hawl i holi ar ddim.

"Dwed i mi, 'ta, os aeth popeth cystal heddiw, pam roedd golwg mor druenus arnat ti pan gyrhaeddais i?"

Hanner gwenodd Alison. "Roeddwn i wedi bwriadu mynd i'r Llew Coch heno. Cymryd dy gyngor di a mentro. Ond pan ddaeth hi i'r pen . . ."

Llyncodd Linda weddillion ei gwin a neidio ar ei thraed.

"Wel, am be rydan ni'n ddisgwyl, 'ta?"

Ond parhau'n wangalon oedd Alison.

"Fedra i dim. Mae gen i ofn gwneud cawl o betha."

Roedd ganddi gystal siawns o ddianc o gasgenaid o uwd.

"Alison, rydw *i*'n bwriadu ymweld â'r Llew Coch heno, ac mi rwyt *ti*'n dod efo fi." Tywalltodd Linda ragor o win i wydr ei chwaer. "Dyna ti, lapia dy hun rownd hwnna tra wyt ti'n newid. Mae un gwydraid yn hen ddigon i mi a finna'n gorfod dreifio adra." Edrychodd ar ei horiawr. "Mi gei di ddeng munud. Ac yna dwi'n mynd hebddot ti."

Roedd hi'n ddistaw yn y Llew Coch wedi bwrlwm y noson cynt. Yn y bar roedd dyrnaid o ffyddloniaid wedi ymgolli yn eu dominos a'u dartiau, ac yn y lolfa prin hanner dwsin o'r byrddau oedd yn llawn. Eisteddodd Alison a Linda ar stolion wrth y bar, ac wedi i Linda gael ei chyflwyno i Gwyneth a Tina buont yn sgwrsio â'r ddwy bob yn ail wrth iddynt fynd a dod wrth eu gwaith. Ceisiodd Linda'i gorau glas i gael eu hanes dan lens Harri, ond ychydig o wybodaeth a gafodd ganddynt, gan fod y ddwy'n teimlo'n swil wrth drafod eu profiadau mewn amgylchiadau mor gyhoeddus.

Teimlai Alison ei hunanhyder yn dychwelyd dan ddylanwad y gwin. Erbyn stop-tap roedd yn fodlon ei byd ac nid oedd am weld y noson yn dod i ben. Roedd am eistedd yno, o fewn hyd braich i'w hanwylyd, yn llowcio pob gair, pob amnaid, pob nodwedd corfforol, a'u storio er mwyn gwledda arnynt drosodd a throsodd.

Yn y man ffarweliodd Linda â'i chwaer â gwasgiad serchog, a chan na fu'n noson brysur, dywedodd Gwyneth wrth Tina am adael yr ychydig wydrau oedd ar y byrddau a'i throi hi am adref.

Doedd fawr o waith perswadio Alison i dderbyn un bach cyn gadael, jyst i gadw cwmni i Gwyneth.

"I griw'r calendar. I ni," cyhoeddodd Gwyneth.

"I ni," cytunodd Alison.

Cymerodd y ddwy lwnc o'u gwydrau cyn i Alison estyn ar draws y bar a mwytho boch Gwyneth yn dyner â'i bys.

Curodd Alison yn ysgafn ar ddrws cefn y Llew Coch cyn camu i'r gegin. Roedd ganddi doreth o waith papur i ddelio â fo, ond gyda dadorchuddiad answyddogol y campwaith yn digwydd ymhen ychydig oriau, roedd hi'n methu'n glir â chanolbwyntio, felly penderfynodd fynd draw i holi a oedd angen help llaw ar Gwyneth gyda'r mân ddanteithion roedd yn eu darparu ar gyfer y criw.

Roedd y gegin yn wag, ond roedd olion paratoadau ar y bwrdd a sŵn mwmian siarad i'w glywed yn y bar. Eisteddodd Alison; byddai'n aros nes i Gwyneth ymddangos cyn gwneud dim, rhag ofn iddi wneud smonach. Ymestynnodd ei breichiau uwch ei phen a dylyfu gên. Roedd hi wedi bod ar ei thraed trwy'r nos wrth ei gwaith yn aml yn ddiweddar a diolchodd fod ychydig ddyddiau o wyliau'n ddyledus iddi cyn hir.

Ond doedd y nosweithiau ddim mor feichus bellach. Yn hytrach na'i herio a'i gwawdio yn ei hunigrwydd, roedd yr oriau dudew'n eu lapio'u hunain o'i hamgylch, yn ei choleddu a'i gwarchod yn ddiogel tan y wawr. Ymddangosai wyneb Gwyneth yn fynych yn ei meddwl megis llusern; gallai estyn bys a llunio pob manylyn o'r pryd a gwedd a oedd mor annwyl iddi, a chanlyn pob modrwy o'r gwallt brown.

Gwyddai pa mor anodd fyddai'r dyddiau a'r wythnosau nesaf i Gwyneth, a doedd hi ddim am ruthro i fynnu ymrwymiad tragwyddol ganddi. Byddai'n rhaid i Gwyneth gydnabod ei theimladau a dygymod â nhw yn ei hamser ei hun. Gwyddai Alison fod Gwyneth wedi priodi, ac yna wedi colli'i gŵr mewn damwain car, a'i bod wedi gweithio'n hynod o galed oddi ar hynny i gadw a chynnal ei bywoliaeth yn y Llew Coch ar ei phen ei hun. Ofnai Alison y byddai unrhyw deimladau datguddiol dieithr yn peri iddi ddadrannu a dyfalu.

A beth am drigolion Nant-yr-Onnen? Beth fyddai eu hymateb hwy petai cyfeillgarwch y ddwy ohonynt yn datblygu i fod yn fwy na hynny? Cynnil gyfeillgar oedd y sylw a roddasai Alison i Gwyneth yn gyhoeddus hyd yma, a hyd yn oed pan oeddynt ar eu pennau eu hunain, ceisiai beidio ag ymddwyn fel gwraig tŷ yng ngŵydd y paced olaf o stwffin ar Noswyl Nadolig, rhag ofn iddi . . .

Clywodd Alison Gwyneth yn ffarwelio â rhywun cyn dod i'r gegin.

"O! Helô, Alison. Be wyt ti'n wneud yma mor gynnar?"

"Croesawus iawn, wir!" Cymerodd Alison arni bwdu. "A finna wedi dod i gynnig help llaw i ti efo'r bwyd."

Ochneidiodd Gwyneth. "Ac mi fydda i'n falch ohono fo hefyd. Mae wedi bod fel ffair yma pnawn 'ma – gohebydd o'r papur lleol ar drywydd stori am y calendar ac yn meddwl y byddai'n cael rhagolwg, a rhyw drafaelwyr jinjabïar a *pork scratchings* wedi penderfynu dod i fwydro 'mhen i. Mi gynigiodd Tina ddod i'm helpu, ond dywedais wrthi am wneud yn fawr o'r cyfle i gael ychydig o seibiant, gan feddwl bod popeth dan reolaeth."

Cododd Alison dan wenu, a gwneud sioe o dorchi ei llewys.

"Reit 'ta, ble wyt ti am i mi ddechra?"

Er gwaethaf pryderon Gwyneth roedd y danteithion i gyd wedi eu gosod yn y lolfa dan orchudd ffoil mewn da bryd, ac roedd digon o amser yn weddill i Alison a hithau gael cyfle i molchi a newid cyn i'r genod gyrraedd.

Roedd Alison yn sefyll ger y sinc yn golchi'r llestri pan ddaeth Gwyneth drwodd o'r bar wedi gosod y platiad olaf o fwydydd gyda'r lleill. "Diolch, diolch o galon i ti," meddai, wrth gasglu'r celfi oedd yn weddill oddi ar y bwrdd ac ymestyn heibio i Alison i'w rhoi yn y sinc.

Sychodd Alison ei dwylo. Trodd i wynebu Gwyneth ac

agorodd ei breichiau. Heb edrych ar ei gilydd safodd y ddwy'n llonydd a'u talcenni'n cyffwrdd; nes i Alison godi un llaw a chosi dan ên Gwyneth yn dringar, cyn codi ei phen yn dyner â'i bys nes bod y ddwy'n edrych i fyw llygaid ei gilydd.

Roedd Alison wedi cael cawod sydyn a rŵan roedd yn ceisio penderfynu beth i'w wisgo ar gyfer y lansiad answyddogol. Roedd ei dillad yn un llanast ar draws y gwely, a dim byd yn plesio – rhy liwgar, rhy ddi-liw, rhy hir, rhy fyr, rhy sidêt, rhy goman. Gwyddai ei bod yn ymddwyn fel rhywun ar ei dêt cyntaf erioed. Ond roedd hi eisiau teimlo ar ei gorau; yn wir, roedd hi *angen* teimlo ar ei gorau er mwyn cadw'r ellyll maleisus a lechai yng nghefn ei meddwl draw, yr ellyll hwnnw a awgrymai mai gwyro o'i llwybr a wnaeth Gwyneth gynnau fach, ac y byddai'n edrych ar Alison â dirmyg didostur o hynny ymlaen.

O'r diwedd dewisodd Alison drowsus du a blows werdd dywyll a weddai i'w gwallt gwinau i'r dim. Ychydig o golur i amlygu ei llygaid llwyd, brwsio'i gwallt nes ei fod fel damasg, dabaid o bersawr, ac roedd hi'n barod.

Roedd Mererid wedi addo y byddai'n galw amdani fel y gallent gydgerdded i'r Llew Coch i gwrdd â gweddill y criw. Gwyddai Alison fod Mererid wedi bod allan gyda Harri sawl gwaith yn ddiweddar, ac roedd yn mawr obeithio y byddai'r berthynas yn ffynnu. Prin bod Mererid wedi crybwyll Berwyn y bradwr ers tro. A chwarae teg i Mererid, doedd hithau heb fusnesu dim ym mywyd Alison, er ei bod yn ymwybodol bod rhyw newid ar y gweill, rhyw benbleth yn ei bywyd, rhyw gynnwrf ar droed. Unwaith byddai'r ddaear yn sadio dan ei thraed, byddai Alison yn rhannu ei chyfrinach â hi. Roedd yn ddigon sicr o'i chyfeillgarwch i fentro gwneud hynny. Ac roedd Linda'n iawn; roedd yn rhaid ymwroli ac achub ar gyfle weithiau. Dal i synfyfyrio oedd Alison pan gurodd Mererid ar y drws a chyhoeddi'n gellweirus ei bod hi'n amser ymweld â ffau'r llewod.

Trodd sgwrs y ddwy at y calendar wrth iddynt gerdded i'r dafarn. Nid oedd Alison wedi rhoi fawr o sylw i'r lansiad ei hun tan y funud honno; roedd amgenach pethau ar ei meddwl. Ond i bob diben, roedd y cynllun wedi bod yn llwyddiant iddi hi'n bersonol, ac yn ddull o wireddu ei gobeithion. Teimlai'n un o'r criw bellach; roedd yn hyderus y câi ei derbyn ganddynt beth bynnag fyddai ffrwyth ei theimladau cariadus tuag at Gwyneth, ac yn bendant roedd y cynllun wedi bod yn gyfrwng iddi nesáu at ei hanwylyd. Wrth i olau croesawgar y Llew Coch ddod i'r golwg, ni allai Alison atal gwên rhag hollti ei hwyneb.

# MEHEFIN

Estynnodd Phyllis y coffr lledr o waelod y cwpwrdd tridarn a etifeddodd gan ei rhieni. Wrth iddi ei agor llanwyd ei ffroenau ag arogl peli camffor. Dechreuodd agor yr haenau o bapur sidan hyfriw a oedd wedi melynu gyda threuliad amser nes ei fod yn lliw cwyr gwenyn, ac yno, yn eu hamwisg brau, gorweddai'r *layette* crandiaf a welodd neb erioed. Byseddodd Phyllis y dillad bychain a greodd hi ei hun â'r gwlân meddal ffein ddeg ar hugain o flynyddoedd yn ôl. Gwyddai y dylai fod wedi cael gwared â nhw ers tro. Ond ni allai, er na fyddai hi'n agor y gist fechan yn aml iawn y dyddiau hyn, dim ond pan fyddai rhywun neu rywbeth yn codi crachen ei cholled drachefn.

Wedi'r cwbl, roedd hi'n gweithio mewn siop dillad babi, ac yn delio â mamau a babanod bob dydd – ac yn cael llawer o bleser wrth wneud hynny. Weithiau byddai'n cael gafael yn y pethau bach a'u magu, neu eu siglo yn eu coetsys tra byddai eu mamau'n pori ymysg yr offer a'r dillad deniadol.

Ond heddiw, pan gyhoeddodd Helen, perchennog y siop, y byddai hi'n nain cyn diwedd y flwyddyn, fe deimlodd Phyllis yr ergyd megis dyrnod.

Roedd Brenda'n nain, ac yn mwynhau pob munud o gwmni ei hwyres fach ddireidus (er nad oedd Sandra, yn nhyb Phyllis, yn malio fawr ddim am les ei merch). Ac roedd hyd yn oed Elsi-busnes-pawb yn nain, er nad oedd honno chwaith yn gwerthfawrogi ei hwyrion a'i hwyres nwyfus. A rŵan, roedd Helen am fod yn nain hefyd.

Lapiodd Phyllis ei hatgofion a'i hiraeth yn eu hamdo bregus a'u diogelu yng ngwaelod y cwpwrdd drachefn.

Trodd yn ôl at ei gwaith, a'r bwlch bythol yn ei bywyd yn peri iddi deimlo fel petai wedi ei diberfeddu.

Y tro cyntaf y gorfododd Tom ei hun arni fe guddiodd Phyllis yn y tŷ am dridiau, yn methu wynebu'r byd. Ar ben y tridiau mentrodd allan i mofyn negesau angenrheidiol, gan fod yr olion corfforol yn dechrau cilio. Ond am y creithiau meddyliol . . .

Ac fel unig blentyn a oedd wedi colli ei rieni, at bwy y gallai droi?

Pan briododd Phyllis â Tom Gruffudd, roedd hi ar ben ei digon. Roedd hi wedi colli ei mam pan oedd hi yn ei harddegau, ac yna ei thad yn fuan wedi iddi adael yr ysgol a dechrau gweithio. Roedd Phyllis yn meddwl y byd o'i thad, a bu ond y dim i Brenda fethu â'i pherswadio i ddod gyda hi i ddawns yn neuadd y pentref rai misoedd ar ôl ei golli. Y noson honno, fel arfer, amgylchynwyd Phyllis gan fintai o wŷr ifanc a swynwyd gan ei gwallt melyn llaes a'i llygaid llwydlas. A chyn diwedd y noson roedd Tom Gruffudd, a oedd ychydig yn dalach, ychydig yn fwy golygus, ychydig yn hŷn, a fesur yn fwy hyderus na gweddill y criw, wedi hawlio Phyllis iddo fo'i hun.

Wedi cyfnod gweddus, ond nid maith, priodwyd Phyllis a Tom. A doedd Phyllis yn disgwyl dim llai na byw'n hapus byth mwy. Parhaodd wrth ei gwaith yn siop ddillad Liverpool Stores yn Nant-yr-Onnen, siop a werthai amrywiaeth o ddillad i bobl a phlant, ynghyd â gwlân a manion gwnïo. Ond pasio'r amser oedd Phyllis yno bellach, gan fod Tom yn dymuno dechrau teulu ar fyrder. Roedd ganddo frawd a chwaer hŷn a oedd yn prysur lenwi eu cartrefi â phlant heini, ac roedd yntau'n awyddus i'w brofi ei hun trwy lenwi ei gartref ef a Phyllis â phitran traed plantos. Tybiai Tom, o ystyried ei bryd golygus ei hun a dengarwch Phyllis, y byddai eu plant hwy yn werth eu gweld. A chan nad oedd gan Phyllis deulu ei hun, a bod y cwpwl ifanc a oedd, wedi'r cyfan, dal dan ddylanwad mis yr afiaith, yn profi cymaint o bleser yn eu hymdrechion i'r perwyl hwnnw, doedd ganddi hi ddim gwrthwynebiad i'r cynllun.

A dyna pryd y surodd melystra eu serch.

Credodd Phyllis yn daer sawl tro ei bod yn disgwyl, dim ond iddi gael ei siomi yn y man. Yn wir, y tro cyntaf roedd wedi gwau casgliad o ddillad babi heb ei ail cyn ei dadrithiad creulon. Siomwyd Tom yn arw, ond roedd yn llawn cydymdeimlad. Y tro nesaf doedd ei gydymdeimlad ddim cweit mor ddiffuant. Ac yna, wedi misoedd o simsanu rhwng gorfoledd a gwewyr, cyngor y meddyg iddynt oedd y dylai'r ddau ohonynt gael profion rhag ofn bod rhyw anghydnawsedd rhyngddynt. Ond gwrthododd Tom ar ei ben, gan ddewis gosod y bai am y sefyllfa yn gyfan gwbl ar Phyllis, ac wedi hynny, wedi'r hyn a ystyriai Tom yn sarhad personol ar ei wrywdod, diflannodd eu cariad i lawr y llethr llithrig a arweiniai at bydew diwaelod dialedd a diflastod.

Ac wedi'r tro cyntaf hwnnw, pan halogwyd hi yn ei chartref, yn ei gwely ei hun, daeth Phyllis i ofni, ac i gasáu ei gŵr. Roedd yr achlysuron pan fyddai'n rhaid iddynt ymweld â'i deulu'n atgas ganddi, gan y byddai hi'n destun cellwair awgrymog, ac roedd yn gas ganddi ymweliadau Tom â'r dafarn wedi diwrnod anodd yn y gwaith, oherwydd mai'r un fyddai cost pob dibrisiad – ei gormesu'n frwnt ac yn ddideimlad.

Roedd gan Phyllis ormod o gywilydd i rannu ei phoen â neb. Hyd yn oed â Brenda. Wedi'r cyfan, er bod Brenda'n cwyno am ddifaterwch Wil, roedd ganddi hi ddau o blant perffaith. Doedd hi ddim fel Phyllis, yn fethiant llwyr.

A phan gredodd Phyllis na allai pethau waethygu, darganfu bod Tom wedi dechrau cadw cylchgronau budr yn y llofft fechan a ddylai fod wedi ei llunio'n llofft i'w babi. Y babi na chafodd Phyllis erioed. Erioed wedi ei gario, erioed wedi ei deimlo'n symud oddi mewn iddi, erioed wedi ei eni na'i gofleidio na'i garu. Ac fel na phetai'n ddigon iddo halogi ei chorff, roedd Tom hefyd am ddifwyno ei dyheadau, ei hatgofion a'i hiraeth. Dyna pryd y plygodd Phyllis y dillad bychain a greodd a'u pacio'n ofalus mewn cas lledr a'u gosod yng ngwaelod y cwpwrdd yn ei llofft rhag iddynt gael eu llygru gan aflendid ei gŵr.

Unwaith eto ni allai Phyllis dros ei chrogi drafod ei thrallod â neb, er iddi holi Brenda'n betrus a oedd Wil yn edrych ar luniau o ferched eraill. Ceisiodd Brenda dynnu Phyllis o'i chragen a mynd at y gwir, oherwydd roedd wedi gwylio'i ffrind yn mynd yn fwyfwy isel ei hysbryd wedi dyddiau cynnar gwynfydedig ei phriodas. Ond ni allai Phyllis ddechrau esbonio hyd a lled ei halaeth, ddim hyd yn oed wrth ei ffrind gorau.

Dros gyfnod, rhoddodd Tom heibio unrhyw gais i guddio'i lenyddiaeth linorog, ac ymhen amser dechreuodd Phyllis ddiolch am gyfrwng ei ddihangfa, gan ei fod yn tynnu ei sylw, am ryw hyd, oddi arni hi.

Bellach roeddynt fel dau estron, wedi eu cipio oddi ar y stryd i fyw o dan yr un to.

Ac ymestynnai gweddill bywyd Phyllis o'i blaen fel twnnel tywyll diderfyn.

". . . ac o ganlyniad fydda i ddim yn adnewyddu'r les ar y siop ddiwedd y mis. Mae'n ddrwg iawn gen i, Phyllis."

Nodiodd Phyllis.

"Ond os bydd angen geirda arnoch ar gyfer swydd arall, byddaf yn falch iawn o'i roi i chi. Mae'ch gwasanaeth wedi bod yn gaffaeliad mawr i'r busnes. Ond gyda'r archfarchnad newydd wedi dechrau gwerthu dillad, a'r trethi wedi codi cymaint yn ddiweddar, roedd dyfodol y siop yn y fantol cyn i mi glywed fod Eleri'n disgwyl. Ac wrth gwrs mae hithau, fel cymaint o ferched y dyddiau yma, yn awyddus i ddychwelyd i'w gwaith wedi cael y babi, felly rydw i wedi penderfynu dirwyn y busnes i ben a rhoi help llaw iddi hi yn hytrach na phydru 'mlaen a gwylio gweddillion y busnes yn diflannu lawr y draen."

Nodiodd Phyllis ar ei chyflogwraig drachefn. Ac er bod

Helen wedi treulio sawl noson ddi-gwsg cyn cyrraedd yr unig benderfyniad ymarferol ynglŷn â dyfodol y siop, ar yr union eiliad honno byddai wedi bodloni ar redeg y busnes ar golled am byth bythoedd amen yn hytrach na wynebu'r olwg o anobaith a welai yn llygaid Phyllis. Gwyddai y byddai'n ffodus iawn i gael swydd arall yn ei hoed a'i hamser, a chymaint o bobl ifanc ar drywydd pob swydd wag.

"Ylwch, Phyllis, cymerwch baned a seibiant bach fan hyn; rydw i'n gwybod bod y newydd wedi bod yn siom fawr i chi, ac mae'n wir ddrwg gen i."

Aeth Helen drwodd i warchod y siop gan adael Phyllis yn eistedd yn y storfa fechan yn y cefn a oedd yn gwneud y tro fel ystafell staff. Eisteddodd yn ei hunfan, yn rhy ddiymadferth i godi a llenwi'r tecell hyd yn oed.

Wedi cau siop Liverpool Stores yn Nant-yr-Onnen, bu Phyllis yn gweithio mewn amryw o siopau yn Nhresarn cyn cael ei swydd bresennol yn Siop Pethau Bychain. Roedd y gwaith a'r cwsmeriaid wrth ei bodd, a chynigient ddihangfa iddi rhag diffrwythdra ei bywyd ar yr aelwyd. Byddai canmol ei chwsmeriaid bychain wrth fodd ei chalon, a chan fod ganddi lygad dda am liw a chynllun a phatrwm, gallai gynghori pob nain a thaid a modryb ac ewythr a ddeuai i'r siop ynglŷn â'r dewis gorau i'w trysor bach arbennig hwy.

A nawr? Nawr byddai'n ôl ar drugaredd Tom, wedi ei chaethiwo yn y tŷ, heb ei gwaith i lenwi rhywfaint ar ei dyddiau hysbion.

Ac fel pe na bai hynny'n ddigon, roedd Brenda wedi cipio braich Phyllis a'i chodi gyda'i braich hithau yng nghyfarfod Merched y Wawr neithiwr i gefnogi cynllun ynfyd Sara i achub neuadd y pentref. Roedd y syniad yn gwbl hurt, wrth gwrs, ac mae'n siŵr y byddai pawb wedi sylweddoli hynny erbyn bore 'ma. Calendar bronnoeth, wir! Onid oedd Phyllis wedi gorfod dioddef presenoldeb pethau felly yn ei thŷ ers blynyddoedd? A phethau gwaeth na hynny. Ni fyddai'n mentro i'r llofft sbâr bellach, ond yn aml iawn deuai ar draws rhai o gylchgronau mochaidd Tom wedi eu gadael yn ddiofal o amgylch y tŷ. Nid bod gan Sara ddim byd cignoeth felly mewn golwg, wrth gwrs. Ond calendar bronnoeth!

Ochneidiodd Phyllis.

Yna clywodd sŵn lleisiau yn y siop a chododd i ddychwelyd at ei gwaith.

". . . yn addas ar gyfer rhywun sydd newydd adael yr ysgol . . .", ". . . yn addas ar gyfer myfyriwr/wraig . . .", ". . . profiad ddim yn angenrheidiol, hyfforddiant ar gael i'r ymgeisydd cymwys, gyda chyfle am ddyrchafiad . . .", ". . . yn chwilio am werthwr(aig) deinamig, brwd-frydig . . .", ". . . yn chwilio am berson aeddfed, profiadol, oddeutu 35–45 oed . . ."

Edrychodd Phyllis o'i chwmpas. Roedd bron pawb yn y lle'n iau na hi. Syllodd ar y merched a eisteddai y tu ôl i'w desgiau'n cyfweld darpar ymgeiswyr ac yn eu cynghori ynglŷn â cheisiadau. Na, doedd hi ddim yn mynd i ymuno â'r ciw i gael ei darostwng a'i labelu'n "rhy hen" i ymgeisio am unrhyw swydd werth ei halen. Edrychodd ar ei horiawr. Roedd Helen wedi caniatáu iddi gymryd awr ginio hwy nag arfer, er mwyn iddi gael cyfle i ymweld â chanolfan waith Tresarn a phori trwy'r cardiau gwynion oeraidd a lenwai'r silffoedd yno. Wel, roedd ganddi bron awr yn weddill o'i hawr ginio estynedig, a doedd hi ddim yn bwriadu gwastraffu eiliad yn rhagor ohoni yn fan'ma, ble roedd yn amlwg mai'r unig le iddi hi a'i thebyg oedd ar y domen.

Roedd Phyllis newydd setlo gyda phaned o goffi, brechdan a thafell o gacen lemwn pan glywodd lais cyfarwydd yn gofyn, "Ga i ymuno efo chi, Phyllis?"

"Cewch siŵr, Sara. Â chroeso."

Wedi i'r ddwy orffen eu cinio a siarad am hyn a'r llall, trodd y sgwrs yn anorfod at gynllun chwyldroadol Sara.

"Dydach chi ddim yn hapus efo'r syniad, Phyllis,"

meddai Sara. "Mi welais i Brenda'n codi'ch llaw chi yn erbyn eich ewyllys yn y cyfarfod."

"Nac ydw, a deud y gwir," cyfaddefodd Phyllis yn ffwndrus. "Cofiwch chi, tydw i ddim yn orlednais," ychwanegodd.

Sut allwn i fod, dan yr amgylchiadau, meddyliodd yn chwerw. Ond ni allai rannu'r baich hwnnw â Sara, na neb arall.

"Peidiwch â phoeni, Phyllis," meddai Sara gan gyffwrdd â'i llaw. "Ddim dod draw i geisio dwyn perswâd arnoch chi i gymryd rhan wnes i. Fyddwn i ddim yn ceisio gorfodi neb i wneud 'run dim. Profais i rym gorfodaeth fy hun pan gollais fy ngwaith gyda chau'r ysgol, a fynnwn i ddim gwthio neb i wneud dim yn erbyn ei ewyllys."

Gwelodd Phyllis gyfle i newid trywydd y sgwrs, yn ogystal â rhannu un o'i phryderon.

"Rydw inna'n colli 'ngwaith ddiwedd y mis," meddai. "Mae Helen yn cau'r siop ac am ymddeol – costau rhedeg busnes yn rhy uchel a gormod o gystadleuaeth gan y siopa mawrion. Ac mae ei merch yn disgwyl babi," ychwanegodd.

Rhoddodd Sara ei llaw ar law Phyllis drachefn.

"Mae'n wir ddrwg gen i, Phyllis. Dwi'n gwybod eich bod chi wrth eich bodd yn y siop, ac ychydig iawn o swyddi sydd ar agor i ferched o'n hoedran ni mewn ardal fel hon, gwaetha'r modd. Dwi'n credu y byddai cymdeithas yn berffaith fodlon ein gweld ni'n cael ein halltudio i ryw ynys bellennig ble byddwn ni'n cael baglu'n ffordd tuag at ein henwendid heb achosi embaras i neb."

Gwenodd Phyllis. Byddai angen rhywun a oedd ynteu'n ddewr iawn neu'n ffôl iawn i geisio gwthio Sara i ymylon cymdeithas.

"Ylwch," parhaodd Sara, "oes gennych chi amser am baned arall?"

Taflodd Phyllis gip ar ei horiawr.

"Oes," meddai. "Alla i ddim aros yn rhy hir ond cefais awr ginio hwy nag arfer gan Helen heddiw, i mi gael cyfle i ymweld â'r ganolfan waith, ac yn union fel y dywedsoch, Sara, mi ddes i o'no'n teimlo mor ddefnyddiol â sosban siocled."

Cilwenodd Sara mewn cydymdeimlad wrth godi i nôl dwy baned arall o goffi.

Tra oedd hi'n aros wrth y cownter ceisiodd Phyllis ei rhoi ei hun yn esgidiau Sara – wedi ei thaflu ar y clwt ar ôl blynyddoedd o wasanaeth i ysgol y pentref, heb sôn am weld y capel yn cau, gan dorri calon ei thad. Doedd syndod felly ei bod mor gadarn ei hymroddiad i gadw'r neuadd ar agor er mwyn sicrhau man cyfarfod i'r hyn oedd yn weddill o rwydwaith cymdeithasol y pentref.

Wedi i Sara ddychwelyd at y bwrdd gyda'r coffi bu'r ddwy'n mwynhau rhyw fân siarad, heb drafod y calendar ymhellach, nes ei bod hi'n amser i Phyllis ddychwelyd i'r siop. Wrth godi, teimlai Phyllis ei hun yn cael ei meddiannu gan bwl o fyrbwylltra gwallgof.

"Mi wna i ymuno efo chi," meddai'n frysiog. "Mi wna i gymryd rhan yng nghynllun y calendar os bydd pawb arall yn fodlon."

"Da iawn, Phyllis," canmolodd Sara hi. "Mi fydd y gwaith i gyd yn cael ei wneud yn chwaethus iawn, cofiwch. Awgrym o'n rhinwedda fydd mewn golwg," ychwanegodd yn smala, "ddim y gwyddoniadur cyfan."

Cerddodd Phyllis yn ei hôl i'r siop. Roedd wedi anghofio'n llwyr am ddiflastod ei phrofiad yn y ganolfan waith, a theimlai hyder ei hymateb i her Sara'n ymchwyddo ynddi. Wedi blynyddoedd o fod yn destun gwawd ei gŵr, efallai mai dyma'i chyfle i dalu'r pwyth yn ôl; yn wir, cyfle i dalu'n ôl am waradwydd blynyddoedd mewn ffordd gwbl briodol. Wedi'r cyfan, ystyriodd Phyllis wrth gyrraedd Siop Pethau Bychain, beth oedd ganddi i'w golli?

Dim byd.

Dim byd o gwbl.

* * *

Roedd ymddangosiad Tom a Wil wedi difetha'r noson allan.

Erbyn iddi gyrraedd y tŷ roedd penfedd-dod Phyllis yn sgil ei safiad gwrthryfelgar hi a Brenda wedi mynd i'r gwynt. Eisteddai yn y gegin yn crogi mŷg o de i lonyddu cryndod ei dwylo, ac yn syllu'n ddwys i'r gwacter, heb sylwi bod y stêm o'r ddiod boeth wedi pylu gwydrau ei sbectol. Beth ddaeth dros ei phen hi i ymddwyn yn y fath ffordd? Beth ddaeth dros ei phen hi i gredu na fyddai Tom yn clywed am y calendar? Beth ddaeth dros ei phen hi pan gytunodd i gymryd rhan yn y cynllun? Rhoddodd ei chwpan i lawr ar fwrdd y gegin, tynnu ei sbectol niwlog a rhoi ei phen yn ei dwylo.

Mae'n debyg y byddai'n talu'n ddrud am ei haerllug-rwydd. Er, ni fu pethau cynddrwg yn ddiweddar. Yn raddol dros y blynyddoedd aeth y cyfnodau rhwng ymosodiadau ymwthiol Tom yn hwy. Un ai roedd yn ei fodloni ei hun â diddanwch ei ddeunydd damniol, neu roedd yn achub ar gyfle i gymryd ei bleser yn rhywle arall.

Pleser!

Ochneidiodd Phyllis cyn estyn am ei sbectol drachefn a'i gosod ar ei thrwyn. Rhedodd ei dwylo trwy'i gwallt brith cwta. Oedd, mi roedd o'n bleser ar y cychwyn, yn yr un modd ag y bu i gymaint o gyplau ifanc nwydus. Ond ers blynyddoedd bellach doedd yn ddim byd ond deheubrawf dideimlad. Ac os oedd Tom yn claddu ei gacwn yn rhywle arall, pob lwc iddo.

Ymlwybrodd Phyllis o'r gegin i'r stafell molchi, ac yna i'w gwely.

Fel rheol, wedi gweld ffilm neu ddarllen llyfr rhamantus, byddai wedi bod yn berffaith fodlon gorwedd yn ei gwely'n ail-fyw'r stori, gan ei dychmygu ei hun yn diosg diflastod ei bywyd ac yn cael ei chofleidio gan yr arwr golygus a fyddai'n tyngu llw o gariad pur iddi yn y fan a'r lle.

Ond heno gorweddai Phyllis yn ei gwely ar binnau, wedi mynd i'w chroen megis draenog. Gyda phob smic credai fod Tom wedi cyrraedd y tŷ. Teimlai fel petai pob pared a phob dodrefnyn yn y lle'n cynllwynio i'w phoenydio â'u clecian a'u crecian, eu gwichian a'u griddfan.

Yna, pan oedd ei nerfau'n gyrbibion a'i chalon yn llamu o'i brest i'w chorn gwddw ac yn ôl gyda phob anadl a gymerai, clywodd Phyllis sŵn goriad yn cael ei osod yng nghlo'r drws ffrynt. Clywodd y drws yn cael ei agor, a'i gau drachefn. Daeth sŵn camau troetrwm Tom i fyny'r grisiau. Peidiodd y sŵn traed y tu allan i'r llofft. Clywodd Phyllis ddolen y drws yn cael ei throi. Yna gollyngwyd y ddolen a chlywodd Phyllis y camau'n symud ar hyd y landin, a drws y llofft sbâr yn cael ei agor.

". . . mae hi'n werth ei gweld, Phyllis. Mi fydd Lowri wrth ei bodd efo hi. Ac mi fydd hi'n gynnes braf iddi dros y gaea. Sbia, Lowri, sbia beth mae Anti Phyllis wedi'i wneud i ti."

Roedd y fechan wedi gwirioni ar y siwmper liwgar, a mynnodd ei thrio'n syth bin a rhoi dawns fach o amgylch y gegin.

"Diolch, Anti Phyllis."

"Ia, diolch yn fawr i ti."

Gwridodd Phyllis. Doedd hi ddim wedi arfer â derbyn y fath glod.

"Wel, gan 'mod i'n treulio mwy o amser yn y tŷ rŵan, waeth i mi drio gwneud rhywbeth defnyddiol efo f'amser," meddai'n ddiymhongar.

Ar ôl i Brenda dynnu'r siwmper oddi ar Lowri a'i phlygu'n ofalus, dychwelodd y fechan i'r stafell fyw gan adael Brenda a Phyllis yn y gegin.

"Yli, gymerwn i banad, Phyl. Dwi newydd fod yn chwythu balŵns dros Gymru, ac mae 'ngheg i'n sych fel sach."

"Wyt ti isio i mi wneud rhywbeth i helpu, Bren?"

"Nagoes, diolch i ti. Mi fydda i'n falch o'r esgus i gael panad a phum munud, a deud y gwir wrthat ti. Mi ddechreuais i ar y balŵns 'na'n syth ar ôl dod o'r siop tra oedd Lowri'n cael ei chinio, a dwi bron â thagu erbyn hyn. Ac mae Sandra wedi addo y bydd hi'n ôl mewn da bryd i helpu efo'r bwyd. Mae hi'n gorfod codi a thendiad at Lowri yn y borea rŵan, a hithau'n wylia 'rysgol, felly does 'na ddim llawer o hwyl ar meiledi, a deud y gwir. Ond mi fydd hi'n siŵr o landio i helpu efo parti Lowri."

Paid â dal dy wynt, Brenda fach, meddyliodd Phyllis, gan eistedd wrth fwrdd y gegin. Mae gen ti gystal siawns o gael help llaw gan Sandra â sydd gan Wil o gael ei ddyrchafu'n Bab.

"Dyna welliant," cyhoeddodd Brenda ar ôl eistedd a chymryd llwnc o'i phaned. "Panad a sgwrs, ac yna, gyda lwc, mi fydda i wedi cael ail wynt erbyn i Sandra gyrraedd 'nôl."

"Oes 'na griw mawr yn dod?" holodd Phyllis.

"Dwsin, os daw pawb," atebodd Brenda. "Ar ddiwrnoda fel hyn mi fydda i'n teimlo f'oed, Phyl."

Gwenodd Phyllis. "Ers i mi fod yn chwilio am swydd, mi dwi wedi mynd i deimlo'n sobor o oedrannus. Mae pawb isio pobl ifanc deniadol, rhad eu cadw, i weithio iddynt, nid rhyw hen goes fel fi."

"Mi ddaw 'na rywbeth, Phyl, mi gei di weld."

Rhoddodd Brenda ei chwpan lawr a chodi'r siwmper i'w hedmygu drachefn.

"Mi roeddet ti'n gwau tipyn erstalwm, toeddet ti?"

"Oeddwn. Ond collais i fynedd efo fo ar ôl . . . wel . . . ti'n gwybod sut mae petha weithia . . ."

"Wn i, Phyl. Ond mi ddylet ti ailafael ynddi. Mi fydda pobol gefnog yn talu pris da am waith crefftus fel hyn yn un o'u siopa mawrion. Aros funud, dwi newydd gofio rhywbeth." Neidiodd Brenda ar ei thraed ac ymbalfalu ymysg y pentwr papurau newydd a orweddai ar ben ei hoergell. "A! Dyma fo. Sbia, mae 'na rywun yn Nhresarn yn gwerthu peiriant gwau. Un go newydd hefyd. Ac am y nesa peth i ddim."

"Ond tydw i 'rioed wedi defnyddio peiriant gwau, Bren."

"Mi fedri di ddysgu. Ac efo dy lygad di am liw a phatrwm byddet ti'n gallu gwneud llond trol o jyrsis mewn dim o dro."

"Ond sut fyswn i'n eu gwerthu nhw?"

"O, croesa'r gamfa honno pan ddoi di ati. Dechra'n y pentre 'ma, yntê? Ac mae rhywun yn bownd o nabod rhywun sy'n nabod rhywun arall."

Roedd Phyllis yn dal i fod yn amheus.

"Yli, mi dorra i'r hysbyseb allan i ti, yna mi gei di gyfle i feddwl am y peth. Ond rŵan a thitha efo amser ar dy ddwylo . . ."

Ar ôl i Phyllis adael, a'r hysbyseb yn ei phoced, dechreuodd Brenda ar y gwaith o ddarparu bwyd ar gyfer y parti gan ddisgwyl gweld Sandra'n ymddangos unrhyw funud i roi help llaw iddi. Wrth daenu menyn ar y pentwr bara o'i blaen, pryderai am ei ffrind. Gwyddai fod gan Phyllis amheuon dybryd ynglŷn â chynllun y calendar, a'r rheini'n amheuon digon dilys hefyd a chysidro'i thrallodion. Gwyddai yn ogystal fod colli ei swydd wedi bod yn loes calon iddi, yn enwedig wrth ystyried anghyflawnder ei bywyd ar yr aelwyd. Er na fyddai Phyllis yn dangos pa mor anodd oedd ei sefyllfa, roedd Brenda'n amau ei bod hi'n wynebu anawsterau a wnâi i'w bywyd hi ei hun ymddangos fel tocyn tymor i Butlins.

Roedd Phyllis wedi cynnig cadw cwmni i Brenda yn ystod yr orchwyl ond gwrthododd, felly doedd hi ddim yn lecio gofyn i Brenda ddod draw i ddal dani yn ystod ei hordîl hithau, gan y tybiai y byddai Brenda'n siŵr o fod yn brysur gyda Lowri. Erbyn y diwrnod tyngedfennol roedd bwriad Phyllis i ddial ar Tom ag un weithred herfeiddiol wedi diflannu. Yr unig reidrwydd bellach oedd peidio siomi gweddill y merched, gan obeithio'n daer y byddai'r profiad erchyll drosodd cyn gynted â phosib.

Trwy lwc, roedd hi wedi nôl y peiriant gwau newydd rhyw ddeg diwrnod cyn y dyddiad hollbwysig (neu yn hytrach, roedd Alun wedi ei nôl a'i ddanfon iddi yn ei fan lyfrgell, gan ddenu llawer o dynnu coes gan ei gwsmeriaid), ac ar hyn o bryd safai ar ganol llawr y llofft sbâr. Doedd Phyllis ddim yn fodlon o gwbl ar ei leoliad yno, yng nghanol llyfrgell lysnafeddog Tom, ond am y tro doedd ganddi'r unlle arall i'w osod, ac o leiaf roedd ei hymdrechion i feistroli'r peiriant wedi bod yn gyfrwng i gadw'r bwganod beirniadol draw tan y funud olaf.

Pan oedd Phyllis ynghanol rhyw bincio llugoer, ar ôl cinio na allai yn ei byw ei wthio i lawr ei chorn gwddw – ac nid ar gownt yr hyn a yfodd y noson cynt, chwaith, a oedd ond megis piso dryw bach yn y môr o'i gymharu â dogn ambell un o'r criw – canodd cloch y drws.

Harri wedi cyrraedd yn gynt na'r disgwyl!

Dechreuodd calon Phyllis ddyrnu yn erbyn ei hasennau fel curiadau'r pen gwely yn erbyn y pared yn ystod un o gyrchoedd meddiannol Tom, ac erbyn iddi gyrraedd y drws ffrynt roedd yn chwys oer drosti.

Ar y stepen safai Brenda, yn wên i gyd.

"Meddwl ella y byddet ti isio rhywun i afael yn dy law di. Ond," ychwanegodd, "os ydi hi'n well gen ti fod ar dy ben dy hun efo Harri, jyst dwed."

"O, na, Bren, dwi'n falch o dy weld ti. Tyrd i mewn."

Arweiniodd Phyllis ei ffrind i'r stafell fyw a suddodd i gadair freichiau cyn i'w choesau roi oddi tani.

"Sut aeth hi efo chdi?" gofynnodd. "A ble mae Lowri?"

Gwenodd Brenda. "Yn well nag oeddwn i wedi'i ddisgwyl, a deud y gwir, Phyl. Mae Harri'n ŵr bonheddig ac yn gwbl ddi-lol; mi fyddi di'n iawn efo fo, mi gei di weld. Ac mae Lowri wedi mynd am dro efo Alun."

Nodiodd Phyllis, heb fod yn siŵr iawn beth i'w ddweud.

Gwenodd Brenda drachefn. "Paid â sbio arna i fel'na, Phyl. Dwi'n gwybod bod pobol y pentre 'ma'n bownd o fod yn clebran am fy nghyfeillgarwch efo Alun. Ond ti'n gwybod sut mae petha wedi bod acw erstalwm iawn . . . Ac mae o'n gwneud i mi deimlo'n hapus."

Syllodd Phyllis ar ei ffrind, a oedd yn wên o glust i glust

ac yn amlwg yn fodlon ei byd.

"Paid â cholli dy gyfle, Bren," meddai'n ddistaw. "Dos amdani, ac os ydi pobol isio clebran, gad iddyn nhw."

Synnwyd Brenda. Wedi'r cyfan, er y gwyddai nad oedd priodas Phyllis wedi bod yn un hapus, fe fyddai hi wastad yn cymryd arni nad oedd pethau'n rhy ddrwg, hyd yn oed gyda Brenda.

"Phyl," mentrodd, gan geisio dewis ei geiriau'n ofalus, "sut mae petha wedi bod yn ddiweddar? A thitha wedi colli dy swydd aballu. Ydi popeth yn iawn?"

Cymerodd Phyllis anadl ddofn ac edrych i fyw llygaid ei ffrind.

"Nac ydi, Bren," atebodd. "Tydi petha heb fod yn iawn ers tro, ers blynyddoedd. A deud y gwir, tydw i ddim hyd yn oed isio bod yma mwyach. A tydw i heb ddod ar draws marchog ar geffyl gwyn fel Alun fyddai'n fodlon fy nghipio i o'ma i borfeydd glasach," ychwanegodd dan gilwenu. "A bod yn onest, tydw i ddim isio un. Mi fodlona i ar ddynion seliwloid a'r rheini sy'n ymwroli rhwng cloria llyfra – mi wn i ble dwi'n sefyll efo rheini."

Cododd Brenda a rhoi ei braich o amgylch sgwyddau Phyllis. Gwyddai fod y datguddiad yma wedi bod yn gam arwyddocaol iddi, yn gam tuag at gydnabod a derbyn ei sefyllfa, a chyda lwc, yn gam tuag at ei goroesi. Wedi saib, gwasgodd Brenda ei hysgwydd.

"Sut mae'r gwau'n dod yn ei flaen?" gofynnodd yn ysgafn. "Wyt ti wedi mentro defnyddio'r peiriant eto?"

Roedd brwdfrydedd amlwg yn llais Phyllis wrth iddi ateb.

"Do, do wir. Ac mae o'n ddigon hawdd ei ddeall unwaith rwyt ti'n eistedd lawr a dilyn y llawlyfr."

"Wel, ydw i am gael gweld yr wythfed rhyfeddod 'ma, 'ta?"

Doedd Phyllis ddim am i Brenda na neb arall weld cynnwys ffiaidd y llofft sbâr. Roedd rhai pethau'n saffach heb eu gweld a heb eu dweud. Am y tro, beth bynnag. Ac roedd un cyffes yn ddigon am un diwrnod, neu pwy a ŵyr sawl deryn corff fyddai'n cael gwynt dan ei adain cyn glanio'n ysglyfaethus ar ei hatgofion chwilfriw?

Arbedwyd Phyllis rhag meddwl am esgus, oherwydd y funud honno canodd cloch y drws drachefn.

Dyna welliant!

Eisteddai Phyllis ar ei gwely yn ei gŵn gwisgo.

Roedd ei theimladau'n un cawdel – ofn ynghylch canlyniad ei gweithred feiddgar; balchder gorfoleddus oherwydd ei bod wedi cydio yn y danadl; ac yn anad unpeth arall, difaterwch ysgubol. Petai Tom yn ei chosbi'n hallt am yr hyn a wnaeth, yna byddai'n rhaid iddi dderbyn hynny, ond bellach hi oedd piau ei meddwl a'i meddyliau, ei hysbryd a'i henaid; roedd hi wedi eu hawlio.

Bu Phyllis wrthi ers ben bore. Roedd wedi cario llond bocseidiau o gylchgronau o'r llofft sbâr i'r sièd, ac yna wedi sgwrio a sgrwbio nes bod nid yn unig pob gronyn o lwch a baw wedi diflannu o'r stafell, ond pob arlliw o ffieidd-dra Tom a'i arferion hefyd. Roedd hi wedi gosod llenni newydd ac wedi llusgo silffoedd o'r stafell fyw i fyny'r grisiau i ddal ei hoffer. Galwodd Alun yn ystod y prynhawn i ddanfon cadair wely oedd ganddo yn ei gartref ac nad oedd ganddo ddefnydd iddi, a haliodd y ddau ohonynt honno i fyny'r grisiau hefyd. Yna gosododd Phyllis y peiriant gwau a'i fân drugareddau yn ei briod le ger y ffenest, a rhoi bollt ar y drws.

Wedi bodloni ar ei hymdrechion, aeth Phyllis i'r bath i sgwrio a sgrwbio drachefn – sgwrio a sgrwbio olion gwaith y dydd, sgwrio a sgrwbio staeniau blynyddoedd o dorcalon ac o ddiffyg, o ddarostyngiad ac o dwyll.

A rŵan wrth eistedd ar y gwely, sylweddolodd Phyllis ei bod yn edrych ymlaen at ymuno â'r criw ar gyfer y lansiad answyddogol yn y man. Doedd hi ddim yn siŵr beth fyddai ei theimladau pan welai'r lluniau, ond, meddyliodd yn chwerw, byddai wedi gweld llawer gwaeth, roedd hynny'n sicr. A châi gyfle i rannu ei newydd da gyda

Brenda, ac efallai un neu ddwy o'r lleill.

Roedd Phyllis wedi rhoi un o'r dilladau bychain a greodd ar ei pheiriant newydd yn anrheg i Eleri wedi dyfodiad y bychan, ac roedd Eleri a Helen wedi gwirioni'n lân ar ei chynllunio a'i chrefftwaith. Yn union fel y proffwydodd Brenda, roedd Eleri'n adnabod rhywun a oedd yn adnabod rhywun . . . ac roedd siawns wirioneddol y câi Phyllis fawd ei throed dros drothwy un o'r siopau dethol hynny a werthai ddillad i blant enwogion a oedd yn fodlon talu unrhyw bris i sicrhau bod eu hepil hwy yn gwisgo dilledyn unigryw.

Oedd, cysidrodd Phyllis wrth newid a phincio, roedd y rhod yn dechrau troi o'r diwedd.

Serch hynny, diolchodd fod Tom wedi cymryd at ddiflannu i rywle ar ôl gorffen ei waith yn ddiweddar, heb ddychwelyd i'r tŷ tan yn hwyr gyda'r nos. Er ei hunanhyder newydd, doedd Phyllis ddim am wynebu ffrae a fyddai'n dwyn y sglein oddi ar noson allan arall. A ble bynnag roedd Tom, a chyda phwy bynnag oedd o, gwynt teg ar ei ôl o, meddyliodd Phyllis, wrth gerdded at y gornel i gyfarfod Brenda.

# GORFFENNAF

". . . ac maen nhw wedi gofyn i mi weithio bora Sadwrn hefyd. Ac mi fysa'n anodd i mi wrthod . . . sori, Mam."

"Paid â phoeni, Beth. Dwi'n dallt yn iawn. Mi ddaw 'na gyfle i ni gael dy weld ti cyn hir, dwi'n siŵr. Ac mi fydd Griff yn gorffen yn y coleg ymhen ychydig wsnosa. Mi welwn ni'r ddau ohonoch chi bryd hynny."

"Dwi wedi siarad efo Griff, Mam. Mae o'n sôn am ddod yma i aros am dipyn dros yr haf. I chwilio am waith, medda fo."

"O . . . dyna ni, 'ta . . . wel, mi welwn ni'r ddau ohonoch chi pan fyddwch chi ddim yn rhy brysur."

"Ia. Mi fyddwn ni'n siŵr o landio fel dwy geiniog ddrwg. A sori am y penwsnos nesa 'ma, Mam."

Ffarweliodd Marian â'i merch cyn rhoi'r ffôn yn ôl yn ei grud a dychwelyd i'r gegin ble roedd Cledwyn wrthi'n ddyfal yn golchi'r llestri swper. Gwelodd yr olwg ddigalon ar wyneb ei wraig.

"Tydi hi ddim yn dod adra?"

Ysgydwodd Marian ei phen. "Nac ydi. Maen nhw wedi gofyn iddi weithio bora Sadwrn. Ac mi fysa hi'n rhy hwyr wedyn i feddwl cychwyn."

"Wel, fedrith hi ddim gwrthod yn rhwydd iawn," rhesymodd Cledwyn, "neu buan y ffeindian nhw rywun arall sy'n fodlon gweithio ar benwsnos."

Nodiodd Marian. "Dwi'n gwybod. Dwi *yn* dallt ei bod hi'n anodd arni, ond anaml iawn rydan ni'n cael cyfle i'w gweld hi dyddia 'ma. A rŵan mae Griff yn meddwl chwilio am waith lawr 'na hefyd."

"Fedri di ddim gweld bai ar yr hogyn, Marian. Duwcs, does 'na ddim byd iddyn nhw ffordd hyn. Tyrd, gafael yn y lliain sychu llestri 'na, wir, i ni gael rhoi'n traed i fyny am dipyn. Ac os rhoddi di wên i mi yn lle sbio mor ddigalon, mi wna i banad i ti."

Yn y man roedd y ddau'n ymlacio yn y stafell fyw pan wafftiodd arogl llosgi i'w ffroenau.

Neidiodd Marian ar ei thraed.

"Fy nhorth dêt a walnyt!" ebychodd gan ruthro i'r gegin.

Dychwelodd i'r stafell fyw ymhen ychydig funudau gan gario torth oedd wedi'i chrasu'n ulw.

Gwelodd Cledwyn y dagrau'n cronni yn ei llygaid glas. Gwyddai fod Marian yn hynod siomedig na fyddai eu merch yn dod adref dros y penwythnos, a sylweddolodd fod perygl y byddai amlosgiad y dorth dêt a walnyt a wnaethai'n arbennig i Beth yn rhoi'r farwol i gydbwysedd rhadlon arferol ei wraig. Cododd a rhoi ei fraich am ei hysgwyddau. Dewisodd ei eiriau'n ofalus.

"Duwcs," meddai'n gysurlon gan bwyntio at y dorth, "fuodd 'na 'rioed ddrwg na fu'n ddaioni. Petai rhywun yn trio dwgyd pres o'r post, mi fedren ni ddefnyddio honna i roi swadan go effeithiol iddyn nhw."

"Sbia, Cledwyn. Maen nhw'n chwilio am rywun i redeg Swyddfa'r Post ym mhentref Gorslas, yr ochr draw i Dresarn. A sbia faint maen nhw'n ofyn am y busnes! Faint o bobol leol fyddai'n medru ystyried buddsoddiad felly?"

Daeth Cledwyn draw o'r tu ôl i'r cownter a darllen yr hysbyseb dros ysgwydd Marian, tra cosai ei chyrls brown cwta ei drwyn. Roedd hi bron yn amser cau am y noson, a'r siop a'r swyddfa bost wedi bod yn reit ddistaw ers i blant a phobl ifanc Nant-yr-Onnen alw mewn am dda-da a chreision ar eu ffordd adref o'r ysgol.

"Duwcs, mi fyddai wedi canu arnon ni petaen ni'n trio cychwyn busnes heddiw," meddai Cledwyn gan ysgwyd ei ben mewn anghrediniaeth wedi darllen y manylion yn y papur. "Fydden ni byth wedi llwyddo i gael ein dwylo ar y math yna o arian."

Wedi gwneud yn siŵr bod popeth yn drefnus yn y siop, gadawodd Marian ei gŵr i gloi'r llythyrdy ac i daro llygad ar gyfrifon y dydd, ac aeth drwodd i'r gegin i ddechrau paratoi swper. Sosej a thatws stwnsh.

Gwenodd Marian wrth fynd ati i blicio'r tatws. Faint o ddraenogod a greodd hi â sosej a thatws stwnsh pan oedd Beth a Griff yn blant? Ugeiniau, mae'n siŵr. Roedd sosej a thatws stwnsh yn un o ffefrynnau'r ddau, a byddai Griff yn dal i fynnu eu llunio'n ddraenog pan oedd yn . . . yn faint? . . . yn ddeg oed a mwy, siŵr o fod.

Dyddiau da oedd y rheini, cysidrodd Marian, wrth sefyll ger y sinc yn syllu drwy'r ffenest a'r crafwr tatws yn segur yn ei llaw. Roedd hi a Cledwyn wedi gweithio'n galed i gael y busnes ar ei draed. Ond er eu bod yn gweithio oriau maith, doedd dim cymaint o bwysau ariannol arnynt ag oedd ar bobl ifanc mewn sefyllfaoedd cyffelyb heddiw. A phan gyrhaeddodd Beth, ac wedyn Griff, llwyddwyd i gyflogi merch ifanc o'r pentref i helpu yn y siop am gyfnod.

Bu'r plant yn ffodus i gael eu haddysg yn ysgol gynradd y pentref. Cofiai Marian sut y byddai'n eu hebrwng adref bob prynhawn, nes i Beth gychwyn yn Ysgol Uwchradd Tresarn, ac i Griff benderfynu ei fod yn rhy fawr i fod gael ei hebrwng adref mwyach. Byddai'r ddau'n anelu'n syth am y siop i chwilio am rywbeth da i'w fwyta, ac er i Marian geisio'u denu â ffrwyth neu damaid iachusol arall, rhyw dda-da lliwgar neu fisgeden wedi ei gorchuddio â siocled moethus melys oedd dewis y plant bob dydd. Gwenodd Marian drachefn. Doedd rhai pethau'n newid dim.

Cofiai hefyd am frwdfrydedd Beth a Griff wrth iddynt roi help llaw yn y siop yn ystod gwyliau'r ysgol, a hynny pan oeddent yn rhy fach i weld dros y cownter heb sefyll ar grât llefrith gwag wedi ei droi ben i waered. Yna, wrth iddynt dyfu'n hŷn a dechrau dallt y dalltings, hawliai'r ddau "gyflog" am eu gwaith. Ymhen amser, wrth gwrs, cyrhaeddodd y ddau'r oedran hwnnw pan oedd y syniad o helpu yn siop y teulu'n codi embaras mawr arnynt. A dyna gaead ar biser cyfnod arall.

Dyddiau da yn wir.

Ysgydwodd Marian ei phen. Trueni bod cymaint o bobl ifanc yr ardal yn gorfod gadael i fynd i'r coleg ac i chwilio am waith. A ddim yn dod yn ôl. Byddai efeilliaid Sioned yn gadael am y brifysgol ddiwedd yr haf. Faint o siawns oedd yna iddyn nhw ddychwelyd a dilyn gyrfa yn eu cynefin? Ynteu a fydden nhw, fel Beth a Griff, yn gorfod mudo i berfeddion Lloegr i sicrhau swyddi?

Edrychodd Marian yn ddigalon ar y tatws yn y sinc. Doedd ganddi fawr o awydd sosej a thatws stwnsh wedi'r cyfan.

"Duwcs, be wyt ti wedi bod yn ei wneud? Dwi wedi gorffen y cownts i gyd tra wyt ti ond wedi plicio dwy daten. Mae'r cwcs clyfar 'na ar y teledu'n creu campweithiau mewn llai o amser na hynny." Roedd Cledwyn wedi gorffen ei waith am y dydd ac yn awyddus i roi rhywbeth yn ei fol.

Trodd Marian. Gafaelodd yn y cadach llestri gwlyb a'i daflu at ei gŵr.

". . . a'r *Guardian* a hanner pwys o'r *chocolate brazils,* os gwelwch yn dda," cwblhaodd Caryl ei negesau. "I Marc," eglurodd wrth i Marian bwyso'r da-da. "Mae o'n gweithio mor galed; wrthi a'i ben yn ei lyfrau bob nos."

"Ydi, mae'n siŵr. Mae'n gyfnod prysur i'r bobl ifanc 'ma rŵan, a'r arholiada ar eu gwarthaf," cytunodd Marian. "Bora rhydd heddiw, Caryl?"

"Ychydig ddyddia o wylia'n ddyledus i mi, Marian. Dim ond gobeithio bod popeth yn rhedeg fel y dylai yn f'absenoldeb."

"Neis iawn. Saith bunt a thri deg ceiniog, plîs."

Roedd Brenda wrthi'n llenwi'r oergell tra oedd Marian yn delio â'r cwsmeriaid. Digon hamddenol oedd pethau wedi bod yn ystod y bore, ac roedd hi'n tynnu am amser paned. Roedd Brenda ar fin diflannu i'r cefn pan ddaeth Nia i mewn gyda Glesni, a oedd yn ei dweud hi'n hallt wrth ei mam.

"Dwi isio mynd i chwara efo Anti Sioned ac Anti Tina. Dwi'm isio mynd i siop, dwi'm isio mynd am dro, a DWI'M ISIO AFAL COCH NEIS!"

Wedi llwyddo i ddenu sylw pawb yn y siop, safodd Glesni'n stond yng nghanol y llawr, croesi ei breichiau, a sgyrnygu.

Ni fedrai Marian a Brenda lai na chwerthin wrth weld ei hosgo penderfynol, ynghyd â'r olwg o anobaith ar wyneb Nia.

"Be ti wedi'i neud rŵan?" holodd Marian Nia'n gellweirus. "Rwyt ti wedi ypsetio Glesni bach, mae'n amlwg."

"Glesni bach, wir!" ebychodd Nia. "Mae Glesni bach wedi bod ar ei thraed trwy'r nos yn cwyno gyda phoen yn ei bol cyn taflu fyny dros bawb a phopeth o fewn cyrraedd. Dyna pam y cadwais i hi adra o'r cylch meithrin bora 'ma. A sbïwch arni rŵan. Croten y fall!"

"Wel, mynd i ddeud oeddwn i, ei bod hi'n edrych yn debyg iawn i Elsi, yn sefyll yn fan'na a'i breichiau wedi'u croesi," meddai Marian.

Rhoddodd Nia ei dwylo dros ei llygaid ac ysgwyd ei phen; roedd y fath syniad yn rhy erchyll i'w ystyried.

Ceisiodd Brenda a Marian seboni tipyn ar Glesni, ond heb unrhyw lwyddiant, nes i Maldwyn y postmon ymddangos. Galwai ef yn y llythyrdy'n ddyddiol i ddanfon llythyrau Cledwyn a Marian, cyn casglu sachaid o lythyrau o Swyddfa'r Post. Rhedodd Glesni ato dan ganu "Postman Pat, Postman Pat" a pheri i Nia ddymuno cuddio y tu ôl i bentwr o focsys powdwr golchi mewn cywilydd. Ond roedd Maldwyn yn hen gyfarwydd â phlant y pentref, a gadawodd i Glesni roi llythyrau Cledwyn iddo. Yna cafodd y fechan helpu'r postmon i wagio cynnwys y blwch llythyrau i'r sach.

Yn y man gadawodd Maldwyn gyda'i lwyth, wedi prynu

lolipop piws i'w gynorthwywraig, a oedd bellach â gwên angylaidd ar ei hwyneb, ac addo y byddai'n trio'i orau glas i ddanfon rhywbeth i'r fferm cyn diwedd yr wythnos.

Llwyddodd Nia i gasglu ei negesau, a chan obeithio bod stumog Glesni wedi setlo, ac na fyddai'r lolipop piws yn ailymddangos yn nes ymlaen, cychwynnodd am adre, gyda Glesni'n dal i fwmian canu "Postman Pat" yn llawen trwy'i lolipop.

Manteisiodd Cledwyn ar ychydig o lonyddwch i agor ei lythyrau tra diflannodd Brenda i'r cefn i wneud te. Agorodd amlen grand a logo newydd sbon Cyngor Sir Cedog arno, a oedd wedi ei gyfeirio ato yn rhinwedd ei swydd fel Cadeirydd y Cyngor Cymuned. Gwelodd Marian ei gŵr yn ffromi dros y llythyr, ac aeth draw ato. Estynnodd Cledwyn y llythyr iddi heb ddweud gair, a darllenodd Marian ef yn frysiog. Safodd y ddau yno gan edrych ar ei gilydd yn fud, nes i Brenda ddod drwodd efo'r te.

". . . a tydi Harri ddim yn rhy brysur yn ystod mis Medi. Ddim eto, beth bynnag. Ond gora po gynta y gwnawn ni drefniada pendant efo fo. Felly os fysach chi a Sara a Sioned yn medru penderfynu ar ddyddiad a chadarnhau hwnnw efo Harri?"

"Iawn, Mererid. Mi gysyllta i efo'r ddwy dros y Sul fel ein bod ni'n gallu gwneud trefniada efo Harri ddechra'r wsnos. A diolch yn fawr iti."

Talodd Mererid am ei phapur a diflannu drwy'r drws, gan adael Marian mewn ffrwcs. Ar y naill law roedd yn falch bod Harri ar gael, a bod y cynllun yn cael gwynt dan ei adain. Wedi'r cyfan, roedd yn rhaid i rywun yn rhywle wneud safiad i warchod cymunedau cefn gwlad, a dywedai Cledwyn ei hun nad oedd 'na fawr o weledigaeth ymysg aelodau'r Cyngor Cymuned. Ac os oedd Cledwyn, a oedd yn hynod amyneddgar, wedi syrffedu ar yr hen gonos ac Elsi Morgan yn mynd trwy eu pethau am ddigwyddiadau cyfrin a gymerodd le ymhell bell yn ôl cyn i'r aelodau diweddaraf ddod o'u clytiau, wel . . . Ac fel y dywedodd Sara, pwy arall yn Nant-yr-Onnen a allai wneud safiad? A faint o bentrefi oedd wedi colli eu canolfannau a'u cyfundrefnau cymdeithasol yn barod? A faint mwy o golledion y byddai'n rhaid i bentrefi cefn gwlad eu dioddef? Oedd, mi roedd hi'n dda o beth bod Harri ar gael.

Ond ar y llaw arall, a fyddai hi, Marian, yn ddigon dewr i roi ei phen yn y ddolen, i sefyll . . .

"Duwcs, gwena, tydi'r byd ddim ar ben, siawns gen i," galwodd Cledwyn o'r tu ôl i'w gownter. "Mi fyddi di'n dychryn y cwsmeriaid fel'na."

Ond gwên ddigon sur a gafodd gan ei wraig, felly gan nad oedd angen ei wasanaeth ar neb y foment honno, daeth draw i'r siop ati.

"Dwyt ti ddim yn hapus efo cynllun Sara?" holodd.

"Ddim yn hapus iawn, a deud y gwir," cyfaddefodd Marian. "Rydw i isio cefnogi'r lleill, ond . . ."

"Yli, gwna di beth bynnag rŵyt ti'n feddwl sy ora," meddai Cledwyn. "Mi dwi'n meddwl eich bod chi'n ddewr iawn i hyd yn oed ystyried y fath sgiâm. Ac os ydi o'n unrhyw gysur i ti," ychwanegodd, gan roi'i fraich o amgylch ei wraig a'i gwasgu'n dynn, "rwyt ti wedi bod ar fy nalen flaen i 'rioed."

". . . mi fedra i ddal y bws, Mam."

"Ond tydi o ddim trafferth, siŵr. Mi fydd dy dad a Brenda'n iawn yma tra 'mod i'n rhoi pàs i ti i Dresarn i'r orsaf fysys. Rhag i ti orfod stryffaglio efo'r bag mawr 'na."

"Mi fydda i'n iawn, Mam. Wir rŵan."

O'r diwedd llwyddodd Griff i ddarbwyllo Marian ei fod yn gwbl tebol o ddal y bws lleol i Dresarn, cyn teithio i gwrdd ag un o'i ffrindiau o'r coleg. Roedd rhieni ei ffrind ar fin gadael y wlad am bythefnos o wyliau, ac roedd Griff wedi derbyn gwahoddiad ei gyfaill i dreulio'r cyfnod hwnnw'n cadw cwmni iddo tra oedd yn gwarchod y tŷ.

Felly, ar ôl ei ymweliad â Beth, roedd Griff wedi treulio rhyw bythefnos gyda'i rieni yn Nant-yr-Onnen, yn bwyta popeth nad oedd wedi ei hoelio i'r llawr, ac yn trin y gegin fel londri, cyn hel ei draed drachefn. Ac er i Marian gymryd arni ddwrdio a dweud y drefn wrtho, roedd hi wedi bod wrth ei bodd yn cael cyfle i dendio ar ei chyw fel mamiar.

Trawodd Griff gusan sydyn ar dalcen ei fam wrth i'r bws ddod i'r golwg. Yna brasgamodd ar draws y lôn cyn i Marian gael cyfle i'w gofleidio. Arhosodd hi y tu allan i'r llythyrdy nes bod y bws wedi symud yn ei flaen i lawr y stryd fawr.

Dychwelodd Marian i'r siop a'i chalon wedi suddo cymaint nes y teimlai fel petai'n llusgo ar y llawr y tu ôl iddi fel hual. Trwy lwc doedd yna ond un neu ddau o gwsmeriaid yn y siop; gallai Brenda ddelio â'r rheini. Amneidiodd Marian arni ei bod yn mynd drwodd i'r cefn i wneud paned.

Wrth ddisgwyl i'r tecell ferwi, ceisiodd Marian resymu â hi ei hun. Fe ddylai fod yn falch o'r newydd a dderbyniodd Griff trwy'r post y bore hwnnw. Fe ddylai fod yn falch ei fod wedi sicrhau swydd ar ôl treulio pedair blynedd yn y coleg. Fe ddylai fod yn falch bod drws ar agor iddo, a chyfle i ennill cyflog da. Ond roedd meddwl am ei mab hefyd yn cefnu ar ei ardal enedigol – nid trwy ddewis, mae'n deg dweud, ond trwy reidrwydd – yn cloffi calon Marian. Ei mab a'i merch yn byw ac yn gweithio ym mhen arall y wlad! Pan edrychai'n synhwyrol ar y sefyllfa, gwyddai fod hynny'n well na'u gweld ar y dôl a heb unrhyw obaith o swydd werth chweil o fewn eu milltir sgwâr. Serch hynny . . .

Roedd Marian wedi bod yn gyndyn iawn o esbonio cynllun Merched y Wawr wrth Griff. Doedd hi ddim yn beth hawdd esbonio i'ch plentyn eich hun, hyd yn oed os oedd o'n ddwy ar hugain oed a dwywaith eich maint (o leiaf!), eich bod yn bwriadu cymryd rhan mewn rhyw hen ffwlbri, beth bynnag eich rhesymeg. Ond chwarae teg i Griff, bu'n hael iawn ei ganmoliaeth i'r criw i gyd am eu safiad. Ac roedd Beth wedi mopio'n lân â'r hyn a alwai'n "mawredd mawr Mam", ac yn holi hynt a helynt y datblygiadau bob tro y byddai'n ffonio. Roedd hi wedi newid ei barn yn llwyr am Ferched y Wawr! Yn hytrach na'u galw'n haid o hen ieir, roedd hi bellach yn eu hystyried yn arloeswyr – criw a oedd yn barod i arwain y gad mewn argyfwng.

O'r diwedd fe ferwodd y tecell a gosododd Marian y mygiau ar hambwrdd yn barod i fynd trwodd i'r siop. Daria! Roedd y llestr siwgr yn wag. Aeth drwodd i'w chegin i'w lenwi, ac yno ar y bwrdd gorweddai tusw o flodau, a nodyn yn dweud: "Mam, diolch am bob dim. Ta-ta tan toc. Griff."

Erbyn i Marian ei hailfeddiannu ei hun roedd y te wedi oeri, a safai Cledwyn y tu ôl i gownter Swyddfa'r Post yn mwmian cwyno am y sychder oedd wedi taro'r pentref.

"Dolly Mixtures, Mint Imperials, Bon-bons a jiw-jiws." Ticiodd Marian yr eitemau ar ei rhestr wrth iddi eu henwi. Dyna ni – fe ddylai fod ganddyn nhw ddigon rŵan i bara nes y deuai'r lori o'r warws ddechrau'r wythnos. Roedd hi'n syndod bod gan blant a phobl ifanc Nant-yr-Onnen yr un dant yn weddill yn eu pennau wrth ystyried y tunelli o dda-da roeddynt yn eu claddu yn ystod gwyliau'r haf.

Rowliodd Marian ei throli tuag at y til ym mlaen y *cash & carry* – sgubor o siop a safai ar gyrion Tresarn. Roedd newydd rowndio'r gornel olaf, gan fanwfro'n ddeheuig heibio'r tyrrau o bapur lle chwech amryliw, pan fu ond y dim iddi daro troli arall yn glep.

"Mae'n ddrwg gen . . ." cychwynnodd y ddwy yrwraig ar yr un gwynt.

"O, helô, Caryl. Doeddwn i ddim yn disgwyl eich gweld chi 'ma," meddai Marian yn syn.

"O, Marian bach, tydi'r byd wedi mynd â'i ben iddo, deudwch. Mae'r lle 'ma'n derbyn archeb gen i bob wythnos ar gyfer y gwasanaeth pryd ar glud – bob wythnos, cofiwch

– ac erbyn i'r genethod ddechrau coginio bore 'ma roedd hanner y cynhwysion yn eisiau. Felly roedd yn rhaid newid y fwydlen ar fyr rybudd – y fwydlen gytbwys y bydda i'n pori drosti bob wythnos er mwyn i'r hen bobol gael deiet amrywiol, ddim rhyw arlwy ddiddychymyg o sbrowts a semolina, naci wir – ac roedd yn rhaid i minna ddod i fan'ma i chwilio am yr eitema coll."

"Wel, rydan ni gyd yn gallu gwneud . . ."

Roedd ymdrechion Marian i dawelu Caryl megis ceisio diffodd llosgfynydd â llond gwniadur o ddŵr.

"A rhyw sbrigyn o hogyn – fawr hŷn na Marc ni, dwi'n siŵr – yn y swyddfa, yn deud wrtha i am fynd â throli a chasglu'r eitema coll fy hun. Y fi, Marian, yn disgwyl i mi wthio troli rownd y . . . y sièd 'ma, fel petawn i yn Marks and Spencers, a finna'n gydlynydd y gwasanaeth pryd ar glud."

"Chwara teg i chi am eich iselhau'ch hun dros yr hen bobol, Caryl. Mi fyddan nhw'n dragwyddol ddiolchgar i chi, dwi'n siŵr."

Ond gwastraff llwyr oedd ymdrech Marian ar goegni.

"Rydach chi'n berffaith iawn, Marian bach," meddai Caryl gan ymfalchïo yng nghyflawniad ei dyletswydd ddirodres. "Mae'n bwysig ein bod ni'n gweithredu er lles y rhai sy'n llai ffodus na ni'n hunain, tydi?"

Cymerodd Marian arni nodio'n ddifrifol. "Fel Sara," meddai, gan yrru'r gwch ymhellach i'r dŵr yn ddrygionus.

"Wel . . . ia, debyg," cytunodd Caryl yn anfoddog, "er tydw i ddim yn siŵr ai dyna'r ddelwedd y dylen ni fod yn ei chyflwyno i'r cyhoedd, chwaith. Wedi'r cyfan, Marian, rydan ni'n wragedd aeddfed, ddim fel Mererid ac Alison a gweddill y to ifanc sy ddim yn gwybod yn well. Ac mi roeddwn i wedi synnu at Sara."

"Ond fel 'dach chi newydd ddeud, Caryl," meddai Marian yn serchog, gan wthio'i hamheuon anesmwyth ei hun i gefn ei meddwl am y tro, "weithia mae'n rhaid ei mentro hi'n arw i warchod pobol sy'n llai tebol na ni'n hunain."

"Wrth gwrs," atebodd Caryl yn siort, "a chan mai dim ond dwsin ohonon ni sy 'na, fyddwn i ddim am i neb ddeud mai fi oedd yr un a ddifethodd y cynllun. Ond rhaid i mi gario 'mlaen i gasglu'r nwyddau 'ma rŵan, Marian; mae 'na bobol yn dibynnu arna i. Bore da."

Trodd Caryl ar ei sawdl. Ceisiodd yrru ei throli'n urddasol i gyfeiriad y tuniau pys, ond gan ei bod hi'n anelu un ffordd, a'r droli'n mynnu trio'i gwneud hi am y cyfeiriad arall, methiant fu ei hymdrech.

Ysgydwodd Marian ei phen. Roedd Caryl yn un ryfedd. Roedd yn benbleth barhaol i weddill y genod pam roedd Caryl yn ei hiselhau ei hun i fynychu Merched y Wawr o gwbl. Wedi'r cyfan, ei chadw ei hun ati ei hun fyddai hi heblaw am hynny. Ac er nad oedd hi cweit mor fawreddog ag Elsi, roedd yn ail teilwng i honno pan ddeuai hi'n fater o dynnu blewyn o drwyn rhywun. Roedd y ffaith bod Caryl wedi cytuno i ymddangos yn y calendar wedi peri syndod syfrdanol ymysg y criw.

Ysgydwodd Marian ei phen eto wrth iddi gychwyn am y til drachefn. Er ei bod hi'n gweithio yn eu mysg bob dydd, roedd hi'n amau, hyd yn oed petai hi'n byw i fod yn gant a hanner, a fyddai hi byth yn deall pobl.

". . . ac yna, ar ôl i Tina agor y drws i ti, mi sefaist ti dan ffenest y llofft yn canu 'Oes Gafr Eto?'."

Roedd Marian wedi gosod stôl ger y cownter, ac eisteddai yno a'i phen yn ei dwylo, tra ceisiai Cledwyn lenwi rhywfaint ar y gagendor yn ei chof ynglŷn â sbloet y noson cynt. Ochneidiodd Marian, a llaciodd rhyw ychydig ar ei bysedd fel y gallai sbecian ar ei gŵr rhyngddynt.

"Oedd 'na rywun arall allan?"

Chwarddodd Cledwyn. "Neb o'r Cwmni Opera Cenedlaethol i werthfawrogi dy berfformiad, yn anffodus; dim ond un neu ddwy arall o'r criw oedd yn baglu eu ffordd am adra mewn cyflwr cynddrwg, os nad gwaeth, na ti."

Ochneidiodd Marian eto.

"Lwcus i ti fod cadw llygad barcud ar Nia'n siŵr o fod wedi cadw gwarchodwraig y cynllun Gwarchod Cymdogaeth yn brysur, neu byddai peryg iddi fynd â'i chwsmeriaeth amhrisiadwy i rywle arall," pryfociodd Cledwyn hi.

Ond doedd hyd yn oed y syniad o gael gwared ag Elsi unwaith ac am byth ddim yn ddigon i ddwyn Marian yn ôl i dir y rhai byw.

"Yli," meddai Cledwyn wedyn, "mi a' i i wneud panad arall i ti tra'i bod hi'n ddistaw 'ma. Fyddi di'n iawn am funud bach?"

Tynnodd Marian ei dwylo oddi ar ei hwyneb a cheisio ffocysu ar ei gŵr trwy lygaid oedd fel olion bysedd plentyn mewn eisin. Nodiodd arno cyn griddfan drachefn pan ddechreuodd ei phen fowndian wrth iddi'i symud. Diflannodd Cledwyn i'r cefn dan wenu.

Yn y man dychwelodd gyda mŷg o de a dwy dabled wen. "Dyna ti, gad i ni weld a fedrwn ni atgyfodi rhywfaint arnat ti cyn i Harri alw amser cinio."

Llowciodd Marian y te a'r ddwy dabled yn ddiolchgar. Sôn am gyflog pechod, wir! Doedd hi byth yn mynd i yfed eto – byth bythoedd. Wir yr.

Ymhen tipyn, a'r boen yn ei phen yn dechrau lleddfu, a'i cheg ddim yn teimlo cweit cyn syched â llith pregethwr cynorthwyol bellach, cofiodd Marian fod ganddi asgwrn i'w grafu gyda'i gŵr.

"Cledwyn Tomos, dwi newydd gofio – dy fai di ydi hyn i gyd."

Syllodd Cledwyn yn syn ar ei wraig o'r tu ôl i'w gownter.

"Ti brynodd y rownd gynta," cyhuddodd Marian ef, "a'n harwain ni i gyd ar gyfeiliorn."

Chwarddodd Cledwyn. "Duwcs, doedd 'na ddim gwaith arwain arnoch chi; roeddech chi'n slochian dros Gymru, y cwbwl lot ohonoch chi, pan adewais i'r Llew Coch."

Er mai bore cymharol ddistaw fu hi, roedd hi bron yn amser cinio cyn i Marian gael cyfle i gael ei gwynt ati gan fod Brenda wedi cael bore rhydd. Cafodd ddigon o gwsmeriaid i gadw'i meddwl oddi ar yr hyn oedd i ddod yn ystod yr awr ginio, ac i'w rhwystro rhag gwyntyllu cyflwr sigledig, nid yn unig ei hargyhoeddiad ond ei chorpws hefyd, yn dilyn gormodaeth y noson cynt.

"Wyt ti am fentro rhywbeth bach i'w fwyta cyn i Harri gyrraedd?" galwodd Cledwyn ychydig funudau cyn eu hawr ginio swyddogol. "Rydw i am gau rŵan hyn, dwi'n meddwl, gan ei bod hi'n wag yma."

Cyn i Marian orffen pwyso a mesur doethineb mentro darn o dost rhag ofn i'w stumog ddechrau cwyno yng ngŵydd Harri, stompiodd Elsi i mewn i'r siop a golwg gynddeiriog arni.

"Hm. Dim Brenda bore 'ma, Marian? Mae'n siŵr ei bod hithau, fel y gweddill ohonoch chi, yn cymryd rhan yn y dichellwaith diraddiol 'ma."

Agorodd Marian ei cheg, ond dal i ymchwyddo fel y seithfed don oedd llid Elsi.

"Ac rwyf wedi fy siomi'n arw yn Nia. Dim parch tuag at fy marn i ynglŷn â'i hymddangosiad yn y calendar coman 'ma. A tydi Ifan fawr gwell. Gadael iddi wneud fel y mynnith yn lle rhoi'i droed lawr. Mae'n syndod 'mod i cystal, wir, dan yr amgylchiadau. Wel, dwi'n gobeithio nad ydych *chi'n* mynd i'm siomi i hefyd, Marian. Dwi'n cymryd eich bod chi wedi sicrhau cyflenwad o hadog melyn i mi bellach?"

Hadog melyn!

Roedd darn o dost yn un peth. Ond hadog melyn?

Rhoddodd Marian ei llaw dros ei cheg a rhedeg i'r cefn.

Roedd Marian wedi llyncu swper brysiog, wedi newid a molchi a phincio, ac roedd ar ei ffordd allan trwy'r drws pan ganodd y ffôn. Dili-daliodd am funud, gan obeithio y byddai Cledwyn yn dod i'w ateb, cyn troi'n ei hôl a chodi'r derbynnydd.

"Helô, Mam, fi sy 'ma. 'Dach chi'n iawn?"

"Helô, Beth. Ydw, dwi'n iawn, diolch. A dy dad hefyd."

"Ylwch, fedra i ddim siarad am hir, dwi ar fy ffordd i nôl Griff o'i waith, ond mae'r ddau ohonon ni'n dod adra fory. Mi fyddwn ni yno erbyn y lansiad swyddogol, ac yn aros dros y penwsnos os ydi hynny'n iawn."

Trawyd Marian yn fud. Beth a Griff yn dod i fyny ar gyfer y lansiad!

"Mam, 'dach chi dal yna?"

"Ydw, ydw, Beth. Mae hynny'n newyddion gwych. Ydach chi'n siŵr y medrwch chi gael amser yn rhydd o'r gwaith?"

"Mae arna'r cwmni 'ma oria i mi, a finna wedi gweithio bron bob bora Sadwrn ers misoedd. Ac mae Griff am ffonio mewn yn sâl bora fory, medda fo. A deud y gwir, Mam, mae Griff wedi cael llond bol yma'n barod; synnwn i ddim petai o'n rhoi'r gora iddi cyn hir, a chwilio am rywbeth yn nes at adra. Person ei filltir sgwâr ydi Griff yn y bôn, dwi'n meddwl. Ac mae o'n colli'ch cael chi'n dawnsio tendans arno fo, Mam!"

Wedi ffarwelio â'i merch, teimlai Marian fel petai wedi diosg pais ddur na wyddai oedd amdani. Aeth drwodd i'r stafell fyw i adrodd y newyddion da wrth Cledwyn a eisteddai yno'n smalio darllen y papur lleol yn gwbl ddiniwed (er ei fod, mewn gwirionedd, yn gwybod am gynlluniau'r pobl ifanc ers rhai dyddiau). Roedd wrth ei fodd yn gweld Marian mor llawen, a gwasgodd hi'n dynn. Yna gyrrodd hi'n ôl i'r stafell molchi i bincio drachefn gan fod ei cholur wedi ei strempio'n llwyr gan ei dagrau.

Yn y man, ailgychwynnodd Marian am y Llew Coch i ymuno â gweddill y criw.

# AWST

Damia! Damia! Damia!

Ble roedd y papurau 'na wedi mynd?

Roedd Gwyneth yn meddwl siŵr ei bod wedi eu cadw yn y drôr efo'r gweddill. Ond doedd dim golwg ohonynt yno. Nac yn 'run o'r drorsys eraill. Nac yn unman arall.

Eisteddodd ar fraich y soffa ac ymbwyllo. Ceisiodd ddwyn i gof pryd y gwelodd y gwaith papur ddiwethaf. Roedd hi wedi taro llygad arno ryw bythefnos yn ôl, gan ystyried yr adeg honno ei bod hi'n hen bryd iddi feddwl am ei gasglu at ei gilydd yn barod i'w gyflwyno i'r cyfrifydd. Rhaid ei bod hi wedi ei gadw'n saff wedi hynny. Ond ymhle? Dyna'r cwestiwn.

Cododd a mynd drwodd i'r gegin i wneud paned.

Yn y man dychwelodd i'r stafell fyw a setlo ar y soffa drachefn. Sigodd ei hysgwyddau. Ar adegau fel hyn byddai pwysau cyfrifoldeb yn ei gormesu, a byddai wedi bod yn fodlon rhoi unrhyw beth i gael rhywun i rannu'r baich. Ond dyna ni, ei phenderfyniad hi oedd aros yma ym mhentref Nant-yr-Onnen wedi colli Robin. Dyn a ŵyr, roedd ei mam wedi erfyn digon arni i ddod yn ôl adref ati hi i fyw, ac i chwilio am waith llai llafurus. Ond byddai cydymdeimlad syrffedus ei mam wedi ei mygu'n lân, ac roedd arni angen profi, iddi ei hun yn bennaf, y gallai ei chynnal ei hun, y gallai fod yn annibynnol, y gallai redeg busnes llwyddiannus.

Ond ar noson fel heno, yn llonyddwch llethol yr oriau mân, wedi diwrnod maith ar ei thraed, heb sôn am ddiflaniad y papurau bondigrybwyll, roedd Gwyneth yn ysu am rywun; rhywun i'w chefnogi a'i chysuro, i'w chofleidio a'i charu.

Safai'r plismon ifanc yn stond, wedi ei arswydo, tra pwysai Gwyneth yn erbyn wal y Llew Coch yn ei dyblau, yn chwydu ar y palmant, a'i gwallt melyn yn cuddio'i hwyneb. Syllodd y llanc yn syn ar y diferion yn tasgu dros ei sgidiau du gloyw. Yna trodd ei ben wrth deimlo'r brecwast helaeth a ddarparwyd iddo gan ei fam yn dechrau corddi yn ei stumog.

Yn y man, stopiodd y sŵn cyfogi a throdd yr heddwas yn ôl at Gwyneth. Roedd ei chefn yn erbyn wal y dafarn, ei hwyneb cyn wynned ag amdo, a'i llygaid ar gau.

Roedd cryndod afreolus wedi meddiannu corff y bachgen o blismon. Doedden nhw ddim yn dweud wrthych am bethau fel hyn pan oeddech yn ymuno â'r heddlu ac yn torri'ch enw yn y man priodol; ddim yn dweud wrthych mai chi fyddai cyfrwng dymchwel bywyd gwraig ifanc yn deilchion o'i hamgylch wrth dorri newyddion drwg. Na, doedd hyn ddim yn cyd-fynd â'i ddelwedd ddisglair, na'r darluniau dychmygus a goleddai ohono'i hun, darluniau ohono'n gwarchod ac yn gwasanaethu, gan lorio drwgweithredwyr â thacliadau godidog.

Estynnodd law grynedig a'i phwyso'n betrus ar fraich Gwyneth. Agorodd hithau ei llygaid glas tywyll. Treiddiodd ysgryd o'i chorun i'w sawdl. Yna gadawodd i'r cyw cwnstabl ei harwain i mewn i'r Llew Coch.

Doedd Olwen, mam Gwyneth, ddim yn hapus iawn pan esboniodd ei merch a'i mab-yng-nghyfraith newydd sbon eu bod yn mynd i gadw tŷ tafarn ym mhentref Nant-yr-Onnen. A dweud y gwir, ers marwolaeth tad Gwyneth sawl blwyddyn ynghynt (digwyddiad a gadarnhaodd farn ei weddw mai dyn cwbl anystyriol ydoedd), fu mam

Gwyneth ddim yn hapus iawn â'r un dim, yn enwedig trefniant a fyddai'n rhyddhau ei merch o afael ei chrafangau chwannog. Ond derbyn y sefyllfa fu rhaid, a chyfyngwyd cwyno hirwyntog, hunandosturiol Olwen i alwadau ffôn aml a maith.

Dan ddylanwad ei gŵr, ac allan o gyrraedd grwgnach gorthrymus ei mam, fe ffynnodd Gwyneth. Roedd Robin yn fachgennaidd o ran pryd a gwedd, yn frwdfrydig ac yn fywiog. Roedd wedi trio'i orau glas i blesio'i fam-yng-nghyfraith na fyddai, mewn gwirionedd, byth yn cael ei phlesio ond wrth chwilio am, a darganfod, ochr dywyll pob sefyllfa. Pan sylweddolodd Robin ei fod yn gwastraffu ei amser yn llwyr, penderfynodd mai'r cynllun callaf oedd sicrhau bod nifer sylweddol o filltiroedd rhwng Gwyneth a'i mam, rhag iddo orfod gwylio'i wraig yn cael ei manglo gan gastiau cyfrwys Olwen.

Cymerodd Gwyneth at ei chartref newydd ac at y gwaith dieithr fel hwyaden at ddŵr. Gydag anogaeth Robin cynyddodd ei hunanhyder, ac er iddi barhau'n fwy distaw a phwyllog na'i gŵr, roedd personoliaethau'r ddau'n cyfannu'r naill a'r llall i'r dim.

Roedd Robin am gyflwyno newidiadau chwyldroadol yn y Llew Coch bron cyn gosod ei frwsh dannedd yn y stafell molchi, ond perswadiodd Gwyneth ef i bwyllo, gan y byddai hynny'n tynnu blewyn o drwyn y selogion, heb sôn am godi gwrychyn y pentref i gyd. A sut bynnag, doedd dim pwrpas iddynt geisio cystadlu â themtasiynau llwydolau'r dref. Wedi'r cyfan, rhywle i ymlacio wedi diwrnod o waith ac i sgwrsio a hel straeon roedd ar bentrefwyr Nant-yr-Onnen ei eisiau, nid clwb nos lle nad oedd modd i'r cwsmeriaid glywed ei gilydd yn sgwrsio, a lle byddent yn bopio nes iddynt syrthio'n swp.

Bu cyfaddawd. Cafodd y Llew Coch ei beintio nes ei fod yn sgleinio fel swllt; prynwyd dodrefn twt a theidi ar gyfer y lolfa, ac estynnwyd y bar fel bod yno le i fwrdd pŵl i ychwanegu at y dartiau a oedd eisoes yn boblogaidd. Yn wir, gweithiodd y pâr ifanc fel lladd nadroedd nes eu bod wedi talu am y newidiadau a sicrhau bod y tafarndy'n fan dymunol i bobl y pentref ymgynnull ynddo. Wedi sawl blwyddyn o waith caled, teimlai Robin a Gwyneth eu bod wedi llwyddo i ddenu busnes sylweddol, a'u bod mewn sefyllfa ariannol eithaf boddhaol.

Nid oedd byw a gweithio ym mhocedi'r naill a'r llall wedi creu unrhyw broblemau iddynt; roedd y ddau'n dal yn gwirioni'n llwyr ar ei gilydd. O dro i dro byddai Robin yn ymweld â chanolfan hamdden Tresarn i nofio neu chwarae sboncen, a pherswadiodd Gwyneth i ymuno â changen y pentref o Ferched y Wawr, er mwyn iddi hithau gael rhyw gymaint o newid. Ond yn gyffredinol roedd y ddau'n fodlon braf yng nghwmni ei gilydd – a'u cwsmeriaid, wrth gwrs.

Ar ôl cyfnod o ffyniant, dechreuodd meddyliau'r pâr ifanc droi tuag at fagu teulu. Ond cyn iddynt gael cyfle i wireddu eu dyhead, fe wawriodd y bore erchyll hwnnw; bore pan nad oedd angen ailstocio'r silffoedd na newid casgen, nac archebu stoc na chyfri cownts, na hwfro na glanhau'r tai bach. Bore rhydd am unwaith; bore y penderfynodd Gwyneth ei dreulio'n gwneud tartenni ar gyfer y rhewgell â basgedaid o afalau a dderbyniasai gan Nia; bore y gwelodd Robin gyfle i fynd i nofio i Dresarn; bore a chwalodd fywyd Gwyneth yn chwilfriw.

Brith gof oedd ganddi o'r oriau, y dyddiau, yr wythnosau a'r misoedd canlynol.

Gallai gofio'i mam yn dod i aros am gyfnod, cyfnod a ymddangosai'n hir iawn, ac Olwen yn ei galarwisg ffasiynol yn edrych yn fwy fel petai newydd gael ei gadael yn weddw na Gwyneth ei hun. Gallai gofio pobl o'i hamgylch yn gyson, fel cadwyn amddiffynnol. Ac aelodau Merched y Wawr yn arbennig – y genod yn ei chofleidio a'i chysuro, yn gwrando ac yn gwarchod.

Ar ôl yr angladd, a Gwyneth wedi llwyddo i gael gwared â'i mam – a oedd yn meddu ar ddawn i wneud sefyllfa dorcalonnus ddengwaith gwaeth – fe redodd y tŷ fel watsh. Tybiai Gwyneth na fyddai byth yn gallu ad-dalu cymwynasgarwch y pentref; roedd y cymorth a'r cynhaliaeth a dderbyniodd yn ystod y misoedd coll hynny yn amhrisiadwy. Ac oedd, mi roedd amser wedi lleddfu'r boen. Roedd hi wedi llwyddo i gael ei thraed dani a dod â'r busnes i'r lan, ac wedi llwyddo hefyd i wrthsefyll swnian diderfyn ei mam arni i roi'r ffidil yn y to a dychwelyd i fyw ati hi.

Roedd Gwyneth yn dal i fynychu cyfarfodydd Merched y Wawr ac yn mwynhau cwmni'r criw yno. Roedd hyd yn oed yn gallu ymweld â'r ganolfan hamdden bellach. Ond er ei bod yn arddangos wyneb llon i'r byd – wedi'r cyfan, ni allai fforddio diflasu ei chwsmeriaid, ac roedd arni ofn am ei bywyd droi mewn i'w mam ei hun – teimlai Gwyneth mai rhygnu byw oedd hi, a bod bywyd bob dydd wedi troi'n rhigol rhwystredig, yn rhibyn rhwyfus, yn ddim ond gwely a gwaith.

". . . a bydd hi fawr hwyrach na naw arna i'n ôl, Jac."

"Popeth yn iawn, Gwyneth. Does arna i ddim brys."

"Diolch. Go brin y bydd hi'n brysur . . . O, helô, Tina."

"Helô, Gwyneth. Haia, Jac. Na, tydw i ddim yn ffwndro; dwi'n gwybod mai noson Merched y Wawr ydi hi. Wedi cael hyd i'r llyfr 'na wnes i addo'i fenthyg i ti, Gwyneth, a meddwl y byswn i'n picio draw efo fo rŵan, cyn iddo ddiflannu yng nghanol anialwch yr hogia 'cw eto."

"Diolch, Tina, dwi wedi bod yn edrych ymlaen at ddarllen hwn."

Aeth Gwyneth â'r llyfr drwodd i'r cefn gan adael Tina a Jac yn siarad ar hyd ac ar led wrth y bar.

"Reit, Tina," meddai Gwyneth wedi iddi ddychwelyd, "awn ni?"

Nodiodd Tina.

"Hwyl, Jac."

"Hwyl, genod."

Ysmiciodd y ddwy eu llygaid wrth ddod allan o far y Llew Coch i heulwen yr hwyrddydd.

"Wyt ti'n siŵr dy fod ti'n hapus yn gadael Jac i weithio ar ei ben ei hun?" holodd Tina.

Cododd Gwyneth ei hysgwyddau a thynnu stumiau ar ei ffrind. "Ydw, dwi'n meddwl 'mod i. Dwi'n dal i ama bod yna ambell i beint yn cael ei dywallt am bris hanner, ac ambell i siortyn dwbl yn cael ei roi am bris un, cofia. Dim ond i'r detholedig rai ymysg ei fêts, yntê?"

"Dwyn 'di'r enw am beth felly," meddai Tina.

"Ond fedra i ddim bod yn berffaith siŵr. Ac mae o'n andros o weithiwr da – ti'n gwybod hynny dy hun. Ac yn barod iawn i helpu allan ar nosweithia fel heno er mwyn i ti a fi gael noson rydd 'run pryd. Heblaw am nosweithia Merched y Wawr, anaml iawn mae o yno ar ei ben ei hun."

"M-m-m."

Nid oedd Tina wedi ei hargyhoeddi. Roedd yn gas ganddi annhegwch. Ac er bod dogn go lew o hwnnw wedi lliwio'i bywyd hi ei hun (er nad fyddai Tina'n edrych arni felly), byddai gweld eraill yn cael eu trin yn annheg yn ei digio'n arw.

"Beth bynnag, mae'n braf cael dod allan o'r tu ôl i'r bar am ychydig oria," cyhoeddodd Gwyneth, "a tydw i ddim yn bwriadu difetha'r noson yn pryderu am Jac a'i fân fisdimanars, gwir neu ddychmygol."

Gwenodd Tina. "Ti'n iawn. Mi gefais inna fora gwyllt yn y cylch meithrin; roedd hi fel syrcas yno. Dwi'n siŵr bod y plant i gyd wedi cael Smarties neu rywbeth arall yn llawn lliwia E i frecwast. Dwi'm yn meddwl bod 'run ohonynt wedi ista lawr am fwy na hanner munud drwy'r bora."

Chwarddodd Gwyneth. "A minna'n cael trafferth i berswadio 'nghwsmeriaid i godi o'u seddi a'i throi hi am adra ambell i noson. Gwell ti na fi, serch hynny," ychwanegodd.

"Ac yna, wedi mynd adra," aeth Tina yn ei blaen, "mi benderfynais i olchi popeth y medrwn i'i wasgu trwy ddrws y peiriant golchi, gan fod y dyn ar y teledu'n deud bod peryg iddi droi tywydd fory. Dwi'n reit falch o esgus i ista lawr erbyn hyn."

"Wel, gyda lwc," meddai Gwyneth, wrth i'r ddwy droi i mewn i neuadd y pentref, "mi fydd hi'n gyfarfod digon hamddenol heno; dim byd mwy beichus na phenderfynu oes yna ddigon o bres yn y coffra i ni gael mynd am drip i rywle 'rha 'ma."

. . . ac mae digon o waith ar gael yma. Mae yna lawer o westai lleol yn hysbysebu yn y wasg am weithwyr – a'r rheini'n westai o safon, rhai *classy*. Gyda dy brofiad di, byddai gen ti siawns dda o gael dy benodi i swydd yn un ohonynt. Mae nifer o'r hysbysebion yn gofyn am staff i ymuno â thimau rheoli. Mewn swydd felly fyddai dim rhaid i ti wneud yr holl lanhau a chario rwyt ti'n ei wneud yn y Llew Coch, a byddwn i'n gallu dod efo ti i ddewis dwy neu dair siwt addas ar gyfer y gwaith. Byddai costau byw'n llawer is i ti hefyd petaet ti'n byw yma efo fi. Rwyt ti'n gwybod bod d'ystafell yn barod bob amser.

Edrych ar y manylion, Gwyneth, a meddwl o ddifrif a wyt ti eisiau parhau i ymdrechu i gael dau ben llinyn ynghyd yn Nant-yr-Onnen ar dy ben dy hun, pan fyddai bywyd yn gallu bod cymaint haws i ti yma, a'r ddwy ohonom yn gwmni i'n gilydd; mae'n unig iawn yma, a'r hen dŷ'n rhy fawr i mi bellach.

Mi fydd yn ddrwg gennyt glywed fy mod yn dal i gael trafferth gyda'r cwpwl ifanc sydd wedi symud i mewn drws nesaf. Maent yn mynnu cynnal rhyw hen farbeciws yn yr ardd er fy mod i wedi dweud wrthyn nhw bod y mwg a'r oglau saim yn amharu'n ofnadwy arna i. Ac mi fydd y plant allan yn yr ardd drwy'r dydd unwaith bydd yr ysgol wedi cau, yn gweiddi a sgrechian fel anwariaid.

Wyt ti'n cofio'r siwt Viyella a brynais yn ddiweddar – yr un *salmon pink*? Roedd yn rhaid i mi fynd â'r sgert i Mari i gael ei chwteuo. Pan ddaeth hi'n ei hôl roedd yr hem bob sut a'r leining yn dangos. Mi es â hi'n ei hôl yn syth bin a chael ar ddeall bod gan Mari ryw lefren o eneth o'r ysgol uwchradd yno ar ryw gynllun *work experience*, a hithau wedi'i gadael yn rhydd ar fy siwt Viyella newydd sbon i! Bu'n rhaid i mi siarad yn reit siarp gyda Mari, wir! A tydw i ddim yn siŵr iawn . . .

Ochneidiodd Gwyneth a rhoi llythyr ei mam naill ochr;

byddai'n mynd trwy weddill y catalog cwynion eto. Ond tra oedd yn gorffen ei phaned cyn dechrau ar waith y dydd, treuliodd ychydig funudau'n bwrw golwg dros yr amryw hysbysebion am swyddi roedd ei mam wedi eu cynnwys yn yr amlen.

Wedi deuddydd, neu dridiau, neu wythnos pan fyddai gwaith y dafarn yn ei diffygio, cyfrifoldeb yn ei llethu ac unigrwydd yn ei threchu, byddai Gwyneth yn cael ei themtio am ychydig gan gynigion gwyngalchog ei mam. Ond gwyddai, petai'n ildio i'r hyn a ymddangosai'n ddewis call, mai dan fawd gorthrymus Olwen y byddai hyd dragwyddoldeb. A beth am bentrefwyr Nant-yr-Onnen? Onid oeddent hwy, pob un wan jac ohonynt, wedi bod yn gefn ac yn gynhaliaeth iddi hi yn ystod y blynyddoedd anodd diwethaf? Oedd hi'n barod i hel ei phac a chefnu ar eu caredigrwydd a'u cefnogaeth? A beth am y calendar arfaethedig? Gyda dim ond dwsin ohonynt, doedd wiw i neb dynnu'n ôl. Byddai ei heglu hi i weithio yn un o'r gwestai *classy* roedd ei mam mor hoff o sôn amdanynt yn esgus – rheswm – nage, esgus gwych dros beidio â chymryd rhan yng nghynllun Sara. Ond allai hi fyth fod mor llwfr â hynny; ddim ar ôl popeth roedd y pentref wedi ei wneud drosti hi.

Sodrodd Gwyneth yr hysbysebion ar fwrdd y gegin, nesaf at lythyr ei mam. Byddai'n eu ffeilio yn y man briodol gyda gweddill y llythyrau sgrwtsh a dderbyniai drwy'r post cyn i'r lori bins alw heibio bore fory. Ond byddai'n rhaid ateb llythyr ei mam. Yn y man. Neu ffonio. Na, byddai sgwennu'n haws, yna fyddai dim rhaid iddi wrando ar y tinc edliwgar, truenus yn llais Olwen pan fyddai'n cadarnhau, unwaith eto, ei bod am barhau i redeg y Llew Coch, a'i bod yn berffaith hapus, diolch yn fawr.

Celwydd golau fyddai dweud ei bod yn berffaith hapus, wrth gwrs. Nid oedd Gwyneth wedi teimlo'n "berffaith hapus" ers colli Robin; dal blawd wyneb oedd hi yn ei bywyd personol hyd heddiw. Ond twyll bach digon diniwed oedd rhyw ddweud felly, a digon teg, dan yr amgylchiadau. A byddai'n haws esbonio cynllun y calendar i'w mam mewn llythyr na dros y ffôn. Byddai Olwen yn cael modd i fyw unwaith y deuai i wybod am hwnnw. Testun chwilio a chwalu a chwyno werth ei gael. A hwnnw wedi glanio fel manna.

Cododd Gwyneth. Wrth roi ei mỳg yn y sinc a sgubo'r manylion am y swyddi naill ochr, hanner gwenodd wrthi ei hun. Petai goblygiadau sgiâm y calendar yn hoelio sylw ei mam am gyfnod, efallai byddai'n atal rhywfaint ar ei hymdrechion i ffaldio Gwyneth yn ei chorlan gysurus gaeth. Dros dro, beth bynnag.

Roedd hwyliau rhyfeddol o dda ar Gwyneth ar ôl sesiwn ym mhwll nofio'r ganolfan hamdden. Am fisoedd wedi'r ddamwain ni allai feddwl am deithio ar hyd y lôn i Dresarn; byddai'n teithio i Gaercedog, tref gryn bellter o Nant-yr-Onnen, i mofyn unrhyw beth nad oedd ar gael yn siop y pentref. Ond bellach gallai yrru yno heb adael darn bach o'i chalon yn y llecyn tawel, twyllodrus ar bwys y ffordd ble cipiwyd Robin oddi wrthi, fel y gwnaethai gyda phob siwrnai cyn hynny. Camodd i'w char gan daflu bag y taclau nofio i'r sedd gefn. Roedd yn dal i wenu wrthi ei hun wrth geisio dychmygu beth fyddai ymateb Elsi petai'n derbyn cais i lamu i'r adwy gan griw'r calendar. Byddai'n ddigon iddi!

Roedd wedi bod yn brynhawn dymunol iawn, ystyriodd Gwyneth wrth dynnu allan o'r maes parcio a throi trwyn y car am Nant-yr-Onnen. Chwarae teg i Tina am gynnig gofalu am y siop – er fe wyddai Gwyneth y byddai Tina'n falch o'r arian ychwanegol. Roedd hi'n rhyddhad cael diosg ei chyfrifoldebau am ychydig oriau, ac roedd cael paned yng nghwmni Mererid ac Alison wedi bod yn fonws. Mererid druan, wedi cael amser go galed, yn ôl pob sôn. Ac Alison . . . doedd Gwyneth ddim yn adnabod Alison yn dda iawn. Byddai'n ei gweld yng nghyfarfodydd Merched y Wawr, wrth gwrs, a byddai Alison hefyd yn ymuno â tho ifanc y gangen a fyddai'n ymgynnull yn y Llew Coch weithiau. Ambell dro byddai'r ddwy'n taro i mewn i'w gilydd yn y pentref ac yn cael rhyw bwt o sgwrs, ond hyd yma, ymddangos yn swil a thawedog a wnâi Alison yng nghanol y criw.

Wrth gyrraedd y gylchfan ar gyrion Tresarn, roedd Gwyneth yn dal i synfyfyrio pan sylwodd ar gornel yr amlen frown a roddodd o'r golwg dan y sedd yn ei hymyl, yn sbecian o'i chuddfan.

Damia! Roedd angen mynd i swyddfa'r cyfrifydd arni; dyna sut y cyfiawnhaodd y prynhawn rhydd yn y lle cyntaf.

Sgrialodd reit rownd y gylchfan gan serio'i llygaid ar y lôn o'i blaen wrth anelu'r car yn ôl am Dresarn, rhag ofn i unrhyw dystion i'w chylchdaith dynnu stumiau arni ac awgrymu nad oedd yn ei hiawn bwyll.

Ymhen cwta ugain munud roedd yn powlio lawr stepiau swyddfa'r cyfrifydd ar frys gwyllt gan ei bod wedi mentro gadael y car ar ddwy linell felen am yr ychydig funudau roedd arni eu hangen i ddanfon yr amlen. Dim golwg o blismon na warden traffig, diolch byth. Brasgamodd Gwyneth ar hyd y palmant a'i llygaid ar ei char nes iddi fwrw'n glep i mewn i rywun oedd yn dili-dalio ar erchwyn y palmant.

"O! Mae'n ddrwg gen . . . Phyllis! Sudach chi?"

"O, helô, Gwyneth. Popeth yn iawn. Fi oedd ar fai'n llusgo 'nhraed yn fan'ma. Trio gweld a oes 'na bobol yn disgwyl tu allan i'r post am y bws oeddwn i, ynteu ydi o wedi mynd."

"Ar eich ffordd adra ydach chi, Phyllis?"

"Ia, ond 'mod i wedi galw yn y llyfrgell i weld oedd ganddyn nhw lyfra patryma ar gyfer peirianna gwau, ac wedi colli cyfrif o'r amser yn llwyr," atebodd Phyllis yn gynhyrfus.

Gwenodd Gwyneth. "Dwi wedi cychwyn am adra unwaith yn barod, ac wedi gorfod troi'n ôl am fy mod wedi anghofio'r neges a ddaeth â mi yma'n y lle cynta. Ylwch, os ydach chi'n barod, mi gewch chi bàs efo fi – mae'r car yn fan'cw."

Wedi pacio Phyllis a'i phecynnau i'r car, cychwynnodd Gwyneth am Nant-yr-Onnen drachefn.

Roedd smotyn pinc ar fochau Phyllis ac roedd hi'n amlwg bron â byrstio â rhyw newydd.

"Gawsoch chi lwyddiant gyda'ch ymchwiliad yn y llyfrgell?" Mentrodd Gwyneth ddyfalu achos y cynnwrf. "Wyddwn i ddim eich bod yn berchen peiriant gwau."

"Newydd brynu un. Siop Pethau Bychain wedi cau, fel y gwyddost ti, mae'n siŵr, a minna'n methu'n glir â chael swydd arall. A dyma Brenda'n gweld yr hysbyseb am y peiriant 'ma yn y papur lleol, a 'mherswadio i i roi cynnig arni."

"Da iawn chi. A hei lwc efo'r fenter. Mi welais i'r siwmper roeddech chi wedi'i gwau i Lowri ar ei phen-blwydd. Mi alwodd heibio efo Brenda un gyda'r nos i ddanfon neges, ac roedd hi'n ddigon o sioe. Mi fydda dillada o'r ansawdd yna'n mynd fel slecs, ond i chi gael bawd eich troed i'r farchnad iawn."

Roedd Phyllis wrth ei bodd â'r ganmoliaeth, a threuliodd ran helaethaf y siwrnai'n anghyffredin o siaradus, yn egluro'i chynlluniau ac yn esbonio'i hymdrechion i fynd i'r afael â'r peiriant.

Roeddent ar gyrion Nant-yr-Onnen pan gollodd Phyllis stêm yn sydyn bwt, a sylweddoli ei bod hi wedi llwyr feddiannu'r sgwrs gydol y ffordd.

"Sori, Gwyneth, dwi wedi parablu'n ddi-baid, a phrin wedi gofyn sut wyt ti."

"Peidiwch â phoeni dim. Dwi'n edmygu'ch dyfeisgarwch chi; mae angen dipyn o blwc i gychwyn eich busnes eich hun fel'na. A dwi'n genfigennus o'ch dawn greadigol. Fe driodd Nain fy nysgu i wau, ond roeddwn i mor drychinebus o wael fe dorrodd ei chalon gyda'r dasg. Yr unig beth greais i 'rioed oedd sgarff i Tedi, a hwnnw'n ddim ond stribyn cnotiog o glyma a thylla, fel y llunia 'na fydda ar y wal yn y dosbarth bioleg, a chitha'n sbio arnyn nhw rhag gwneud eich gwaith – llunia tu mewn i goesa pobl, yn dangos y cyhyra a'r gewynna, ac ati."

Chwarddodd Phyllis. "Chefais inna fawr o grap ar *biology* yn 'rysgol chwaith. Ond falla bod yr holl wersi *home economics* 'na'n mynd i ddechra talu'u ffordd o'r diwedd."

Wrth i'r car fynd heibio neuadd y pentref, roedd yn anorfod y byddai'r ddwy yn holi ei gilydd ynglŷn â'u teimladau am y calendar, a'r diwrnod tyngedfennol yn prysur nesáu.

"Doeddech chitha, fwy na minna, yn fodlon iawn ar y dechra, Phyllis."

"Nac oeddwn, a deud y gwir wrthat ti. Roedd y syniad reit wrthun i mi, ond bod Brenda wedi codi'n llaw i 'radeg y bleidlais."

"Digon cyndyn oeddwn inna hefyd. Ond 'tydi pawb yn Nant-yr-Onnen wedi gwneud cymaint drosta i yn ystod y blynyddoedd diwetha 'ma? A finna'n teimlo bod hwn yn gyfle i mi dalu rhyw ddarn bach o'r gymwynas honno'n ôl."

"Ia, fel'na gwelais inna hi yn y pen draw hefyd," cyhoeddodd Phyllis yn llon, wrth i Gwyneth ei helpu at ddrws y tŷ gyda'i bagiau, "fel rhyw fath o dalu'n ôl."

Roedd Gwyneth yn falch ei bod wedi llwyddo i ddal yn dynn yn y mymryn argyhoeddiad a feddai. Doedd hi ddim wedi siomi'r criw, beth bynnag.

Ond er gwaetha'i hymdrechion i gadw'n brysur, roedd y prawf a'i hwynebai wedi bod yn flaenllaw yn ei meddwl yn ystod y dyddiau blaenorol. Gan ei bod hi'n ben tymor, roedd y Llew Coch bellach ar gau tan amser te yn ystod yr wythnos, ac yn agor trwy'r dydd ar benwythnos yn unig, nes y byddai pethau'n dechrau prysuro eto ym mis Rhagfyr. Digon o gyfle i hel bwganod felly. Byddai rhannu ei hofnau a'i hamheuon gyda rhywun wedi lliniaru rhywfaint ar anesmwythyd Gwyneth; byddai wedi gwerthfawrogi llais cynnes, cyfarwydd yn holi hynt a helynt ei phrofiad o flaen y camera; byddai wedi suddo i swcr cesail glud a gynigiai loches iddi wedi i'w beiddgarwch byrhoedlog ballu. Ond dyna ni, roedd hi wedi goroesi, ac wedi ymdopi â'i chyrch i fyd modelu bronnoeth. Gallai geisio anghofio am y peth rŵan nes y deuai'n amser lansio'r calendar yn y man.

Doedd Gwyneth heb weld 'run o'r criw yn ystod y dydd i weld sut hwyl gawson nhw arni. Efallai y byddai un neu ddwy o'r to ifanc yn galw heibio nes ymlaen, meddyliodd wrth roi sglein munud olaf i'r bar. Ar y llaw arall, cilwenodd wrth fynd draw i dynnu'r bollt yn ôl, roedd hi'n bur debyg bod y rhan fwyaf ohonynt wedi cael mwy na'u siâr neithiwr, tra oeddent yn ymwroli ar gyfer yr awr fawr.

Bu'n noson reit ddistaw ar y cyfan. Cafodd Gwyneth a Tina gyfle i furmur drafod eu treialon gerbron Harri cyn i Alison ymddangos gyda'i chwaer, a rhannu ei hanes hithau. Roedd chwaer Alison, Linda, ar dân eisiau manylion profiadau Gwyneth a Tina, ond cyndyn oedd y ddwy i drafod eu dinoethi gyda rhywun dieithr, chwaer Alison neu beidio. Ac er nad oedd llawer o gwsmeriaid yn y tŷ, gwyddai Gwyneth y byddai hanes eu helyntion trwy'r pentref fel tân gwyllt, dim ond i un hen gant cegog glustfeinio arnynt.

Ar ddiwedd y nos, wedi i Linda gychwyn adref, ac i Gwyneth ryddhau Tina rhag yr ychydig waith clirio oedd yn weddill, cymerodd Gwyneth ac Alison ryw lwnc bach cyn i Alison hithau adael. Ac wrth i'r gwin leddfu'r pryderon fu'n hofran ym meddyliau'r ddwy, a'u llenwi â hyder heriol trwy wydrau rhosliw ôl-ddoethineb, ni theimlodd yr un ohonynt unrhyw chwithdod pan estynnodd Alison ar draws y bar a chyffwrdd wyneb Gwyneth â'i bys, cyn i Gwyneth, yn ei thro, estyn ei llaw hithau a gafael yn dynn yn llaw Alison.

* * *

Beth fyddai Robin yn ei ddweud? Beth fyddai pentrefwyr Nant-yr-Onnen yn ei ddweud? Beth fyddai ei mam yn ei ddweud?

Eisteddai Gwyneth mewn caffi yn Nhresarn yn mwytho mŷg o de, a'r bag dillad newydd a geisiodd ei roi o'r golwg dan y bwrdd yn rhwbio'i choes gyda phob symudiad a wnâi, fel cath yn chwilio am faldod.

Doedd hi ddim yn gall. Dyna'r unig eglurhad. Roedd blynyddoedd o stryffaglio ar ei phen ei hun wedi troi ei hymennydd yn brywes. Ac roedd teimladau dieithr, dryslyd yr wythnosau diwethaf wedi ei gwthio dros y dibyn i bydew breuddwydiol.

Ysgydwodd Gwyneth ei phen ac ochneidio nes creu tonnau mân yn yr hylif claear yn ei mŷg. Os nad oedd hi wedi drysu'n llwyr, yna pam yn y byd roedd hi newydd wario ffortiwn ar ddillad i'w gwisgo yn y lansiad ddiwedd yr wythnos, a hithau â llond wardrob o ddillad gartre oedd yn hen ddigon da?

Pam yn wir?

Cofiai Gwyneth brofi dyddiau fel hyn pan oedd hi'n hogan ysgol. Hi ei hun, a dwy neu dair o'i ffrindiau, y rhai oedd yn ddigon ffodus i gael dêt ar gyfer nos Sadwrn ('radeg hynny credent mai oed gydag un o dduwiau plorynnog y chweched uchaf oedd pinacl bywyd), yn cribinio'r dref trwy'r prynhawn gan chwilota am rywbeth i'w wisgo; chwilio a chwalu i geisio darganfod *y* dilledyn, y dilledyn fyddai'n eu llenwi â chyfaredd byrbwyll, y dilledyn y gallent ei daenu dros eu hunanhyder hyfriw.

A rŵan dyma hi, yn ymddwyn yn union yr un fath.

Dros Alison.

Ers colli Robin doedd neb wedi cyffwrdd calon Gwyneth, neb wedi codi chwant arni, neb wedi gwneud iddi deimlo mai hi oedd y person pwysicaf yn ei fywyd. Gwyddai na fyddai Robin am iddi suro fel rhyw hen brwnsen, neu'n waeth fyth, dynnu ar ôl ei mam. Ond hyn?

Oedd ei theimladau tuag at Alison yn annaturiol? Oeddynt o ddifrif? Ynteu wedi ei hudo am iddi gael ei chyffwrdd a'i choleddu oedd hi? Ac wedi iddi fod mewn diffeithwch emosiynol cyhyd, onid oedd perygl iddi wirioni ar unrhyw un a gynigiai gysur cofl iddi?

Rhoddodd Gwyneth ei mỳg ar y bwrdd. Roedd y te eisoes yn oer ac yn anghynnes fel dŵr pwll.

Na, fyddai hi ddim yn nôl paned arall. Fyddai'n elwa dim o eistedd fan hyn rhagor; roedd ei meddyliau fel y sgarff 'na a luniodd i Tedi druan flynyddoedd yn ôl, yn un gybolfa gymysglyd garfaglog.

Cododd ac ymlwybro tua'r car, a'r bag papur crand yn siffrwd yn gyhuddgar bob cam o'r ffordd.

". . . popeth yn iawn, Sara, peidiwch â phoeni."

"Ond fe ddylwn i fod wedi dod heibio i'ch helpu gyda'r paratoadau, Gwyneth fach. Fedra i ond diolch i chi am eich trafferth."

"Does dim isio i chi ddiolch i mi, wir. Fi ddylai ddiolch i chi, a phawb arall yn y pentra 'ma. Roeddwn i'n falch o'r cyfle i gael gwneud cyfraniad."

"Twt lol, Gwyneth. Os na fedrwn ni helpu'n gilydd ar ein ffordd trwy'r hen fyd 'ma, beth yw'r pwynt? Ond mi wn i'n iawn sut 'dach chi'n teimlo; mi fu'r pentre 'ma'n gefn i minna pan gaeodd yr ysgol a'r capel, a phan gollais fy nhad. Dyna pam roeddwn i'n teimlo'i bod hi'n ddyletswydd arna i i ymladd dros y neuadd."

"Roeddech chi yn llygad eich lle, Sara; weithiau mae'n rhaid gwneud safiad dros yr hyn sy'n bwysig i chi, ac os ydi pobl yn wfftio ar gownt hynny, wel . . ."

"Yn hollol, Gwyneth. Ac os bydd . . ."

Ymhen ychydig funudau dychwelodd Gwyneth y ffôn i'w grud. Cymerodd gipolwg o amgylch y lolfa i sicrhau bod popeth yn ei le, yna rhedodd i fyny'r grisiau, ac roedd ar ei ffordd i'r gawod cyn sylweddoli nad oedd ganddi'r un syniad sut olwg oedd ar bethau yn y lolfa wedi'r cyfan.

Ceisiodd ymbwyllo, a rhoi trefn ar ei meddyliau.

Roedd ei theimladau a'i hemosiynau'n chwyrlïo yn ei

phen fel dail yr hydref yn yr awel. Serch hynny, er gwaethaf ei dryswch a'i hamheuon, gwyddai Gwyneth ei bod yn hapus, yn hapusach nag y bu ers colli Robin.

Wrth newid i'r dillad newydd teimlai y gallai droedio'n gadarn tua'r dyfodol; bod ganddi nerth i wynebu pob anhawster a rhagfarn a ddeuai i'w rhan; bod ei henaid yn rhodio'r uchelfannau yn ysgafn droed.

Allai hi ddim llai na chyfaddef wrthi ei hun fod y ffaith ei bod yn wrthrych gwên gariadus Alison yn toddi ei thu mewn fel menyn melyn ar dost poeth; bod y ffordd ddidwyll y mynegai Alison ei thynerwch iddi yn dwyn dagrau i'w llygaid; bod caru, a chael ei charu, yn ddigon – yn ormod bron – wedi treulio cyfnod mor hir mewn diffeithwch dideimlad diderfyn, ble claddwyd ei hanfod hoffus dan amwisg galar a gwacter.

Wrth ddychwelyd i lawr y grisiau i groesawu'r genod, doedd dim amheuaeth ym meddwl Gwyneth nad yma oedd ei dyfodol; yma roedd hi'n ffitio'n glyd i batrwm bywyd; yma roedd ei lle.

# MEDI

. . . a theimlaf nad yw ond yn deg dy rybuddio fy mod yn bwriadau gwerthu'r tŷ pan fydd Marc yn gadael am y brifysgol, ac y byddaf hefyd yn cymryd camau cyfreithiol i ddwyn ein priodas i ben yn swyddogol.

Rwy'n siŵr na fydd hyn yn sioc fawr i ti, gan dy fod, fel finna, yn gwbl ymwybodol mai sioe fu'n perthynas ers tro byd. Yn wir, mae'n debyg dy fod yn edrych ymlaen at gael dy draed yn rhydd.

Am y tro bydd popeth yn parhau fel y mae - rwyf am i fywyd Marc fod mor gyfforddus a sefydlog â phosib am weddill ei gyfnod yn yr ysgol. Byddwn yn falch pe na baet yn sôn am gynnwys y llythyr hwn wrtho, ac yntau â'i arholiadau TGAU ar ei feddwl.

Mae'n wir ddrwg gen i mai fel hyn y trodd pethau allan.

Cofion,
Morris

Cododd yr ychydig lymeidiau o goffi roedd Caryl wedi eu hyfed i'w chorn gwddw, ac edrychodd Llinos, swyddog gweinyddol y swyddfa, yn syn arni pan gododd oddi wrth ei desg a rhuthro o'r stafell.

Er ei bod hi'n prysur dynnu am amser te, roedd yr haul yn dal i befrio yn yr awyr a Caryl yn gwywo yn y gwres. Aeth rownd cornel ei chartref i chwynnu rhagor ar y borderi blodau trefnus a thestlus a orweddai ger tair ochr o'r tŷ, gan ddiolch am glaearwch cysgod. Plygodd i chwynnu'r twffiau tila a fentrodd i ganol y rhengoedd disgybledig dihoenllyd, a hwythau'n pendrymu yn sgil y tes.

Trwy ffenestri agored Annedd Hedd gallai Caryl glywed

sŵn sgwrsio a smaldod, chwerthin a chyfeillach, ynghyd â chlindarddach llestri a llais llon rhyw gyflwynydd radio neu deledu'n parablu'n ddi-baid. Meddyliodd am yr hambwrdd roedd wedi ei osod yn barod iddi ei hun yn y gegin. Roedd arno liain gwyn wedi ei frodio'n gywrain, a les o'i amgylch; roedd arno lestri o'r tsieina gorau; roedd yno ddail te o'r siop arbenigol yn Nhresarn yn barod i'w rhoi yn y tebot; roedd yno gyffrwyth danteithiol mewn potyn pwrpasol, a llwy arian i'w godi; roedd yno deisen foethus yn barod i'w thorri. Ond a Marc yn aros gyda'i dad dros y penwythnos – ei seibiant olaf cyn ei arholiadau, mae'n debyg – ni fyddai yna neb i rannu mân glecs y dydd dros yr arlwy amheuthun. Gyda phwy arall y gallai hi rannu sgwrs dros baned? Er ei bod yn byw yn Nant-yr-Onnen ers bron i bymtheng mlynedd, doedd ganddi'r un ffrind mynwesol yno. Am funud gwelodd Caryl ei hun fel plentyn eto, yn darparu te bach i'w theganau. Ond roedd hi'n rhy hen i eistedd a smalio gyda Tedi a Doli bellach. Ac wrth bwyso'i llaw yn erbyn carreg oer y tŷ i'w sadio'i hun wrth symud ymlaen â'i thasg, mynnodd wadu drachefn y gwirionedd a lechai yn ei hisymwybod, sef mai muriau a adeiladodd hi ei hun oedd yn ei chaethiwo, ac yn cadw pawb arall draw.

Roedd Caryl Jenkins wedi ymgymryd â nifer o swyddi gweinyddol cyn mynd i weithio i gwmni Morris Harper. Fel un o griw o weinyddesau yn ei swyddfa yn y ddinas, roedd Caryl yn llawn parchedig ofn tuag at y dyn ei hun. Am gyfnod, beth bynnag. Er mai cwmni cymharol fychan ydoedd ar y pryd, roedd hi'n amlwg bod hwn yn ddyn â'i fys ar y pŷls, yn ddyn oedd am fynd i rywle, a gorau po gyntaf y cyrhaeddai.

Wedi cyfnod o sylwi ar Morris yn slei, ac edmygu ei chwaeth mewn dillad a'i graffter ym myd busnes, heb sôn am degwch ei bryd a gwedd, bachodd Caryl ar y cyfle pan ddaeth swydd fel cynorthwywraig bersonol y bòs yn wag. Ac unwaith yr oedd wedi ei hymsefydlu ei hun yn sancteiddrwydd ei swyddfa, doedd dim yn ormod o drafferth iddi. Yn wir, byddai bob amser yn fwy na pharod i wirfoddoli i aros yn y swyddfa ar ôl yr oriau swyddogol er mwyn helpu Morris i gryfhau ac ymestyn ei ymerodraeth.

Byddai Caryl yn gwneud ymdrech i wisgo'n ddeniadol ar gyfer ei gwaith, gan ddewis dillad a amlygai ei chorff siapus. Byddai'n gosod ei gwallt melyn yn fwndel ar ei phen, gan adael i ambell gudyn ddianc yn bryfoclyd, a byddai'n codi'n blygeiniol gan dreulio oriau'n creu effaith ffwrdd-â-hi â'i cholur. A chan ei bod mor barod i weithio oriau ychwanegol, a Morris a hithau'n byw ar eu pennau eu hunain mewn fflatiau digon dilewyrch, beth oedd yn fwy naturiol na'r ddau ohonynt yn rhannu ambell swper ar derfyn diwrnod diwyd? Ac er bod rhywbeth afiach amcanus ynglŷn â dull Caryl o fynd ar sodlau Morris, mi roedd hi'n rhyw lun o'i garu, mae'n debyg – dim cweit cymaint ag yr oedd hi'n caru ei arian a'i ddelwedd a'i safon, mae'n wir, ond yn ei garu ddigon (neu wedi'i pherswadio'i hun ei bod hi, beth bynnag) i dderbyn ar unwaith pan ofynnodd iddi ei briodi.

Roedd Caryl â'i bryd ar briodas gyda'r trimins i gyd, fel y rheini y rhoddir lle blaenllaw iddynt mewn cylchgronau sgandal a sglein. Ond nid oedd ganddi hi a Morris ddigon o deulu rhyngddynt i lenwi cwpwrdd crasu, heb sôn am eglwys neu gapel, felly trefnwyd gwyliau moethus ar ynys nefolaidd a fyddai'n cyfuno'r briodas a'r mis mêl, gan gadw pawb gartref yn hapus gyda dathliad ysblennydd ar ôl dychwelyd.

Symudodd Caryl i fyw i fflat Morris, ond cyn hir prynodd y ddau fflat fwy, mewn rhan fwy ffasiynol o'r ddinas. Perswadiodd Caryl ei gŵr y byddai'r ffaith bod y ddau ohonynt yn gweithio gyda'i gilydd yn rhoi gweddill y staff mewn lle cas, felly gwell fyddai iddi hi aros gartref a chanolbwyntio ar gydlynu'r gweithwyr oedd am droi'r fflat yn balas.

Ac yna, cyn i Caryl gael cyfle i eistedd yn ôl ac edmygu

canlyniadau ei chynllunio chwaethus, cyrhaeddodd Marc. Nid oedd Caryl wedi bwriadu cael plentyn am sbel wedi priodi (os o gwbl, a dweud y gwir). Rhyw bethau bach swnllyd, stomplyd oedd plant yn ei barn hi, heb unrhyw barch tuag at ei hangen beunosol am wyth awr di-dor o gwsg, na thuag at ei phapur wal drudfawr. Ond unwaith y ganwyd Marc, roedd Caryl a Morris wedi gwirioni'n lân arno. Roedd ganddo wallt euraid ei fam a llygaid glas ei dad, a thrwy lwc fe setlodd i gysgu'r nos yn ddidrafferth wedi ychydig wythnosau. Er i Caryl orfod gweithio'n galetach nag a wnaethai ers tro (roedd wedi perswadio Morris y byddai dynes lanhau yn hanfodol cyn i'r paent sglein sychu ar y sgertins newydd), roedd wrth ei bodd pan fyddai pobl yn eu stopio ar y stryd ac yn canmol y ceriwb bach yn ei bram.

Wrth weld ei fab yn dechrau bustachu o amgylch y fflat, penderfynodd Morris fod angen mwy ar y bychan na chael ei baredio trwy strydoedd swnllyd, llychlyd y ddinas. Roedd arno angen awyr iach i roi gwrid ar ei fochau; roedd arno angen blewyn glas i chwarae pêl; roedd arno angen cyfle i gymysgu â phlant bach eraill a dysgu dadlau dros deganau; roedd arno angen cynhaliaeth cymdeithas glòs, yn hytrach nag awyrgylch amhersonol, pawb drosto'i hun y ddinas; roedd arno angen gwreiddiau.

Nid oedd ar Caryl unrhyw awydd gadael y ddinas a chyfleusterau ei chartref moethus, ond am unwaith dangosodd Morris beth o'r penderfyniad unplyg a gludai ei gwmni o nerth i nerth. Trwy gysylltiadau busnes clywodd am y llain tir ar bwys capel Bethesda ym mhentref Nant-yr-Onnen, a thrwy ddirgel ffyrdd dynion busnes cefnog, roedd y tir wedi ei brynu a'r tŷ wedi ei adeiladu mewn da bryd i Marc gael sefyll ar flaenau ei draed a cheisio llywio'i feic tair olwyn newydd sbon dros y lawnt dilychwyn.

Ac wrth gwrs, roedd y syniad o fod yn feistres ar ddau gartref yn apelio at Caryl. Cafodd foddio'i holl fympwyon unwaith eto wrth gynllunio sut i addurno'r tŷ newydd. Ac fel yr âi Marc yn hŷn, fe gâi ddigon o gyfle i aros yn y fflat yn y ddinas a manteisio ar yr holl atyniadau yno.

Treuliai Morris bob penwythnos gyda nhw, a byddai'n dotio at ddatblygiad ei fab a'i ymateb brwd i ryfeddodau'r byd o'i gwmpas. Treuliai'r ddau oriau gyda'i gilydd, yn chwarae yn yr ardd neu'n cerdded y llwybrau o amgylch y pentref. A phan fyddai'r tywydd yn wael, byddent yn codi gwrychyn Caryl drwy orchuddio llawr y lolfa â thai a chledrau a threnau bach, cyn symud ymlaen i adeiladu wythfed rhyfeddod o Lego ar fwrdd y stafell fwyta.

Ond yn naturiol, wrth i Marc dyfu dechreuodd dreulio mwy a mwy o'i amser rhydd gyda'i ffrindiau ysgol. A dechreuodd Morris dreulio mwy a mwy o'i benwythnosau yn y fflat yn y ddinas. Roedd y sefyllfa'n siwtio Caryl i'r dim, ac yn ei rhyddhau i ffidlan â'i ffigiarîns mewn unigedd ysblennydd, a neb dan draed i hel llwch a chreu llanast.

Pan gychwynnodd Marc ar ei yrfa yn Ysgol Uwchradd Tresarn derbyniodd Caryl swydd fel cydlynydd gwasanaeth pryd ar glud yr ardal, gan ei bod bellach wedi polisio holl ddodrefn a thrugareddau'r tŷ hyd drwch blewyn o'u tranc. Adlewyrchai'r swydd hon y ddelwedd oedd ganddi ohoni ei hun. Ystyriai ei hun yn gymwynas-wraig i'r gymdeithas leol, ac ymunodd â changen Merched y Wawr y pentref gan gredu y byddai ei phresenoldeb yn ychwanegu rhyw gymaint o goethni at weithgareddau'r criw yno (rheswm llawer mwy dilys ym meddwl hunandwyllodrus Caryl na chyfaddef mai cwmni tila oedd ornaments a thlysau).

Methu, neu wrthod gweld oedd Caryl, mai byw mewn tŵr ifori ydoedd, a'r gagendor rhyngddi hi a Morris wedi tyfu'n ddibyn diwaelod. Bellach byddai ei gŵr yn ymfalchïo yng nghwmni ysbeidiol ei fab rhwng cyfnodau o ymgolli yn ei waith gyda difaterwch di-droi'n-ôl, tra treuliai Caryl ei dyddiau yn y pentref yn tendiad ei safle a'i delwedd, heb feddwl am eiliad fod yna berygl i'w chadarnle ddymchwel o'i hamgylch.

* * *

Nonsens! Nonsens llwyr!

Roedd Caryl ar ei ffordd i'r ganolfan gymdeithasol yn Nhresarn y bore wedi'r cyfarfod tyngedfennol. Yn y ganolfan y darperid prydau pryd ar glud yr ardal, ac yno hefyd roedd ei swyddfa.

Beth oedd wedi dros ben Sara? Yn disgwyl iddynt wneud *exhibition* ohonyn nhw eu hunain fel'na. Ac roedd neuadd y pentref mewn cyflwr ofnadwy, beth bynnag. Llawn gwell iddi gael ei dymchwel, ac i'r Cyngor ddatblygu'r safle. Wedi'r cyfan, roedd y gwaith addasu a wnaeth Albert Parry yng nghapel Bethesda yn chwaethus dros ben, a'r tai roedd wrthi'n eu hadeiladau rŵan, heb fod nepell o'i dŷ ei hun ar safle'r ysgol, yn ddigon del, er mai tai bychain oeddynt. Ond ychydig iawn o dai uwchraddol oedd yna yn y pentref. A chan mai dolur llygad oedd y neuadd erbyn hyn, byddai'n well o lawer ei chwalu ac adeiladu stad fechan o *executive homes* yno, i ddenu'r math iawn o bobl i'r ardal. Piti bod y safle mor agos i'r hen dai cyngor ble roedd Phyllis a Brenda a Tina'n byw. Ond wedi plannu tipyn o goed a llwyni . . .

Wrth gwrs, roedd Caryl wedi codi ei braich adeg y bleidlais y noson cynt, rhag i'r gweddill gael cyfle i'w chyhuddo o fod yn hen Jeremeia. Ond credai y byddent i gyd wedi callio erbyn y bore, ar ôl iddynt sylweddoli cynllun mor ddiraddiol oedd o.

Dal i dwt-twtian wrthi ei hun oedd Caryl wrth iddi barcio'r car a cherdded tuag at ei swyddfa. Yn y cyntedd yn disgwyl amdani roedd Angela, un o hoelion wyth y criw o wirfoddolwyr a ddanfonai'r prydau parod i'r cwsmeriaid yn feunyddiol.

"Bore da, Mrs Harper. Mae'n ddrwg gen i'ch poeni chi, ond mae rhywun wedi trio torri mewn i un o'r fania dros nos. Does dim byd wedi'i gymryd, hyd y gwela i, ond mae un o'r ffenestri wedi'i malu, a sgriffiada ar y drysa. Mae'n rhaid bod pwy bynnag fu wrthi wedi cael ei styrbio cyn mynd ati i wneud rhagor o ddifrod."

"Ydach chi wedi cysylltu â'r heddlu, Angela?"

"Naddo, ddim eto, Mrs Harper. Newydd ddarganfod y difrod oeddwn i, a meddwl y byddai'n well rhoi gwybod i chi yn gynta."

Ochneidiodd Caryl yn ddramatig. "Gadewch o efo fi, Angela. Mi gysyllta i â'r heddlu. Yna mi ddo i draw i'r gegin i weld sut mae petha yno bore 'ma."

Beth fyddai'n digwydd i'r gwasanaeth hebddi hi, meddyliodd Caryl. Ysgydwodd ei phen wrth dip-tapian tua'r gegin yn ei siwt ddrudfawr a'i sgidiau sodlau main, wedi iddi siarsio'r plismon ifanc a atebodd y ffôn ei bod yn disgwyl swyddog gydag awdurdod i alw arni *ar fyrder*.

Erbyn amser paned roedd Caryl wedi colli'r ddadl â Meri, a oedd yn teyrnasu yn y gegin ac wedi dyfarnu yn erbyn dewis y dydd ar fwydlen Caryl, sef *spaghetti Bolognese,* gan farnu y byddai pensiynwyr yr ardal yn hapusach gyda phlatiaid o bastai bugail. Yn ogystal, roedd Caryl wedi gorfod siarad yn reit siarp â'r sarjant a yrrwyd i ymchwilio i'r difrod i'r fan. Roedd hi'n amlwg bod hwnnw wedi penderfynu ymlaen llaw nad oedd gan yr heddlu lleol na'r amser na'r adnoddau i ddelio â rhyw fân droseddau felly, a hyd yn oed petaent yn rhoi pob heddwas yn y sir i weithio ar yr achos, byddai ganddynt gystal siawns o ddal y troseddwyr ag oedd gan hen feic peniffardding fy nhaid o ennill y Tour de France.

Wrth dderbyn ei choffi gan Llinos yn ddiolchgar, cofiodd Caryl ei bod wedi taro'r post yn ei bag wrth adael y tŷ. Bodiodd trwy'r dyrnaid amlenni nes gweld llawysgrifen Morris ar un ohonynt. Lledodd ei llygaid llwyd. Ffonio byddai Morris yn ei wneud fel arfer os oedd am gysylltu â hi neu Marc.

Gwthiodd Caryl ewin hir gwaetgoch ei bys bach dan fflap yr amlen i'w hagor.

Petai cath neu ddraenog, twrch neu fochyn daear, gwdihŵ neu ystlum, wedi bod mor haerllug ag ymweld â gardd gymen gysáct Caryl y noson honno, yna efallai, tua dau o'r gloch y bore, byddai wedi sylwi ar rimyn gwantan o olau rhwng dellt bleind ffenest y gegin, a chlywed sŵn ochneidio dolefus yn dianc o'r stafell.

Yn y man cododd Caryl ei phen oddi ar ei breichiau ac anadlu'n ddwfn. Roedd llewys ei gŵn gwisgo sidan yn soeglyd gan ddagrau. Estynnodd hances boced frodiog gwbl annigonol o'i phoced a cheisio chwythu ei thrwyn heb wneud sŵn fel corn gwlad a deffro Marc. Doedd y neisied ddim yn ddigonol i'r gwaith a chododd Caryl i nôl y rholyn papur cegin a'r patrwm blodau mân arno oedd yn cyd-fynd â'r llestri – byddai'n rhaid iddo wneud y tro. Cydiodd yn y gwpan de a safai'n barchus ar ei soser yng nghanol y bwrdd – roedd bron yn oer bellach. Tywalltodd Caryl yr hylif claear i lawr y sinc a berwi'r tecell drachefn.

Agorodd y cwpwrdd uwch ei phen i estyn cwpan a soser glân. Llithrodd llewys llaith ei gŵn gwisgo lawr ei breichiau noeth. Plygodd Caryl y llewys nes bod y defnydd drud yn un rholyn am ei phenelin ac estyn am y llestri glân drachefn. Stopiodd a'i llaw yn yr awyr am ennyd, yna cydiodd yn un o'r mygiau di-chwaeth y byddai'n dweud y drefn wrth Marc am eu defnyddio gyda'i ffrindiau. Sodrodd y mỳg ger y tecell a mynd drwodd i'r stafell fyw i nôl y botel wisgi o'i chwpwrdd crand pwrpasol.

Yn ôl yn y gegin gwthiodd y cistiau te addurniedig a'r pecynnau coffi naill ochr er mwyn cael gafael ar y jar Nescafé, a gwnaeth baned o goffi parod iddi ei hun. Ychwanegodd joch o wisgi iddo a dychwelyd at y bwrdd. Am ennyd, syllodd Caryl a Homer Simpson yn hurt ar ei gilydd cyn iddi gymryd cegiad sydyn o'r coffi. Llifodd ias trwy ei chorff wrth iddi sawru'r hylif chwerw.

Ar y bwrdd o'i blaen roedd y papur lleol wedi ei daenu ar hyd ac ar led, gyda'r tudalennau a arddangosai dai ar werth a chyhoeddi hysbysebion am swyddi, yn amlwg. Roedd taranfollt Morris wedi chwalu cocŵn cyfforddus, cysgodol Caryl yn chwilfriw, ac i rywun nad oedd wedi sbecian y tu draw i wahanfuriau breintiedig digonedd ers blynyddoedd maith, ymddangosai'r byd yn lle blin a bygythiol.

Sut na welodd hi hyn yn dod? Ynteu dewis dallineb a wnaeth? Ble fyddai'n byw? Beth fyddai'n ei wneud? Beth fyddai pobl yn ei feddwl? Beth fyddai'n ei ddweud wrthynt?

Cymerodd Caryl lwnc arall o'r coffi adfywiol. Ac un arall. Ac un arall.

Gorffennodd y mygiad i gyd. Ac wedi diwrnod heb fwyta prin damaid, dim ond rhyw gogio swpera gyda Marc, doedd ei phen fawr o dro'n dechrau troi dan ddylanwad y ddiod ddieithr. Ond doedd hynny'n ddim o'i gymharu â chwyrlïo gwallgof ei meddyliau gydol y dydd a'r nos.

Wrth ymlwybro'n ôl i'w gwely ar flaenau ei thraed, cynllun Sara oedd y peth olaf ar feddwl Caryl. Ac wrth gau drws ei llofft ddi-fai, ddienaid, roedd hi'n llawn sylweddoli bod muriau'r plas perffaith ar fin cael eu dymchwel. Bellach doedd ei chadarnle chwaethus yn ddim ond tŷ ar y tywod.

". . . ac mae dy fam a finna'n gwybod dy fod wedi gwneud dy ora – does dim isio i ti boeni."

"Tydw i ddim yn poeni am yr arholiada, Dad – wel, ddim mwy na neb arall. Mi aethon nhw'n iawn. Wir yr."

"Ond mae 'na rywbeth ar dy feddwl di, Marc. Rwyt ti wedi bod yn dawedog iawn ers i ti gyrraedd."

"M-m-m . . . wel . . . a deud y gwir . . . dwi'n poeni am Mam."

"Dy fam? Pam? Be mae hi wedi'i ddeud wrthat ti?"

"Tydi hi ddim wedi deud dim byd, ond dwi'n siŵr ei bod hi'n gofidio am rywbeth. Mi fydda i'n ei dal hi weithia'n syllu i nunlla, a golwg . . . wn i ddim . . . golwg unig ofnadwy arni."

"Tydi dy fam 'rioed wedi bod yn un am gymysgu ryw lawar – mae hi reit hapus yn potsian yn y tŷ gan herio pob sbecyn o lwch a phob awgrym o anhrefn."

"Ond dyna beth arall. Mae hi fel petai wedi colli diddordeb yn yr holl drugaredda 'na sy ganddi. Mae'r ddynes llnau'n dod, ond tydi Mam ddim yn cyboli i fynd rownd ar ei hôl hi'n symud hwn a pholisio'r llall . . . a mae hi wedi dechra yfed coffi parod a gwneud te tramp . . . mewn mŷg."

"Na, tydi hynny ddim yn swnio fel dy fam o gwbl. Fyddai hi byth yn . . ."

"*Ac* mae hi wedi cytuno i gymryd rhan yng nghynllun hollol wallgo Sara Watcyn i godi pres i achub neuadd y pentre."

"Cynllun Sara Watcyn?"

"Ia, maen nhw – Merched y Wawr Nant-yr-Onnen – yn mynd i wneud *topless calendar* i godi pres a . . ."

"Dy fam? Mewn *topless calendar*? Blydi hel!"

Roedd ar Caryl eisiau eistedd i lawr yn y fan a'r lle, yng nghanol y maes parcio, a chrio.

Roedd wedi methu'n lân â dofi'r troli oriog; roedd y sbrigyn yn y swyddfa wedi gwenu'n ffals arni wrth iddo adrodd, "Sori, Mrs Harper, wneith o ddim digwydd eto, Mrs Harper," pan fygythiodd hi roi busnes y gwasanaeth pryd ar glud yn nwylo cwmni arall; roedd hi wedi bod yn ddigon piwis gyda Marian; roedd hi'n bwrw glaw mân erbyn iddi ddod allan o'r siop, a rŵan byddai ei gwallt fel nyth brân erbyn iddi gyrraedd y ganolfan; ac roedd yn cael trafferth garw codi'r tuniau a'r pecynnau trwm i gist y car yn ei sgert dynn a'i sodlau pigfain. Cododd y pecyn olaf i'r bŵt a'i wthio'n ddi-hid i ganol y lleill. Sythodd, a sefyll am ennyd i gael ei gwynt ati cyn cau'r gist yn glep.

Ymhen munud neu ddau roedd Caryl yn anelu am Dresarn, a'r troli anystywallt wedi ei adael yn amddifad yng nghanol maes parcio'r *cash & carry* fel arwydd o'i dig. Trawodd swits y radio gan feddwl y byddai tonau tyner Classic FM yn tawelu'r tyndra a lenwai ei chorff.

*"Switch on to Atlantic 252 to hear tomorrow's hits today,"* meddai llais gorfrwdfrydig rhyw ynfytyn arwynebol, cyn i'r car gael ei lenwi â chybolfa o seiniau ansoniarus oedd yn ddigon i ferwino clustiau'r person mwyaf hawdd ei blesio.

Diffoddodd Caryl y radio ar unwaith.

Marc eto. Roedd wedi gofyn ganwaith iddo aildiwnio'r radio i'w hoff orsaf hi ar ôl cael pàs i rywle. Yn enwedig gan ei bod hi'n cymryd pum munud da o ffidlan iddi ddod o hyd iddi ei hun.

Ysgydwodd Caryl ei phen.

Roedd y glaw'n gwaethygu hefyd. Ond roedd rhywbeth cysurlon yn su rhythmig y weipars, ac wrth i'r dafnau glaw gael eu clirio i'r naill ochr a'r llall, gresynai Caryl na allai ymdrin â'i phroblemau yn yr un modd. Slaes ffordd hyn, slaes ffordd acw, a'i holl ddryswch wedi ei sgubo o'r neilltu.

Ceisiai gadw wyneb er mwyn Marc, ond doedd ei chalon ddim ynddi: ddim yn ei gwaith, ddim mewn gofalu am y tŷ, ddim mewn dim byd, a gwyddai fod yr ymdrech i ymddangos fel petai popeth yn iawn yng ngŵydd Marc yn ei gwneud yn fyr ei thymer â phawb y deuai i gysylltiad â nhw. Ni allai rannu ei gofidiau â neb. Wedi'r cyfan, roedd wedi mynnu cadw rhyw bellter rhyngddi a gweddill y pentref ers iddi ddod yno i fyw; wedi hawlio ei lle ris neu ddwy'n uwch na phawb arall ar yr ysgol gymdeithasol, ac wedi cymryd arni ei bod yn gwneud ffafr â gweddill y merched trwy fynychu cyfarfodydd Merched y Wawr.

A dyna sut y daeth cynllun melltigedig Sara i hofran uwch ei phen fel angel angau, tra oedd gweddill ei bywyd yn datod yn raflins o flaen ei llygaid. Doedd yno neb i glytio'r darnau, neb a allai chwipio nodwydd hud ffordd hyn a ffordd acw a gwneud y cwbl yn gyfan drachefn.

Trodd Caryl y car i mewn i faes parcio'r ganolfan. Yna cerddodd draw at yr adeilad a rhoi ei thrwyn heibio drws y gegin. Roedd y merched wrthi'n coginio, yn laddar o chwys; byddai'n rhaid iddi gario'r pecynnau i mewn ei hun. O leiaf doedd hi ddim yn bwrw yn Nhresarn, meddyliodd, wrth agor cist y car ac estyn y pecyn cyntaf yn ddiamynedd. Teimlodd un o'i hewinedd gwritgoch yn plygu ac yn torri gyda'r ymdrech. Dihangodd dau ddeigryn o'i llygaid a rhedeg yn araf i lawr ei gruddiau.

Roedd Caryl yn cael trafferth codi. Wedi treulio cymaint o amser ar ei phengliniau'n lapio trugareddau mewn papur newydd a'u gosod mewn bocsys, roedd ei chymalau wedi cloi'n llwyr. Trwy bwyso ar y gadair agosaf llwyddodd i stryffaglio ar ei thraed a sythu ei choesau.

Paned. Roedd hi siŵr o fod yn amser paned a saib fechan. Ymlwybrodd Caryl tua'r gegin fel petai ganddi ddwy goes bren. Wrth basio'r seidbord stopiodd yn stond pan welodd y ddrychiolaeth a syllai'n ôl arni o'r drych uwchben y seld: dynes â'i gwallt melyn wedi dianc blith draphlith o'i *chignon* – a llygaid fel panda.

Gwenodd yn chwerw ar ei hadlewyrchiad. Wedi ymgolli yn y dasg o bacio'i chymdeithion ffals, roedd Caryl wedi llwyr anghofio am y môr o ddagrau a wylodd ar ôl dychwelyd i'w hysblander ynysig wedi'r profiad arteithiol. Pwysodd yn erbyn cwpwrdd y gegin tra oedd yn disgwyl i'r tecell ferwi, yn dal i gamu o'r naill droed i'r llall i ystwytho cyffni ei choesau.

Ni fu'r ordîl ei hun cynddrwg ag y disgwyliasai, er ei bod yn teimlo'n hurt ac yn hyfriw yn stelcian y tu ôl i bentwr o blatiau ar un o'r trolïau bwyd yng nghegin wag y ganolfan. Ond roedd wedi penderfynu bod Harri'n werth y byd. Roedd yn broffesiynol ac yn fonheddig, a chyflawnwyd y cyfan yn gyflym ac yn ddidrafferth. Mae'n debyg y byddai rhyw hen gant o gwsmer yn gwneud sylwadau am ddwmplenni pan welai'r calendar, a phwy a ŵyr beth fyddai ymateb cyfarwyddwr y gwasanaethau cymdeithasol. Ond hen ddyn bach coman oedd hwnnw.

Sylweddolodd Caryl yn sydyn fod y tecell yn berwi'n grychias y tu ôl iddi, a rhoddodd yr hyn a ystyriai bellach yn ddŵr bywyd yn ei mŷg – llwyaid o goffi parod a joch o wisgi.

Na, nid y tynnu lluniau ei hun oedd wedi ypsetio cymaint arni, ond dod yn ôl i dŷ gwag. Byddai wedi bod yn rhyddhad cael rhannu ei phrofiad gyda rhai o'r merched eraill, ond ni wyddai ble na sut i ddechrau croesi'r gagendor yr oedd hi ei hun wedi ei greu. Roedd wedi ymuno â gweddill y criw yn y Llew Coch neithiwr, ac wedi mwynhau'r noson a'r teimlad anghyfarwydd o fod mewn cwmni hwyliog. Ond synhwyrai fod y merched yn gweithio nerth deg ewin i'w chynnwys yn yr hwyl, yn ymbalfalu am dir comin wrth greu sgwrs. Roedd yn ysu am gael dweud wrthynt am lythyr Morris, ac egluro nad gwynfyd gwastadol oedd ei bywyd er gwaethaf ei holl rodres, a'i bod yn crefu am eu cyfeillgarwch a'u cyngor.

Doedd Marc ddim gartre chwaith; roedd o wedi mynd i aros noson gyda ffrind. Gwyddai Caryl fod y ffaith ei bod yn cymryd rhan yng nghynllun Sara wedi creu embaras enfawr iddo ac yntau rhwng dau oed. Ond byddai'n ymdopi. Byddai'n rhaid iddo. Fel y byddai'n rhaid iddi hithau ymdopi â'r holl newidiadau oedd ar y gorwel.

Dychwelodd i'r stafell fyw gyda'i phaned. Edrychodd ar y papurau newydd a'r bocsys yn un llanast ar y llawr. Roedd wedi gwagio hanner ei chwpwrdd gwydr ond roedd ugeiniau o ffigiarîns i'w pacio eto. Heb sôn am bopeth arall. Ond ni fyddai'n meddwl am hynny heno, neu byddai'r cwbl yn ymddangos fel copa gwbl anorchfygol, a phetai hynny'n digwydd, beth a ddeuai o'i phenderfyniad newydd sgleiniog?

Doedd Caryl ddim am fyw celwydd mwyach, ac yn bendant ddim am ddwy flynedd arall nes bod Marc yn gadael yr ysgol. Wedi wythnosau o boeni, o bendroni, o bryderu yn feunosol a thrwy gydol yr oriau mân, roedd wedi cydnabod hynny heddiw. Yno, o flaen Harri, wrth gyflawni'r weithred ddirdynnol, roedd wedi gwawrio ar Caryl mai dim ond ei dwylo hi ei hun a allai gyweirio ei dyfodol, nid dwylo Morris na neb arall. Doedd hi'n gweld dim bai ar ei gŵr am geisio sicrhau bywyd sefydlog i Marc, ond doedd hi ddim yn deg disgwyl iddi hi fodoli mewn rhyw lun ar ebargofiant emosiynol am fisoedd. Er, erbyn meddwl, efallai mai dyna sut y bu hi i Morris ei hun ers tro, gyda dim ond gwely a gwaith i'w gysuro rhwng ymweliadau Marc. Ond bellach roedd hi'n rhy hwyr, yn rhy hwyr iddi hi a Morris. Doedd dim dewis ond casglu'r tameidiau a cheisio'u hasio at ei gilydd a dechrau drachefn.

Roedd Marc wedi profi llwyddiant yn ei arholiadau ac ar fin cofrestru yn chweched dosbarth Ysgol Uwchradd Tresarn. Fyddai byw mewn tŷ llai'n effeithio nemor ddim arno. Neu efallai y byddai'n manteisio ar y cyfle i fynychu coleg chweched dosbarth yn y ddinas a byw gyda'i dad am gyfnod; wedi'r cyfan, ychydig iawn oedd gan Nant-yr-Onnen i'w gynnig i bobl ifanc.

Wrth yfed gweddill y coffi wisgi llanwyd corff Caryl â gwres adfywiol, anghyffredin. Cydiodd yn un o'r ornaments oddi ar y seidbord a'i daflu at wal bella'r stafell fyw a'i wylio'n malu'n yfflon. Gwnaeth yr un peth ag un arall a syllu ar hwnnw'n syrthio'n deilchion. Wrth bledu'r trugareddau at y pared fesul un, meddiannwyd corff Caryl gan bwl o chwerthin afreolus, ac wrthi iddi ildio i'r ysfa a chwerthin llond ei bol, gallai deimlo'i chorff yn diosg tensiynau blynyddoedd.

". . . a meddwl y byddwn i'n rhoi gwybod i ti eu bod nhw'n dod i osod yr arwydd bore fory."

"Yli, Caryl, does 'na ddim cymaint â hynny o frys – tydw i ddim am dy daflu allan ar y palmant na dim felly. Cymer hynny o amser fynni di."

"Diolch yn fawr i ti, Morris, ond gora po gynta y cawn ni drefn ar betha. A rŵan 'mod i'n fodlon fy meddwl ynglŷn â Marc, does dim diben loetran yn y tŷ mawr 'ma ar fy mhen fy hun. Mae o wedi setlo, tydi?"

"Ydi . . . ydi . . . mae o wrth ei fodd yn y coleg, ac wedi datblygu bywyd cymdeithasol llawer mwy cyffrous na f'un i dros nos . . . ac mae'r ddau ohonon ni'n addasu i fyw bywyd bob dydd gyda'n gilydd, yn hytrach na chyfnodau gwyliau sy'n meithrin rhyw atgofion afreal."

"Dwi'n falch . . ."

"Ond be wnei di? Aros yn Nant-yr-Onnen? Ynteu symud i Dresarn neu rywle arall?"

"Aros yma, dwi'n meddwl. Mae Albert Parry, y dyn a addasodd gapel Bethesda, wyddost ti, wrthi'n adeiladu rhes o dai bychain ar gyrion safle'r hen ysgol. Byddai un o'r rheini'n ddelfrydol i mi."

"Wel, os mai dyna ti'n feddwl fyddai ora . . ."

"Gawn ni weld sut fydd petha'n mynd. Yli, dwed wrth Marc y ffonia i eto . . . a 'mod i'n falch ei fod o'n hapus . . . ac y byddai'n syniad da iddo wneud rhyw fymryn o waith coleg rŵan ac yn y man. Rhaid i mi fynd; dwi'n cyfarfod y genod yn y Llew Coch cyn bo hir."

"Cyfarfod y . . . ? O . . . wel, gobeithio y cei di noson werth . . . Y calendar! Heno 'dach chi'n lansio'r calendar!"

"Ia . . . naci . . . dim ond rhyw lansiad bach answyddogol rhyngon ni'n hunain sy 'na heno. Nos fory bydd y cyhoedd yn cael ein gweld yn ein holl ogoniant."

"Choelia i fyth – Caryl Harper – pryd ar glud yn y cnawd!"

"Paid â rwdlian, Morris."

"Na, Caryl, wir rŵan. Dwi'n edmygu dy safiad di. Dwi'n gwybod bod Marc wedi bod yn ddigon chwithig ynglŷn â'r fenter ar y dechrau, ond mae'r ddau ohonom yn falch ohonot ti, ohonoch chi i gyd. Mwynhewch eich noson."

"Diolch, Morris. Mi ro i wybod i ti am unrhyw ddatblygiada. Hwyl am y tro."

"Hwyl i ti . . . a Caryl . . ."

"Ia?"

"Pob lwc."

Gosododd Caryl y ffôn yn ôl yn ei grud. Gwenodd ar ei hadlewyrchiad yn nrych y cyntedd. Erbyn amser cinio drannoeth, dywedodd wrth y wraig drwsiadus a syllai'n ôl arni, fe fyddai'r pentref i gyd yn gwybod bod y tŷ ar werth. Byddai clywed y postyn yn cael ei forthwylio i'r ddaear fel hoelio caead arch ei gorffennol. Ond heno . . . roedd heno'n gyfle i geisio cyfamod, i geisio cyfeillgarwch, i geisio cyfeiriad – cyfle, yn wir, i osod sylfeini o'r newydd.

# HYDREF

Roedd Tina wrthi'n didoli'r dillad glân yn bentyrrau taclus ar ôl eu smwddio. Ceisiai anwybyddu'r raflins ar waelod breichiau siwmperi ysgol Iwan; byddai'n rhaid iddynt wneud y tro tan wyliau'r haf, raflins neu beidio. Byddent yn rhy fach ar gyfer y tymor newydd, beth bynnag, yna byddai Iwan yn cael siwmperi Gareth, a Gareth yn cael rhai Hywel. Petai ganddi ychydig mwy o arian . . . ond wedyn, petai'r Wyddfa'n gaws . . .

Toedd hi'n beth rhyfedd, meddyliodd Tina, pan fyddai rhieni plant bychain yn clywed rhywun yn dweud bod llawer mwy o drafferth gyda phlant wrth iddynt fynd yn hŷn, roeddynt yn gyndyn iawn, yng nghanol eu dryswch di-gwsg, o gredu'r ddoethineb honno. Ond roedd o'n berffaith wir; roedd llawer mwy o bethau i boeni yn eu cylch wrth i'r hogiau gyrraedd eu harddegau.

Roedd Tina'n poeni am arholiadau TGAU Hywel yn un peth, ac yn poeni llawer mwy na Hywel ei hun yn ôl pob golwg. Roedd ganddo ddigon rhwng ei glustiau, ond ym marn Tina, agwedd hynod ddi-hid oedd ganddo tuag at ei waith ysgol, a hithau'n gobeithio y byddai ei mab hynaf am ddychwelyd i'r chweched dosbarth, er gwaethaf y straen ariannol. Ond roedd am weld y tri yn gwneud yn dda ac yn manteisio ar bob cyfle.

Cael dau ben llinyn ynghyd – dyna'r broblem fwyaf. Roedd ceisio dal ei phen uwchben y dŵr tra oedd yn gwneud dwy swydd yn ogystal â holl waith y cartref yn mynd yn drech na Tina ar brydiau, serch ymdrechion ysbeidiol yr hogiau – yn enwedig Hywel – i'w helpu. Ac am Robart . . .

Wrth gychwyn am y grisiau gyda'r dillad, stopiodd Tina yn y cyntedd y tu allan i ddrws y stafell fyw a ffromi. Trwy'r drws gallai glywed sŵn offrwm prynhawnol y teledu ar gyfer y rhai oedd un ai'n gaeth i'w cartrefi, yn benwan, neu'n dioddef o syrffed angheuol. Llwyddodd i wrthsefyll yr ysfa i gicio'r drws ar agor a thaflu'r teledu trwy'r ffenest i'r ardd ffrynt; yn hytrach, parhaodd i fyny'r grisiau i ddanfon y dillad glân i stafelloedd yr hogiau. Roedd stafell Iwan a Gareth yn edrych fel petai rhywun wedi ei chodi, ei throi â'i phen i waered, ei hysgwyd yn reit dda, ac yna'i gosod yn ôl yn ei lle. Roedd Tina wedi hen laru swnian arnynt i gadw'u stafell yn daclus, a byddai'n disgwyl nes y byddai'r llanast wedi mynd ar ei nerfau'n llwyr cyn mynd ati i glirio a glanhau'r lle ei hun. Twt-twtiodd dan ei gwynt wrth godi rhai o'r dillad budron a'r trugareddau niferus oddi ar y llawr cyn cadw'r dillad glân yn y drorsys. Ymhen dim roedd wedi casglu llond peiriant a mwy o olch, a hithau'n meddwl ei bod wedi gorffen y gwaith hwnnw am y diwrnod.

Wedi dychwelyd lawr y grisiau â llond ei hafflau, taflodd Tina gipolwg ar y drych yn y cyntedd wrth fynd am y gegin drachefn. Ffromodd wyneb piwis yn ôl arni, wyneb â thrwyn smwt a llygaid gwyrddlas, wedi ei fframio gan wallt brown golau ac aroleuadau cartref ynddo.

Mi fyddi di'n ddeugain cyn diwedd y flwyddyn, meddai'r wraig yn y drych wrth Tina'n reit stowt, a thincial di-ddim y teledu'n diferu trwy ddrws y stafell fyw. Mae blynyddoedd gorau dy fywyd yn gwibio heibio dan dy drwyn di, a beth wyt ti'n bwriadu ei wneud ynglŷn â'r peth?

Caeodd Tina'r drws ffrynt y tu ôl iddi a sefyll yn stond yn y cyntedd.

Distawrwydd.

Moelodd ei chlustiau ger drws y stafell fyw.

Tawelwch.

"Robart?"

Dim smic.

Diolch byth, meddyliodd Tina, o leiaf roedd ei gŵr wedi'i styrio'i hun ddigon i fynd i seinio 'mlaen. Roedd hyd yn oed yr ymdrech honno wedi profi'n drech na fo sawl gwaith yn ddiweddar. Estynnodd ei llaw am ddolen drws y stafell. Dyma'i chyfle i glirio tipyn ar y papurau newydd a'r cwpanau budron y byddai Robart yn eu gadael yno; ac i agor y llenni a gadael golau dydd i mewn; ac i daflu'r ffenestri ar agor led y pen er mwyn i'r awyr iach gael cyfle i buro rhywfaint ar yr aer drwg.

Safodd Tina ger y drws am ennyd ac yna gollyngodd y ddolen. Na, roedd hi wedi cael bore prysur yn yr ysgol feithrin, roedd ganddi lond gwlad o waith i'w wneud cyn i'r hogiau gyrraedd adre o'r ysgol, a shifft yn y Llew Coch yn ei hwynebu gyda'r nos. Os oedd Robart yn fodlon treulio'i ddyddiau'n pydru byw mewn stafell a oedd yn cau fyny o'i amgylch, rhyngddo fo a'i fusnes.

Ymhen ychydig funudau roedd Tina'n eistedd wrth fwrdd y gegin gyda phaned a brechdan caws a thomato. Daliodd ei bawd i fyny a syllu'n fyfyrgar ar y plaster yr oedd newydd ei osod arno – canlyniad ymdrech lew y tomato i geisio'i ryddid ar y funud olaf. Plaster bach diniwed yr olwg fel hwn oedd man cychwyn ei pherthynas â Robart, meddyliodd. Man cychwyn perthynas ddelfrydol, ddilychwyn; nosweithiau rhamantus, rhyfeddol; carwriaeth fel un mewn stori tylwyth teg.

Merch annibynnol oedd Tina. Wedi gadael yr ysgol, sicrhaodd swydd mewn fferyllfa o fewn ychydig wythnosau. Roedd yn mwynhau'r gwaith, ac roedd ganddi law at drin pobl. Ymhen cwta ddwy flynedd, yn sgil ei pharodrwydd i ddysgu popeth a oedd i'w ddysgu ynglŷn â'r swydd, ac ymddeoliad y rheolwr, roedd wedi ei phenodi'n rheolwraig y siop. Dathlodd Tina'r dyrchafiad hwn, a'r codiad yn ei chyflog, trwy symud o'r tŷ roedd yn ei rannu â thair o ferched ifanc eraill, a rhentu ei fflat ei hun. Bu wrthi am fisoedd yn casglu amrywiol drugareddau, ac yn eu gosod a'u hailosod nes ei bod yn gwbl fodlon ar ei chartref newydd.

Gyda'i llygaid gwyrddlas, ei thrwyn smwt, a'i gwallt brown golau, doedd Tina byth yn brin o ddarpar gariadon. Ond er iddi ymgymryd â sawl carwriaeth lugoer, teimlai bob tro mai tarfu ar ddedwyddwch ei chartref twt a theidi y byddai rhyw labwst o lanc pan ddeuai i'r fflat i'w chyrchu i'r pictwrs neu i'r dafarn, megis tarw du Cymreig mewn arddangosfa gwniaduron.

Yn wir, perthynas barhaol oedd y peth olaf ar feddwl Tina pan redodd gŵr golygus, oddeutu deg ar hugain oed i mewn i'r siop un diwrnod braf o haf, ei wyneb yn wynnach nag oferôl claerwen Tina ei hun, ac yn dal ei law dde'n dynn yn ei law chwith, tra diferai dafnau o waed yn hamddenol i lawr ei grys glas golau.

Erbyn i Tina sodro'r gŵr ar stôl yng nghefn y siop (rhag ofn iddo lewygu yn y fan a'r lle), a'i berswadio i adael iddi roi ei law'n ofalus dan y tap dŵr oer, gallai weld mai arwynebol oedd ei anaf, er gwaetha'r alanas ar ei grys, a phrysurodd i'w lanhau cyn ei esmwytho ag eli melyn a gosod plaster arno.

Wrth i'r lliw ddychwelyd i wyneb y claf, ni chymerai ei wrthod pan fynnodd fod ei gymwynaswraig yn deilwng o swper, o leiaf, am ei achub ef, os nad ei grys, o enau trychineb. A thros bryd godidog yn un o fwytai moethus y dref, dechreuodd Tina a Robart ddod i adnabod ei gilydd. Peiriannydd oedd Robart wrth ei alwedigaeth, gartre ar wyliau o'r Dwyrain Canol, ble roedd ganddo gytundeb ffafriol, ac erbyn diwedd y noson roedd yn anodd dweud p'run o'r ddau oedd wedi gwirioni fwyaf. Roedd Robart wedi ei gyfareddu gan y ferch fywiog o'i flaen, a Tina hithau wedi mopio ar Robart, gyda'i wallt brown trwchus, ei lygaid oedd fel pyllau o siocled tywyll twym, a'i ffordd ffug-ddifrifol o sgwrsio tra oedd ei lygaid yn loyw â direidi.

Treuliodd y ddau bob eiliad posib gyda'i gilydd nes y daeth hi'n amser i Robart ddychwelyd i'r Dwyrain Canol, yna bu llythyru brwd a galwadau ffôn di-ri am rai misoedd nes i'w gytundeb ddod i ben. Pan ddaeth adref, ailgydiodd y ddau yn eu carwriaeth gan ddilyn holl draddodiadau ysblennydd straeon tylwyth teg, gyda phriodas ledrithiol a mis mêl hudolus. Prynwyd a dodrefnwyd tŷ sylweddol yn y dref, a derbyniodd Robart swydd o bwys gyda chwmni lleol. Parhaodd Tina â'i gwaith yn y fferyllfa hyd nes y bendithwyd y cwpl ag un, dau, tri o feibion hardd a heini.

Yna, pan oedd Tina a Robart yn disgwyl dim mwy a dim llai na byw yn hapus byth wedyn, talodd ellylles fechan faleisus ymweliad â hwy. Aeth y cwmni oedd yn cyflogi Robart i'r wal yn gwbl ddirybudd. Serch eu siom, doedd Robart na Tina'n poeni'n ormodol ar y pryd; wedi'r cyfan, roedd Robart yn dipyn o arbenigwr yn ei faes – byddai fawr o dro'n dod o hyd i swydd arall. Ond wrth i'r wythnosau o laesu dwylo droi'n fisoedd, cydiodd y felan yn Robart. Byddai iselder yn ei feddiannu am gyfnodau cynyddol; byddai ei hwyliau'n anwadal, a'i gydbwysedd yn ansefydlog; byddai'n troi gyda'r gwynt.

Roedd Tina mewn cyfyng gyngor. A fyddai Robart yn gallu gofalu am yr hogiau petai hi'n dychwelyd i'r gwaith? A fyddai hi'n gallu sicrhau swydd? A fyddai hi'n gallu ennill digon i gynnal y pump ohonynt a'r tŷ?

Yn y pen draw doedd dim dewis. Roedd y gymdeithas adeiladu ar eu sodlau ac yn bygwth ailfeddiannu'r tŷ, felly a Robart yn ei bydew diwaelod personol, bu'n rhaid i Tina, gydag Iwan yn ei drol, a Gareth a Hywel wrth ei chwt, ymweld â stribedi o swyddfeydd sychlyd, digroeso. Yna, wedi methu'n glir â darganfod fformiwla'r swyn a fyddai'n gwneud popeth yn iawn drachefn, fe geisiodd ei gorau glas i drawsnewid y tŷ cyngor a gynigiwyd iddynt yn Nant-yr-Onnen yn gartref clyd, cysurus. Bu wrthi'n sgwrio a sgrwbio a phaentio a phapuro hyd yr oriau mân, tra oedd yr hogiau yn eu gwlâu a Robart yn un swp diymadferth o flaen y teledu.

Unwaith roedd hi wedi llwyddo i gael trefn ar y tŷ, ac yn hapus bod yr hogiau wedi setlo yn yr ysgol, dechreuodd Tina chwilio am waith. Gan nad oedd ganddi gar, a gan fod ei hangen hi ar yr hogiau y tu allan i oriau'r ysgol, roedd yn falch o dderbyn swydd fel cynorthwywraig i Sioned yn y cylch meithrin. O leiaf roedd hi'n teimlo ei bod hi'n gwneud ymdrech, a châi fynd allan o'r tŷ am ychydig o oriau bob dydd. Wrth i'r hogiau dyfu a mynd yn fwyfwy annibynnol, roedd hi'n falch hefyd o'r gwaith a gynigiwyd iddi gan Gwyneth yn y Llew Coch.

Rhwng popeth, pur anaml y byddai gan Tina'r hamdden i gnoi cil ar ei sefyllfa. A Robart yng ngafael digalondid diderfyn, teimlai fod rheidrwydd arni i geisio creu rhyw fath o normalrwydd i'r hogiau. Ond 'leni, a'i phen-blwydd yn ddeugain yn prysur nesáu, roedd rhyw deimlad o anfodlonrwydd wedi dechrau ystwyrian ynddi. Roedd ei meddyliau'n ferw o rwystredigaethau a siomedigaethau blynyddoedd, a choblyn bychan grwgnachllyd y tu mewn iddi'n ailadrodd y rhigwm: dwi wedi blino ar hyn; dwi wedi blino cadw wyneb; dwi wedi blino trio cynnal pawb a phopeth; dwi wedi blino ar yr ymdrech barhaus i gael dau ben llinyn ynghyd; dwi wedi blino rhedeg ffwl-sbîd jyst er mwyn aros yn f'unfan.

". . . a fedra i ddim credu'n bod ni i gyd wedi cytuno efo chdi."

"O paid, wir, Gwyneth . . . dwi'n chwys oer drosta i wrth feddwl 'mod i wedi deud y fath betha."

"Ond y chdi sy'n iawn, Tina. Erbyn i ni ddisgwyl i'r Cyngor Cymuned bwyso a mesur goblygiadau pob agwedd o'r sefyllfa o bob ongl mi fydd holl blant yr ysgol feithrin ar eu pensiwn . . . a'r ddwy ohonon ni'n canu deuawd i Pedr."

"M-m-m . . . ond calendar bronnoeth? A Sara'n gyfrifol am y syniad?"

"Wel, mi wyddost dy hun yr effaith gafodd cau Ysgol Nant-yr-Onnen arni . . . a hynny ar ôl iddi golli'i thad . . ."

"Ia . . . a fedrwn ni ddim fforddio colli'r unig fan cyfarfod sy ar ôl yn y pentra."

"Mi fyddwn i'n gallu gwneud trefniada i bobol ddefnyddio'r lolfa yma o bryd i'w gilydd . . . ond fyddai hynny'n dda i ddim i'r cylch meithrin."

"Mi fyddai'n handi iawn i Sioned a minna petai pawb yn penderfynu cambihafio'r un pryd."

"A byddai sgandal felly'n edrych yn dda ar dudalen flaen y papur lleol, Tina. Mi fedra i weld y pennawd rŵan: Dwy wraig o Nant-yr-Onnen – sy'n ddigon hen i wybod yn well – yn feddw, ac eneidiau diniwed a dilychwyn y cylch meithrin lleol yn eu gofal!"

"Mae gofyn gras, mynadd, blacin gwyn a dos go lew o gaffîn efo ambell un, coelia di fi. Yli, well imi orffen clirio'r byrdda 'ma, neu mi fyddwn ni yma drwy'r nos."

"Ti'n iawn . . . o leia mae awgrym Sara wedi bwrw Jac a'i fisdimanars – os misdimanars hefyd – i ebargofiant."

"Am y tro, Gwyneth."

Er gwaetha'r sterics a'r strancio, byddai Sioned a Tina bob amser yn colli selogion yr ysgol feithrin pan fyddent yn gadael i fynychu'r ysgol gynradd yn Nhresarn. Yn ystod eu cyfnod yng ngofal y ddwy, byddai pob plentyn yn datblygu o fod prin mwy na thwdlyn bychan simsan i fod yn unigolyn diddorol a dihafal.

Ychydig ddiwrnodau cyn diwedd y tymor roedd hi'n ben-blwydd y cythraul yn y cylch meithrin, a Sioned a Tina wedi aros ar ôl yn y neuadd wedi i'r trysorau bychain gael eu tywys adref am ginio. Roedd y ddwy'n wrthi'n brysur yn gwahanu'r trenau bach (a oedd yn byw gyda'r cledrau pren) oddi wrth y ceir bach (a oedd yn byw gyda'r garej), ac yn llusgo'r dynion Duplo o grafangau temtwragedd y tŷ dol.

"Wyddost ti be?" meddai Tina, a oedd yn eistedd yn ei chwman ger un o'r byrddau bychain a'i phengliniau'n sownd oddi tano. "Dwi'n gwybod fod y syniad o annog y plant i'n helpu i dwtio ar ddiwedd pob sesiwn yn un canmoladwy iawn ond, bobol bach, mi fysan ni'n arbed oria o waith trwy glirio'n hunain."

"Dwi'n meddwl dy fod ti'n iawn," chwarddodd Sioned, a oedd hefyd yn eistedd ger un o'r byrddau, yn chwynnu pentwr o deganau a'i phengliniau dan ei gên. "Tydw i 'rioed wedi cofleidio'r *spring cleaning* 'ma, hyd yn oed adra. Efo Dylan a Ceri yn y tŷ, mae fel trio golchi glo'n lân."

"Nabod y teimlad yn dda," cytunodd Tina gan ochneidio. "Dwi'n siŵr fod corwynt wedi ymgartrefu yn stafell Iwan a Gareth. Ac o ran symud dodrefn a llnau y tu ôl iddyn nhw, a rhyw nonsens gorfrwdfrydig felly . . . wel . . . rhywbeth mae rhywun yn ei wneud pan mae 'na 'z' yn y mis ydi peth felly."

"Wel, mi hoffwn i gael tynnu fy llun ar gyfer y mis efo 'z' ynddo fo cyn belled ag y mae'r calendar 'ma yn y cwestiwn," meddai Sioned, "ar ôl i rywun roi *spring clean* iawn i mi."

"Dwi'n dal i drio gweithio allan be ddaeth dros fy mhen i'n cefnogi'r awgrym," meddai Tina gan ysgwyd ei phen. "Mi alwais yn y Llew Coch ar y ffordd i'r cyfarfod i roi llyfr i Gwyneth, ond wnes i ddim cyffwrdd dropyn . . . wir yr."

"Mae'n rhy hwyr i ailfeddwl bellach. Mi fuodd Sara a Marian mewn cysylltiad dros y penwsnos, ac maen nhw wedi penderfynu ar ddau neu dri o ddyddiada ym mis Medi i'w cynnig i Harri, gan obeithio y bydd un ohonyn nhw'n gyfleus iddo fo."

"Does dim troi'n ôl, felly."

"Nagoes . . . a beth bynnag, gan mai'r cylch meithrin sy'n gwneud y defnydd mwya o'r neuadd . . ."

Torrodd sgrech gan Tina ar draws synfyfyrio Sioned; roedd hi wedi darganfod dant wedi pydru yng ngwaelod y bocs teganau.

Ychydig funudau'n ddiweddarach roedd y ddwy'n llymeitian coffi ac yn barod i gladdu clamp o faryn Mars yr un roedd Sioned wedi eu cuddio yn ei bag (dylanwad Dilys, mae'n amlwg). Roedd arweinwyr cylchoedd meithrin i fod i gymell y plant yn eu gofal i fwyta'n iach, yn hytrach na llowcio logiau mawr o siocled pan fyddai'r chwa leiaf o chwithigrwydd yn eu bygwth.

"Dwi'n meddwl y gadawn ni weddill y tegana fel maen nhw," penderfynodd Sioned, "bendramwnwgl neu beidio."

Nodiodd Tina, a'i cheg yn llawn Mars melys.

"Ond mi allwn ni dynnu rhywfaint o'r gwaith oddi ar y walia er mwyn i'r plant gael mynd â'u campweithia adra ddiwedd yr wsnos," parhaodd Sioned.

Nodiodd Tina drachefn, gan ddal i ymrafael â moddion Sioned ar gyfer sioc.

"Mae rhai o'r gweithgaredda 'ma rwyt ti wedi'u cynllunio ar gyfer y plant sy'n mynd i'r ysgol yn arbennig o dda, Tina," meddai Sioned. "Wnest ti 'rioed feddwl am fynd i ddysgu? Nid 'mod i isio cael gwared ohonot ti," ychwanegodd yn frysiog.

"Wel," meddai Tina, wedi llwyddo i ennill y frwydr gyda'r lefiathan siocled o'r diwedd, "mi wnaeth un o'r athrawon yn yr ysgol awgrymu'r peth, ond ar y pryd y cwbl roeddwn i isio'i wneud oedd canu'n iach i'r lle, a chael swydd ac arian ac annibyniaeth."

"Tydi hi byth yn rhy hwyr," awgrymodd Sioned.

"Fedrwn i ddim, siŵr . . . ddim efo'r hogia . . . a Robart . . . a does gen i ddim pwt o gymwystera . . . ac mae'n costio ffortiwn i fynd i'r coleg dyddia 'ma . . . heb sôn am fod fel gwsberan yng nghanol llond lle o betha bach ifanc . . ."

Syllodd Sioned ar ei ffrind. Wedi sawl blwyddyn o gydweithio yn y cylch meithrin fe wyddai Sioned fwy am fywyd Tina nag a dybiai ei chynorthwywraig. Cododd a mynd at y wal i ddechrau tynnu'r arddangosfa liwgar i lawr. Ond cyn cychwyn trodd i edrych ar Tina dros ei hysgwydd.

"Ai hon ydi'r un Tina a'n hysgogodd ni i gyd i gydio yn y danadl ychydig wsnosa'n ôl? Ai hon ydi'r un Tina a wfftiodd ein hamheuon a'n hofna? Be ddeudist ti bryd hynny – y byddi di'n ddeugain cyn diwedd y flwyddyn? Wel, beth amdani, Tina? Os wyt ti am newid petha, rŵan amdani."

* * *

Wedi'r glec, ymledodd y staen coch yn araf dros y palmant. Safodd Tina'n stond, wedi fferru yn ei hunfan. Yna clywodd glonc tun ffa pob yn rowlio linc-di-lonc i'r gwter, a dau dun pys a thun afal pîn i'w ganlyn.

Diwrnod olaf yr ysgol cyn gwyliau'r haf oedd hi; roedd yr ysgol feithrin wedi bod fel ffair trwy'r bore, a'r plant i gyd wedi cynhyrfu'n lân. Roedd Tina wedi cychwyn am Dresarn yn syth o'r cylch er mwyn stocio rhywfaint ar ei chypyrddau cyn i'r hogiau gyrraedd adref a dechrau bwyta dros Gymru, gan obeithio bod gartref i'w cyfarch â phantri llawn, gwên groesawus ac addewid o rywbeth blasus i de.

A ble roedd hi? Yma ar gornel y stryd yn Nhresarn, a phen ôl sgwâr sbeitlyd y bws yn diflannu o'r golwg. Roedd hi'n padlo mewn sôs coch, a'i siopa'n dianc i bob cyfeiriad. Ac roedd arni eisiau crio. A chwerthin. A sgrechian.Yn fwy na dim roedd arni eisiau sgrechian. Sgrechian dros y stryd. Sgrechian ei bod wedi laru, wedi cael llond bol, wedi ffieiddio ar fod yn firiman, wedi digio, diflasu, digalonni, wedi cyrraedd pen ei thennyn.

Cyneuodd Tina'r lamp fechan a safai ar y cwpwrdd wrth y gwely: ugain munud wedi dau – deng munud yn hwyrach na phan edrychodd gynnau fach. Cododd ar ei heistedd a lapio'i breichiau'n dynn am ei phengliniau. Ystyriodd fentro lawr i'r gegin i wneud paned, ond gan fod Robart yn dal yn y stafell fyw penderfynodd beidio. Efallai mai wedi syrthio i gysgu yng nghanol rhyw ffilm hwyr ar y teledu oedd o – byddai hynny'n digwydd o dro i dro.

Tra oedd hi'n gweithio yn y Llew Coch y noson honno doedd Tina ddim wedi cael llawer o gyfle i bendroni ynghylch yr awr fawr a'i hwynebai drannoeth, er mai dyna oedd wedi denu'r criw i gyd i'r dafarn. Ac roedd hi'n falch hefyd nad oedd wedi cael cyfle i lowcio hanner cymaint o blwc potel ag ambell un. Ond wedyn, roedd hi'n annhebygol bod yfwyr y rheng flaen yn gorwedd yn effro yn eu gwlâu fel hithau, a'u meddyliau'n un gybolfa gythryblus.

Yfory, neu yn hytrach heddiw, roedd yn rhaid talu iawn am ei beiddgarwch. Ac er gwaethaf ei geiriau rhyfygus noson y cyfarfod, roedd hi'n llawn arswyd wrth feddwl am y sesiwn gyda Harri. Hi a'i cheg fawr, yn rwdlan am ei phen-blwydd. Pa wahaniaeth fyddai gwneud sioe ohoni ei hun yn ei wneud i'w bywyd go iawn? Dim iot. Heblaw am sicrhau ei swydd yn yr ysgol feithrin, os byddai'n lwcus.

Doedd Robart ddim wedi cymryd arno ei fod wedi clywed am y cynllun, ac roedd hynny'n ddigon posib ac yntau yn hafan ei hunandosturi. A doedd hi ddim wedi crybwyll y calendar wrth yr hogiau – wedi'r cyfan, nid dyna'r pwnc mwyaf hwylus i'w drafod gyda thri o fechgyn yn eu harddegau. Cymerai eu bod nhw wedi medru derbyn y syniad, o leiaf.

Roedd hi wedi bod yn ymdrech cadw trefn ar yr hogiau dros yr haf, fel bob haf, a hynny heb gyfyngu ar eu hannibyniaeth. Gweithiodd hithau shifftiau ychwanegol yn y Llew Coch er mwyn iddynt gael tipyn mwy o bres poced nag arfer, ac er mwyn sicrhau y byddai'r dillad priodol ganddynt ar gyfer y tymor newydd.

Roedd Hywel wedi gwneud yn rhagorol yn ei arholiadau TGAU, er gwaethaf ei phryderon, ac wedi dychwelyd i'r chweched dosbarth ar ôl sawl dadl. Roedd ef am adael yr ysgol a dechrau ennill ei damaid, ond mynnodd Tina y byddai'n well ganddi ymlafnio am dragwyddoldeb a diwrnod na gweld ei mab hynaf yn colli ei gyfle. Erbyn hyn byddai Hywel yn gwneud tipyn o amgylch y tŷ yn ddistaw bach, ac yn cadw llygad ar Iwan a Gareth. Roedd wedi sicrhau swydd dydd Sadwrn iddo'i hun yn un o siopau Tresarn, a byddai'n codi'n blygeiniol i ddal y bws bob wythnos.

Na, doedd ganddi ddim rheswm i boeni'n ormodol am Hywel. Nac am Iwan a Gareth chwaith – er bod eu stafell yn edrych fel petai'n gartref i lond syrcas o fwncwn, a'u bod yn tyfu allan o'u dillad bron cyn iddi gael cyfle i dorri'r labeli oddi arnyn nhw. A doedd hynny'n ddim syndod, o ystyried faint o fwyd roedden nhw wedi ei gladdu dros yr haf.

A Robart? Wel, dros y blynyddoedd roedd hi wedi trio a thrio'i berswadio i dderbyn help at ei gyflwr: wedi seboni, ysgogi, rhesymu; wedi gweiddi, sgrechian, blagardio. Ond bellach roedd hi wedi derbyn mai'r unig un a allai helpu Robart oedd Robart ei hun. Ganddo fo roedd yr hawl a'r gallu i ddewis gwneud hynny.

A hithau? Oedd ganddi hi ddewis? Ers blynyddoedd roedd hi wedi gorfod rhoi'r hogiau'n gyntaf a gwneud ei gorau dan yr amgylchiadau, gan gladdu ei dyheadau a'i hanghenion ei hun. Ac er iddi geisio'i gorau glas gydol yr haf i gladdu geiriau Sioned hefyd, roeddynt yn mynnu codi i'w chystwyo a'i chystuddio. Rŵan bod yr hogiau yn eu harddegau, oedd modd iddi droi ei bywyd ar ei ben? Oedd modd iddi drawsnewid ei sefyllfa? Oedd modd iddi herio'r dyfodol?

Yfory; nage, heddiw. Heddiw roedd y sesiwn gyda Harri – byddai'n delio â'r ordîl hwnnw'n gyntaf. Ac yna . . .

". . . mi roeddwn i'n swp sâl ar y pryd, cofia, ti'n gwybod hynny dy hun . . . rhwng yr ofn a gweddillion y pen mawr . . . ond rŵan . . . rŵan mi fyswn i'n gallu bwrw tin dros ben o un pen i'r pentra i'r llall."

Edrychodd Tina'n amheus ar Sioned ar draws bwrdd y gegin; nid dyma'r un Sioned y bu'n rhaid iddi ei llusgo dros drothwy'r neuadd y bore hwnnw. Roedd Tina ei hun yn teimlo bellach fod ceisio dal ei gafael ar y mymryn hunanhyder oedd yn weddill ganddi fel trio ffrwyno meráng.

Amneidiodd Sioned i gyfeiriad y stafell fyw. "Ydi Robart yma?" holodd.

"Nac ydi," ysgydwodd Tina ei phen. "Mae o wedi cael ei yrru i siopio i Dresarn . . . doedd o ddim yn hapus iawn."

"Ddeudodd o rywbeth am y calendar?" gofynnodd Sioned wedyn.

"Dim gair," atebodd Tina. "Wn i ddim ydw i'n falch nad oedd rhaid i mi gynnig eglurhad, ynteu'n ddigalon am ei fod mor ddi-hid."

Estynnodd Sioned ar draws y bwrdd a rhoi gwasgiad fach i law ei ffrind. Ers i Tina ddechrau gweithio yn yr ysgol feithrin bu'n gynorthwywraig selog a siriol, ond gallai Sioned ddirnad bod ei bywyd yn un llafurus a llesteiriol, er gwaethaf ei gwên barod.

"Ydi petha dal . . .?" mentrodd.

"Ydan, dal yr un fath," atebodd Tina, "ond dwi wedi bod yn meddwl am d'awgrym . . . a chdi sy'n iawn . . . dewis Robart ydi troelli'n ei unfan . . . ac mi fedra inna ddewis . . . dewis trio gwella petha i'r hogia a finna . . . mi fydd hi'n anodd . . . anodd ar y naw, dwi'n sylweddoli hynny . . . anodd mynd 'nôl i'r dechra . . . a sobor o anodd yn ariannol . . . ond os na ddewisa i'r llwybr yma . . . os dewisa i adael petha fel y maen nhw . . . mi fydda i wedi mynd o 'ngho."

Cododd Sioned a cherdded o amgylch y bwrdd i gofleidio'i ffrind; os oedd rhywun yn haeddu cyfle, Tina oedd honno.

"Da iawn chdi, a mi fydd y genod i gyd y tu ôl i chdi; fyddi di ddim ar dy ben dy hun."

"O, dwyt ti ddim yn mynd i gael gwared ohona i eto," sicrhaodd Tina hi. "Dwi'n gobeithio cael lle ar gwrs yng Ngholeg Tresarn ddiwedd y mis, cwrs fydd yn gam tuag at gael fy nerbyn i addysg uwch. Cwrs rhan-amser fydd o, ond o leia mae'n fan cychwyn."

"Mae hynny'n . . ."

Torrodd caniad cloch y drws ar draws Sioned.

Cuddiodd Tina ei hwyneb yn ei dwylo. "O, na," llefodd.

Casglodd Tina'r cardiau pen-blwydd oddi ar silff ffenest y gegin. Wedi'r holl aros ac aflonyddu ac anesmwytho, roedd ei phen-blwydd – y pen-blwydd ych-a-fi a dim yn y rhif – wedi pasio'n ddigon digynnwrf.

Roedd mamau'r cylch meithrin wedi cyflwyno tusw o

flodau iddi, a Sioned wedi dod â chlamp o deisen i'r cylch y bore hwnnw. Fe wfftiwyd y bananas a'r afalau canol bore, a chafodd pawb dafell o gacen. Trueni ei bod wedi cymryd gweddill y bore i ddadlynu'r plant oddi wrth ei gilydd, wedi iddynt lwyddo i'w gorchuddio'u hunain mewn eisin o'u corun i'w sawdl! Yna ras i ddal y bws i'r coleg yn Nhresarn (lle nad oedd neb yn ymwybodol ddiwrnod mor ddirdynnol oedd hi) am y cyntaf o'i sesiynau'r wythnos honno. Roedd mynychu'r coleg ddau brynhawn yr wythnos wedi bod yn agoriad llygad i Tina. Roedd hi wedi sylweddoli bod llawer o bobl mewn sefyllfa debyg iddi hi, yn ceisio ailafael yn eu gyrfaoedd a'u bywydau. Oedolion oedd pawb ar y cwrs – nifer o wragedd yn ceisio dychwelyd i fyd gwaith wedi magu teulu; eraill wedi colli eu swyddi ac am newid cyfeiriad; rhai yn chwilio am her newydd, ac ambell un na wyddai Tina eto beth oedd ei gymhelliad.

Ac wedi'r wythnos gyntaf honno, pan oedd pawb a phopeth yn ddiarth, a Tina'n berchen ar lai na llond llwy de o hunanhyder, roedd wedi cymryd at y cwrs fel hwyaden at ddŵr. Câi foddhad o ymgodymu â thasgau a mwy o her iddynt na cheisio penderfynu beth i'w goginio i de a sut i wneud i gynnwys ei phwrs bara tan ddiwedd yr wythnos, a byddai'n gwneud ei gwaith cartref ar ddydd Sul, 'run fath â'r hogiau.

A chwarae teg i'r hogiau, rhaid eu bod wedi bod yn cynilo ers tro cyn prynu'r bag newydd hwnnw iddi ar ei phen-blwydd; bag ymarferol yn llawn papur a phensiliau a phob math o fân betheuach a fyddai'n ddefnyddiol iddi yn y coleg. Roedd hyd yn oed Robart wedi arwyddo cerdyn (er mae'n siŵr mai Hywel a'i prynodd), ac wedi rhoi cysgod cusan ar ei boch wrth fwmian cyfarchion ar fore ei phen-blwydd cyn diflannu i'r stafell fyw.

Ac mewn ychydig ddiwrnodau byddai'r calendar yn cael ei lansio; bu bron iddi anghofio am hwnnw yng nghanol ei chynnwrf. Roedd Mererid wedi sicrhau'r criw fod y lluniau'n werth eu gweld, a bod pob un ohonynt yn edrych yn bictiwr, ond nes y byddai'n ei gweld ei hun mewn du a gwyn, daliai Tina i amau y byddai'n edrych fel drychiolaeth. Ond roedd hi'n rhy hwyr i boeni bellach.

Ai'r calendar fu'r catalydd? Ynteu ei phen-blwydd? Ynteu rhyw ddewiniaeth ddirgel a grewyd pan ddaeth y ddau at ei gilydd? Ac mewn gwirionedd, a oedd ots? Nag oedd, mae'n debyg, ystyriodd Tina wrth roi'r cardiau mewn amlen i'w cadw yng ngwaelod y wardrob gyda gweddill ei thrysorau; dim ots o gwbl.

# TACHWEDD

Un bach arall. Fyddai un bach arall yn gwneud dim drwg i neb. Ac roedd ganddi hanner awr dda cyn y byddai angen cychwyn am y neuadd i nôl Aled o'r cylch meithrin. Efallai byddai rhyw lymaid bach eto yn pylu tipyn ar y dwndwr a ddeuai o'r safle adeiladu ar gyrion yr ardd. Roedd y dynion eisoes wedi bod yn gweithio yno ers rhai wythnosau, ac roedd yn gas gan Anwen feddwl am yr holl styrbans a ddeuai i'w rhan dros y misoedd nesaf, a'r fath sŵn a stomp yn digwydd o fewn tafliad carreg i'w drws cefn.

Roedd Aled wrth ei fodd, wrth gwrs, ac wedi cael modd i fyw yn sefyll yn ffenest ei lofft yn gwylio'r Jac Codi Baw yn tyllu. Ond pryderai Anwen am y plant, ac roedd wedi siarsio'i mab a'i merch rhag mynd yn agos i'r safle, gan y gwyddai mor chwilfrydig y gallai plant bach fod, ac mor hawdd y byddai iddynt eu niweidio'u hunain.

Go damia Albert a'i gynlluniau. Ym marn Anwen byddai wedi bod yn well o lawer gadael y llain tir fel yr oedd, neu osod tarmac arno i greu maes parcio i bobl y pentref, neu lunio cae chwarae i'r plant. Doedd dim angen mwy o dai yn Nant-yr-Onnen, siŵr. Roeddynt wedi bod yn lwcus gyda mwyafrif preswylwyr Annedd Hedd gan eu bod yn Gymry glân gloyw oedd yn barod i ymroi i fywyd cymdeithasol cymuned glòs y pentref. Ond mwy o dai . . . ?

Mwy o arian i gynnal ymerodraeth bersonol Albert, roedd hynny'n sicr, barnodd Anwen yn chwerw wrth wagio'i gwydr. Roedd hi wedi hen sylweddoli y gallai arian, yn ogystal â rhoi rhyddid i'w berchen, droi'n arf ac yn hual.

* * *

Gyda'i gwallt du sgleiniog yn chwifio am ei hysgwyddau, a'i llygaid brown tywyll, roedd Anwen yn ferch ifanc ddeniadol. Mor ddeniadol nes bod Albert Parry, pan oedd allan yn y dref gyda chriw o wŷr ifanc hwyliog un noson, wedi penderfynu yr eiliad y gwelodd hi ymysg ei ffrindiau mai dyma'r ferch a fynnai ar ei fraich.

Gweithio fel teipyddes yn swyddfa'r Cyngor Sir a wnâi Anwen, a chan ei bod hi'n ben-blwydd un o'i chydweithwyr, roedd y merched yn bwriadu dathlu mewn steil. Ac roeddynt yn fwy na pharod i groesawu'r gwŷr ifanc i ymuno â nhw am ychydig o sbort a sbri.

Er nad oedd Anwen gyda'r mwyaf beiddgar o'r criw – roedd y rheini yn y rheng flaen yn annog y llanciau, yn gwbl ddiangen, i'w dangos eu hunain, â sylwadau awgrymog ac ystumiau pryfoclyd – anelodd Albert yn syth amdani. Glynodd wrthi fel gelen am weddill y noson, ac yn wir, llwyddodd i'w gwahanu oddi wrth ei ffrindiau a'i thywys i dafarn fechan ddistaw, yn ddigon pell o sŵn rhialtwch y lleill. Ac roedd Anwen yn ddigon bodlon cael ei harwain i ffwrdd gan Albert, wedi'i hudo gan ei wallt coch cyrliog, ei lygaid glas gloyw, a'i osgo hunanfeddiannol a phenderfynol.

Profodd Anwen dynnu coes didrugaredd yn y gwaith y bore canlynol, yn enwedig a hithau'n ferch ddiymhongar a chanddi duedd i gochi at ei chlustiau hyd yn oed pan ofynnai rhywun iddi faint oedd hi o'r gloch. A phrofodd lawer mwy o dynnu coes yn ystod yr wythnosau dilynol, wrth i Albert ddod i'w nôl ar ôl mynychu rhyw gyfarfod neu'i gilydd, a mynd â hi allan am bryd bach blasus ganol dydd, neu ei danfon adref o'r gwaith ddiwedd prynhawn wedi sesiwn o bwyllgora. Gan mai merch wylaidd oedd hi ei hun, ni allai Anwen lai nag edmygu egni ac eiddgarwch Albert ynghylch pob agwedd o'i fywyd. Boed o'n trin busnes neu ramant, doedd o ddim yn ddyn i lusgo'i draed.

Byddai Anwen wedi bod yn berffaith hapus symud mewn yn ddistaw bach i'r bwthyn a etifeddodd Albert gan ei nain a'i daid. Ond na, roedd yn rhaid ymweld â'i rhieni yn ei thref enedigol a threfnu priodas ysblennydd – dathliad drudfawr fu'n achos poen meddwl mawr i Anwen am fisoedd ymlaen llaw, a ble profodd Albert i'w theulu y gallai ddenu'r adar o'r coed petai galw am hynny. Byddai Anwen wedi bod yn berffaith hapus gyda phenwythnos rhamantus o fis mêl mewn gwesty cefn gwlad, ac yna cyfnod o botsian yn hamddenol yn y bwthyn. Ond na, roedd yn rhaid trefnu mis mêl moethus, cofiadwy a choeth. Byddai Anwen wedi bod yn berffaith hapus gyda chodi estyniad ar y bwthyn i greu stafell neu ddwy'n ychwanegol i'r plant yn eu tro. Ond na, roedd yn rhaid prynu safle'r hen ysgol, a llunio'r adeilad yn honglad o fyngalo. A swyddfa i'r busnes. Ac yn wir, yn llygaid mwyafrif pentrefwyr Nant-yr-Onnen, roedd Anwen yn wraig oedd ar ben ei digon. Doedd dim diwedd i'r hyn a wnâi Albert Parry dros ei deulu, ac roedd pob moethusrwydd ar flaenau eu bysedd.

Wrth ddanfon Delyth, ac yna Aled, i'r cylch meithrin, daeth Anwen i adnabod Nia, a datblygodd cyfeillgarwch cyfforddus rhwng y ddwy ferch, er bod eu hamgylchiadau yn dra gwahanol. Ac efallai mai Nia oedd yr unig un a ddeallai fod gormodedd yn gallu gormesu lawn cymaint â phrinder.

Teimlai Anwen ei hun fod y trimins a'r trugareddau twyllodrus yn cau amdani. Roedd y cysuron costus yn ei mygu, a'r moethusion mynych yn gwasgu ac yn gwasgu arni nes ei bod hi ei hun bron â throi'n dryloyw. Doedd prin dim ohoni ar ôl; jyst digon i gael ei pharedio fel gwraig ddelfrydol oedd yn byw mewn tŷ delfrydol gyda theulu delfrydol – dim mwy. Yn wir, teimlai Anwen fel doli glwt; doli glwt roedd rhywun wedi gyrru stîm-roler drosti – 'nôl ac ymlaen, 'nôl ac ymlaen – a rŵan, wel, doedd dim angen agor drws iddi fynd drwyddo bellach, oherwydd roedd yr hyn oedd yn weddill ohoni mor ddisylwedd nes y gallai lithro oddi tano.

Heblaw am y plant, byddai wedi . . . Wedi be? Madael? Cael affêr? Ei hyfed ei hun yn llonydd yn y gwter agosaf?

* * *

"Anwen! Anwen!"

Sodrodd Anwen y gwydr yn y meicrodon a chau'r drws arno.

Go damia! Roedd y ffaith bod y swyddfa drws nesa i'r tŷ yn golygu bod Albert yn gallu picio i mewn yn annisgwyl yn rhy aml o lawer. A rŵan bod y gwaith wedi dechrau ar y safle yng ngwaelod yr ardd, doedd dim posib gwybod pryd y byddai'n ymddangos.

Rhoddodd Albert ei ben rownd drws y gegin. "Sut mae'i dallt hi am banad?"

Cymerodd Anwen arni ei bod hi'n brysur. "Mae'r tecell yn fan'cw," amneidiodd, "a mi gymera inna un os wyt ti'n cynnig."

"Be ti'n neud?" holodd Albert tra oedd yn disgwyl i'r tecell ferwi.

"O . . . dim ond twtio ryw chydig ar y cypyrdda 'ma," atebodd Anwen yn ddidaro.

Ffromodd Albert. "Tydi'r ddynas llnau ddim yn gwneud ei gwaith yn iawn? Mi fedra i chwilio am rywun arall i chdi."

"Mae'r ddynas llnau'n berffaith tebol, Albert. Er dwn i ddim pam mae angen dynas llnau arnon ni o gwbl. 'Toes gen i drwy'r dydd bob dydd i gadw trefn ar y tŷ 'ma fy hun?"

Canodd ffôn symudol Albert cyn iddo gael cyfle i ateb, ac aeth allan i'r cyntedd i dderbyn yr alwad.

Agorodd Anwen y meicrodon, estyn y gwydr ac yfed ei gynnwys ar ei dalcen cyn ei roi yn y peiriant golchi llestri.

Be oedd yn bod ar Albert? Onid oedd o'n deall pa mor ddiflas oedd ei bywyd? Onid oedd o'n deall bod byw yn y tŷ dilychwyn yma, a oedd yn edrych fel petai wedi'i godi yn ei grynswth o dudalennau rhyw gylchgrawn crand, yn ei gyrru hi i fyny'r wal a rownd y nenfwd?

Sylweddolodd yn sydyn fod y tecell wedi berwi, a dechreuodd wneud paned.

"Sori," brathodd Albert wrth daro'i ben heibio'r drws,

"rhaid i mi fynd – rhyw broblem gan yr hogia. Wela i chdi nes ymlaen."

Rhoddodd Anwen y mygiau'n ôl yn y cwpwrdd. Doedd hi ddim wedi croesawu ymweliad annisgwyl ei gŵr o bell ffordd, ond o leiaf roedd ei ymddangosiad wedi torri ar undonedd ei diwrnod. Doedd dim angen nôl Aled o'r ysgol feithrin heddiw gan ei fod yn mynd adref gyda Nia i gael cinio a chwarae gyda Glesni, a felly byddai ar ei phen ei hun nes i Delyth ddod adre o'r ysgol.

Edrychodd Anwen o'i chwmpas ar ei chegin ddisglair ddi-fai. At bwy y gallai droi? Wrth ba un o ferched y pentre, a chymaint ohonynt yn ymdrechu'n ddyddiol i gynnal eu teuluoedd a chael dau ben llinyn ynghyd, y gallai ddweud bod y bywyd delfrydol yma'n gwneud iddi deimlo fel petai ar ynys bellennig? Wrth bwy y gallai gyfaddef y byddai'n well ganddi faeddu tipyn ar ei dwylo a theimlo'i bod yn gwneud rhywbeth na chael ei chadw fel rhywogaeth unigryw ymhell o grafangau pobl go iawn a'u problemau.

Agorodd ddrws y cwpwrdd a ddaliai'r llestri gorau, ac ymbalfalodd y tu ôl i'r tŵr tsieina nes cael gafael ar wddw'r botel. Cododd hi'n ofalus o'i chuddfan ac estyn gwydr glân.

Roedd Nia yn llygad ei lle, wrth gwrs. Tendiad ar bobl eraill oedd y ddwy ohonynt yn barhaus, er bod Nia'n gwneud tipyn mwy o'r hyn a fyddai'n cael ei gyfrif yn waith go iawn. Gorchwyl digon diflas oedd cynnal fi fawr Albert, ond doedd o ddim yn waith caled corfforol. Er, ystyriodd Anwen wrth gerdded adre ling-di-long o dŷ Nia drannoeth y cyfarfod, byddai hi'n croesawu cyfle i gael gwared â rhai o'i rhwystredigaethau â dos iawn o . . . o. . . o unrhyw beth, a dweud y gwir – palu'r ardd, sgrwbio'r llawr – unrhyw beth y byddai Albert yn ei gysidro'n waith cwbl israddol ac anaddas iddi hi ymgymryd ag o fel gwraig contractwr adeiladu mwyaf llwyddiannus Nant-yr-Onnen.

Byddai Albert yn cael mynd i'w grogi am unwaith pe

bai'n dechrau cega am y calendar 'ma; wedi'r cyfan, ei gynnig o oedd wedi plannu'r hedyn o obaith ynglŷn â dyfodol y neuadd ym meddwl Sara yn y lle cyntaf – er mai yn y cyfarfod neithiwr y clywodd Anwen ei hun amdano. Wel, doedd hi ddim yn bwriadu dweud yr un gair wrth Albert am gynllun Sara chwaith. Ac yn sicr ni fyddai'n dweud wrtho ei bod hi wedi ymrwymo i fod yn un o'r criw dethol. Wedi'r cwbl, iawn i'r ŵydd, iawn i'r ceiliagwydd.

Wrth nesáu at y tŷ a chlywed y twrw a'r trwst yn dod o'r safle adeiladu gerllaw, teimlai Anwen ei hyder hwyliog, a oedd yn morio'n braf megis cwch bapur ar sieri Nia, yn dechrau suddo. Ac erbyn iddi gyrraedd y drws roedd yr hyn oedd yn weddill o'i hwyliau da wedi diferu trwy wadnau ei thraed i ebargofiant.

". . . ac wn i ddim be ddaeth dros dy ben di i gytuno efo'r fath ffwlbri diraddiol. Wyt ti'n trio codi cywilydd arna i?"

"Fedri di ddim fy nadu i. Ac o leia mi fydd o'n rhywbeth i'w wneud, rhywbeth i basio'r amsar."

"Rhywbeth i basio'r amsar? Be sy'n bod arnat ti, Anwen? Rwyt ti wedi cael popeth roeddet ti'i isio yn y tŷ 'ma. A mwy na hynny hefyd. Mi fyddai hanner merched y pentra wrth eu bodd yn cael byw bywyd fel d'un di. Ac rwyt ti am gymryd rhan yng ngwallgofrwydd Sara Watcyn i basio'r amsar?"

"Ac i warchod y neuadd. Mi ddylet ti fod yn blês; ymateb i dy gynnig di wnaeth Sara."

"Ia, ond doeddwn i ddim yn credu bod yna unrhyw . . . hynny ydi, doeddwn i . . ."

"Doeddet ti ddim yn credu bod yna unrhyw obaith y byddwn ni'n gallu codi'r arian; dyna oeddet ti'n feddwl, yntê? A be oedd am ddigwydd i'r safle wedyn?"

"Wn i ddim. Meddwl wnes i y byddai pawb wedi dod at ei gilydd i drio codi'r arian trwy . . . wel, trwy ddulliau mwy confensiynol."

"A meddwl mai methiant fyddai unrhyw ymgyrch felly? Meddwl y bydden ni wrthi nes byddai . . . nes byddai Jac y Jwc yn cael ei ethol yn Archdderwydd, yn codi rhyw geiniog neu ddwy yma a swlltyn acw? A be fyddai'r cam nesa, Albert?"

"Wn i ddim . . . dim ond trio helpu oeddwn i."

"Helpu pwy? Helpu Nant-yr-Onnen ynteu helpu Albert Parry? Ai caffaeliad newydd sbon i diriogaeth d'ymerodraeth bersonol fyddai safle'r neuadd, heblaw am safiad Sara?"

"Sara Watcyn, wir! Mae'n hen bryd i rywun roi tro yng ngwddw'r hen iâr yna. Treulio'r holl flynyddoedd yn clwydo ar ei phen ei hun yn y tŷ capel sy wedi troi ei hymennydd yn frywes, siŵr iawn. Beth arall sy'n esbonio'r ffaith iddi ddyfeisio cynllun mor lloerig?"

"Wedi cymryd yn erbyn Sara wyt ti am ei bod hi'n gwrthod gwerthu'r tŷ capel i ti. Wel, os na wneith hi gymryd ei bwlio dros werthu hwnnw, yna wneith hi ddim cymryd ei bwlio dros y neuadd, saff i ti. A chei di ddim fy mwlio inna tro 'ma, chwaith. Tydw i ddim am adael gweddill y genod yn y baw. Ac mi *fydda* i'n un o griw'r calendar, embaras neu beidio."

". . . a waeth i ti heb â dechrau troi'r tu min yn fygythiol fel'na efo fi, Prys. Sut gebyst y gallwn i fod wedi dyfalu y bydden nhw'n cael rhyw syniad twp fel hwn i'w penna?"

Ymhen rhai munudau ffarweliodd Albert a Prys â'i gilydd yn ddigon swta, a sodrodd Albert ei ffôn yn nrôr ei ddesg yn ddiamynedd, cyn ei gau yn glep. Hen sinach bach hafin dan din oedd Prys Harris, meddyliodd, wrth droelli ei gadair o'r naill ochr i'r llall ger ei ddesg yn y swyddfa. Roedd Albert wedi neidio at ei gynnig heb lawn ystyried goblygiadau'r cynllun, a rŵan roedd o mewn cyfyng gyngor. Roedd ymgyrch Merched y Wawr i godi'r arian angenrheidiol i brynu amser i safle'r neuadd yn ymddangos fel petai wedi cael gwynt dan ei adain, ac roedd Anwen wedi pwdu'n llwyr. Fu fawr o Gymraeg rhyngddynt ers wythnosau, ac roedd hi'n edrych yn fwyfwy tebygol y byddai'n rhaid iddo weithredu ar ei gynnig a chlirio'r safle a gosod caban yno ar gryn golled. Roedd yn bosibilrwydd cryf y byddai cais y Cyngor Cymuned am arian Loteri i adeiladu neuadd newydd (cais y bu cefnogaeth Albert ei hun iddo'n llugoer iawn) yn dwyn ffrwyth, ac na fyddai'r tir yn cael ei werthu o gwbl. A rŵan, i goroni'r cyfan, dyma'r bwbach hunanbwysig 'na'n bygwth dweud wrth bawb mai fel tai gwyliau y bwriadwyd y tai newydd ar bwys ei dir, ac nid fel tai i bobl leol. Hynny ydi, os nad oedd ef, Albert, yn fodlon gwneud iawn am y cyfran ariannol y byddai Prys wedi ei dderbyn pan fyddai telerau'r cytundeb a fyddai'n trosglwyddo perchnogaeth y safle i ddwylo Albert yn derbyn sêl bendith y Cyngor.

Damia Prys Harris! Damia Sara Watcyn! Damia Merched y Wawr a phob blydi mudiad arall oedd yn cyfarfod yn yr hofal 'na o neuadd! 'Toedden nhw ddim yn gallu gweld mai bendith fyddai dymchwel y lle a chael stad o dai twt ar y safle? Nac oeddent, debyg; roedden nhw'n dal i fyw yn yr oes o'r blaen, oes y seiat a'r sesiwn. Wel, roedd o, Albert Parry, am gamu'n feiddgar i'r unfed ganrif ar hugain, ac am gofleidio pob cyfle a gâi i ennill llewyrch a llwyddiant iddo'i hun. Efallai mai fel 'Bert bach' y byddai'n cael ei adnabod gan ambell hen goes yn y pentre 'ma am byth, ond yn sicr nid fel 'Bert bach' yr oedd o'n ei ystyried ei hun bellach. Naci wir!

Ond byddai'n rhaid iddo drio cymodi efo Anwen. Fel rheol, gwaith digon hawdd oedd ei throi o amgylch ei fys bach, ond roedd hi wedi styfnigo'n llwyr dros y calendar 'ma. Roedd o wedi hen laru ar yr awyrgylch yn y tŷ, a phetai hi'n dod i wybod am ei gynlluniau ar gyfer y tai yng ngwaelod yr ardd, byddai wedi ta-ta domino arno wedyn i drio seboni efo hi. Oedd, roedd o rhwng y diawl a'i gynffon, doedd dim dwywaith amdani.

Cuchiodd Albert ei aeliau wrth roi cic fileinig i'r llawr nes ei fod yn troelli'n wyllt yn ei gadair.

Prys Harris . . . m-m-m . . . byddai'n gadael i Prys Harris stiwio yng nghanol ei fygythiadau am sbel. Mi ddeuai'r cyfle i roi halen ym mhotes hwnnw yn y man. Gobeithio.

"Sws i Mam, Aled."

"A fi."

"A sws i Glesni."

"Pob hwyl i ti, Anwen. A paid â phoeni, mae Harri'n werth y byd."

Tynnodd Anwen wyneb ar ei ffrind wrth droi am adref. Tua hanner ffordd i lawr y llwybr at y giât trodd i godi ei llaw ar Nia a'r plant, ac o gornel ei llygad gwelodd gyrtans Elsi'n stwyrian. Llwyddodd i wrthsefyll yr ysfa i stopio'n stond gyferbyn â'r byngalo bendigaid a thynnu ei thafod ar y wraig fusneslyd. Cilwenodd wrth gyrraedd y giât a arweiniai i'r lôn. Roedd wedi amau ers tro fod y cyfuniad o ymdrechion Albert i'w chadw fel dol tsieina, a chwmni cyson plant ifainc, yn dechrau cael dylanwad andwyol ar ei meddwl. A rŵan dyma'r prawf, petai ei angen. Gwenodd drachefn wrth geisio dyfalu beth fyddai ymateb Albert petai Elsi'n dweud y drefn wrtho oherwydd bod ei wraig wedi cymryd at sefyll yng ngardd y byngalo a thynnu stumiau arni drwy'r ffenest. O leiaf roedd Nia'n edrych yn well, meddyliodd. Roedd golwg sobor o sâl arni ben bore pan gyrhaeddodd Anwen ac Aled yno i nôl Glesni.

Fedrai Anwen ddim dweud â'i llaw ar ei chalon ei bod yn edrych ymlaen at yr orchwyl o'i blaen. Ond am a wyddai, doedd y weithred arfaethedig ddim wedi peri cymaint o loes iddi hi ag a wnaethai i ambell un arall o'r criw. Y syndod mwyaf, meddyliodd, oedd y ffaith ei bod hi wedi dal at ei phethau a herio gwrthwynebiad a grwgnach parhaus Albert. A hithau'n amlwg yn cyfrannu tuag at ymgyrch a fyddai'n tanseilio unrhyw gynlluniau ar gyfer safle'r neuadd, roedd o'n eu cadw o dan big ei gap. Ond byddai wiw iddi adael y genod i lawr. Albert neu beidio.

Doedd dim smic i'w glywed o leoliad y gwaith adeiladu ger y tŷ; y dynion yn manteisio ar y tywydd braf ac yn cymryd awr ginio lawn, mae'n rhaid. Gwae nhw petai Albert yn eu darganfod yn gorffwys ar eu bri am eiliad yn rhagor nag oedd eu hawl, meddyliodd Anwen wrth gamu i'r gegin. Yn reddfol, aeth yn syth i'r cwpwrdd ac estyn potel a gwydr.

Safodd Anwen wrth y sinc yn y gegin a diolch eto fod rhyw broblem oedd yn gysylltiedig â joban reit broffidiol wedi mynd ag amser a sylw Albert ers diwrnod neu ddau bellach. Trodd ac estyn y botel olaf o'i lloches y tu ôl i'r llestri gorau, a'i gosod ar bwys y sinc gyda gweddill y rheng yr oedd wedi eu casglu at ei gilydd o'u cuddfannau o amgylch y tŷ.

Agorodd y botel gyntaf, ac yn araf bach tywalltodd ei chynnwys i lawr y sinc. Dechreuodd ei chalon fowndian yn erbyn ei hasennau fel pêl goch Aled pan fyddai ef a Delyth yn ei chicio hwnt ac yma yn yr ardd, ac ar ôl iddi osod y botel wag yn y bocs wrth ei thraed, sylweddolodd ei bod yn laddar o chwys oer. Pwysodd yn erbyn y sinc gan anadlu'n ddwfn, i geisio adennill ei chydbwysedd. Wrth iddi sefyll yno'n syllu i'r ardd gwelodd y bêl goch, wedi ei gadael yn ddidaro ger y tŷ bach twt, a llanwodd ei llygaid â dagrau parod wrth i euogrwydd ei meddiannu unwaith eto.

Y diwrnod blaenorol, a'r haf bach Mihangel yr oeddynt wedi ei brofi yn ystod yr wythnos yn bygwth tynnu i'w derfyn, roedd Anwen ac Aled wedi mwynhau picnic yn yr ardd amser cinio, ar ôl dychwelyd o'r cylch meithrin. Ond wedi i'r criw ar gyrion yr ardd ailafael yn eu gwaith, ciliodd y ddau ohonynt i'r tŷ, gan fod y mwstwr yn tarfu ar eu hymdrechion i hamddena a chael hwyl. Cyhoeddodd Aled ei fod yn mynd i'w llofft i lunio fferm fel fferm Glesni â'i flociau adeiladu a'i lu o anifeiliaid a pheiriannau plastig lliwgar. Siarsiodd ei fam i wrando am yr alwad i ddod i fwrw golwg dros y stad pan fyddai wedi ei chwblhau.

Bu Anwen yn tindroi yn y gegin am sbel gan chwilio am unrhyw dasgau oedd angen ei sylw. Ond newydd ddod i ben â'i gorchwylion am yr wythnos oedd y lanhawraig, felly roedd pob celficyn a theclyn yn ei le ac yn ddisglair a dilychwin. Yn y man tywalltodd ddiferyn bach iddi ei hun, ac aeth drwodd i eistedd ar y soffa yn y stafell fyw.

Y peth nesaf roedd Anwen yn ei gofio oedd Delyth yn galw arni wedi iddi gyrraedd adre o'r ysgol. Roedd hi wedi syrthio i gysgu. Am bron i ddwyawr! Cyfarchodd Anwen ei merch, ac aeth y ddwy i fyny'r grisiau i weld sut hwyl roedd Aled yn ei gael wrth adeiladu ei faenor. Ond doedd dim golwg ohono yn ei lofft. Roedd y llawr bron o'r golwg o dan doreth o frics llachar ac anifeiliaid o bob lliw a llun. Ond roedd y pensaer wedi diflannu.

Rhuthrodd Anwen o'r naill stafell i'r llall gan alw ar ei mab. Llifodd ton o orffwylltra drosti wrth iddi sylweddoli nad oedd Aled yn unman yn y tŷ. Roedd lled hysteria ei mam wedi peri i Delyth gynhyrfu drwodd, ac roedd hithau bellach yn beichio crio.

Wedi chwilio pob twll a chornel roedd yn rhaid wynebu'r posibilrwydd y gallai'r bachgen bach chwilfrydig fod wedi gosod ei fryd ar ryfeddodau swnllyd y safle adeiladu.

Petai hynny wedi digwydd . . . petai Aled wedi cael damwain . . . petai wedi dioddef niwed difrifol . . .

Ar ôl i'r plant gael eu swper, eisteddodd Anwen ar stepen y drws cefn yn gwylio'r ddau'n chwarae yn yr ardd. Diolchodd fod gwaith Albert yn galw, a'i fod wedi ei ffonio i ddweud wrthi y byddai'n hwyr arno'n cyrraedd adref. Allai hi ddim ei wynebu y funud honno.

Wedi i Delyth ddarganfod Aled yn cysgu'n sownd yn y tŷ bach twt, a'i fawd yn ei geg, roedd Anwen wedi gwasgu ei phlant mor dynn nes y bu ond y dim iddi hollti pob un o'u hesgyrn. Bellach roedd y ddau yng nghanol rhyw gêm ac wedi anghofio popeth am argyfwng y prynhawn, ond er nad oedd Aled yn ddim gwaeth ar ôl yr helbul, roedd y profiad wedi rhoi cryn ysgytwad i Anwen. Oedd, mi roedd hi wedi syrffedu ar ei bywyd; oedd, mi roedd hi wedi diflasu; oedd, mi roedd hi wedi laru. Ond nid ar Delyth ac Aled oedd y bai am hynny. Roedden nhw'n haeddu gwell; yn haeddu gwell ganddi hi. Ac nid ar Albert oedd y bai i gyd chwaith. Nid o bell ffordd. Ei bywyd hi, Anwen, oedd o, ac yn hytrach na chael ei chario gyda lli pobl eraill, roedd yn hen bryd iddi godi hwyl a dilyn ei llwybr ei hun, er na wyddai i ble fyddai'r llwybr hwnnw'n ei harwain. Ddim eto. Ond fyddai hi byth yn codi hwyliau fel hyn. Roedd yn rhaid i bethau newid. Roedd yn rhaid iddi hi ei hun newid . . . derbyn cyfrifoldeb drosti ei hun a hwylio yn nannedd y gwynt petai'n rhaid. Wedi'r cyfan, roedd hi wedi goroesi gwrthwynebiad Albert ac wedi cael ei hanfarwoli fel un o griw'r calendar. A doedd hi ddim am i Delyth ac Aled edrych arni fel rhyw froc môr dynol, yn glynu wrth y botel am bob cymorth a chynhaliaeth.

Gwenodd Anwen wrth weld Aled yn cael y gorau ar ei chwaer trwy eistedd arni a'i chosi'n ddidrugaredd. Cododd, ac aeth draw atynt i ymuno yn yr hwyl . . .

Ffliciodd Anwen y dagrau poethion o'i llygaid yn ddiamynedd. Daliodd bob potel uwchben y sinc yn ei thro, nes bod pob diferyn wedi diflannu i lawr twll y plwg. Byddai heno'n brawf ar ei phenderfyniad newydd, meddyliodd â gwên chwerwfelys. Byddai'n rhaid iddi fodloni ar lemonêd yn y lansiad answyddogol a gwylio gweddill y criw yn disbyddu ffynnon y Llew Coch. Ond nawr amdani neu ddim.

Roedd Albert wedi cael siars ers dyddiau i fod adref mewn da bryd, a byddai gweld y genod heno yn tynnu ei meddwl oddi ar ei gwendidau fel mam, boed y rheini wedi eu gorliwio gan bwysau euogrwydd neu beidio. Efallai y câi gyfle i rannu ei phrofiad poenus a'i phenderfyniad â

Nia, gan wybod ei bod yn siŵr o gael cefnogaeth a chymorth gan ei ffrind – a gweddill y criw hefyd – petai pethau'n mynd yn drech na hi.

Cododd Anwen y bocs a oedd bellach yn llawn o boteli gwag. Wedi cael gwared arnynt unwaith ac am byth, aeth i fyny'r grisiau i molchi a thwtio dipyn arni ei hun cyn mynd draw i'r cylch meithrin i nôl Aled.

# RHAGFYR

Roedd Sara wrthi'n tywallt paned iddi ei hun pan glywodd y post yn syrthio ar y mat ger y drws ffrynt. Yna clywodd lais Meirion y postmon yn cyfarch Gareth, mab Tina, wrth iddo yntau wthio'r papur dyddiol trwy'r blwch llythyrau. Cododd a mynd drwodd i'r cyntedd i weld a oedd unrhyw beth o bwys ymysg yr amrywiol amlenni. Yna tynnodd ei gŵn gwisgo cynnes yn dynnach amdani wrth setlo'n ei hôl wrth fwrdd y gegin i bori drwy arlwy'r bore.

Dau lythyr sgrwtsh yn cynnig benthyciadau helaeth iddi am bris baw (dyna oedd yr addewid a wnaed mewn llythrennau breision, o leiaf), a dwy daflen liwgar yn ceisio'i pherswadio i wario ffrwyth yr haelioni tragwyddol ar lu o drugareddau nad oedd arni eu heisiau na'u hangen.

A dau fil. Nododd Sara'r symiau oedd yn ddyledus a rhoddodd y biliau o'r neilltu am y tro. Yn hwyr neu'n hwyrach, meddyliodd, byddai'n rhaid iddi wynebu'r ffaith ei bod hi'n mynd yn fwyfwy anodd cadw ei phen uwchben y dŵr yn ariannol. Ymddangosai ei phensiwn yn anrhydeddus, ac yn wir, o'i chymharu â llawer un roedd hi'n ffodus iawn, ond roedd y tŷ capel yn llyncu arian, yn union fel y byddai glaslanc yn llyncu lluniaeth ar ôl peidio â bwyta am ryw ddwyawr. Roedd cost cynhesu'r tŷ'n sylweddol; roedd y trethi'n codi'n flynyddol; ac am y gwaith cynnal a chadw roedd ei angen arno . . .

Ond byddai gwerthu'r hen gartref yn gwneud iddi deimlo fel petai'n bradychu holl egwyddorion a gwerthoedd ei thad – ynghyd â llafur ei oes. Ac onid oedd hi wedi gwneud hynny unwaith yn barod?

Tywalltodd Sara baned arall iddi ei hun ac estyn y papur. Yn ôl ei harfer, wedi bwrw golwg dros y penawdau, trodd at dudalen yr hysbysiadau teuluol.

> OWEN, DYFAN, Ebrill 21, 2003, gartref ym Min y Môr, Llanllwyni, wedi gwaeledd byr, yn 70 mlwydd oed. Priod annwyl Julia, tad tyner Gwawr a Guto, taid arbennig Ifan a Deio. Gwasanaeth cyhoeddus yng Nghapel y Cwm, Llanllwyni am 12 o'r gloch, Ebrill 28, ac i ddilyn yn y fynwent. Blodau'r teulu yn unig ond derbynnir . . .

Gorweddai cwpan Sara'n deilchion, ac ymledai gweddillion ei the yn ddiferion dioglyd ar draws y llawr teils.

Fel un haf hirfelys hirhoedlog yr oedd Sara'n meddwl am ei phlentyndod bellach. Wrth ddwyn ei hatgofion i'w chof, teimlai fel petai'n agor caead y bocs tlysau a roddodd ei thad iddi pan oedd yn ferch fach. Ac roedd hi'r un mor sicr o hudoliaeth ei hatgofion heddiw ag y bu o ymddangosiad y ddawnswraig yn ei gwisg fale yn troi'n osgeiddig i dincial y gerddoriaeth wrth iddi agor y blwch. Anodd oedd i'w chof ddethol dyddiau a digwyddiadau neilltuol bellach, ond roedd un peth yn sicr – gyda'i thad y byddai Sara bob amser, ym mhob man, ac yn gwneud pob peth.

Bu mam Sara'n glaf am nifer o flynyddoedd, a datblygodd perthynas arbennig rhwng y ferch fach ddifrifddwys a'i thad, a oedd mor barod i roi o'i amser a'i egni nid yn unig i'w wraig a'i ferch, ond i bob un o'i braidd yn Nant-yr-Onnen yn ogystal. Er mai dyn ei filltir sgwâr oedd Daniel Watcyn, roedd hefyd yn ddyn dysgedig a deallus, ac er ei fod yn pryderu ynglŷn â chyflwr iechyd bregus ei wraig, roedd yn ŵr oedd yn fodlon ei fyd, gŵr yr oedd llewyrch ei ffydd i'w ganfod yn amlwg, gŵr yr oedd ei ymddiriedaeth yn ei Dduw yn ddiamod a diamau.

Erbyn i gyfnod Sara yn yr ysgol ddod i ben, roedd

cystudd ei mam wedi ei llethu'n llwyr, ac roedd Sara'n bwriadu aros gartref i gynorthwyo gyda'r gwaith o ofalu amdani. Roedd hi am geisio ysgafnhau tipyn ar faich ei thad, er ei fod ef yn ymgymryd â phob dyletswydd annymunol yn y clafdy â graslonrwydd diderfyn a diflino, a gwrthododd yn lân â chytuno â'r fath syniad. 'Toedd Sara wedi gweithio'n galed ac ennill canlyniadau ardderchog yn yr ysgol? Ac yn bwriadu llusgo byw yn y pentref 'ma yn lle gwneud defnydd addas o'i gallu? Dod i ben? Wrth gwrs y byddai ef a'i mam yn dod i ben. Doedd Sara ddim i ofidio am hynny. A doedd o ddim am glywed gair arall am y peth. Yng nghanol pobl ifanc llawen a llon oedd ei lle hi rŵan. Diar mi, bwriadu ymgilio yn y tŷ capel, wir! A chyn i Sara gael ei gwynt ati, roedd hi a'i thaclau'n ceisio'u gorau glas i ymgartrefu yn neuadd breswyl y coleg hyfforddi athrawon yn y ddinas.

Patrwm ei bywyd am rai misoedd wedyn oedd treulio diwrnodau'r wythnos yn y ddinas yn astudio, ac yna ymgymryd â'r daith lafurus i Nant-yr-Onnen ar nos Wener i fwrw'r Sul gyda'i rhieni. Anaml y byddai'n aros yn y neuadd dros y penwythnos, heblaw bod rhyw achlysur arbennig yn hawlio'i phresenoldeb, gan y byddai mwyafrif ei chyfeillion newydd yn y coleg hefyd yn ei throi hi am adref, ac roedd gwawr o euogrwydd am adael ei rhieni yn dal i liwio'r mwynhad a brofai yn ei bywyd newydd – nes iddi dderbyn lleoliad mewn ysgol nid nepell o'r coleg ar gyfer ei chyfnod o ymarfer dysgu, cyfnod a fyddai'n cwmpasu ei thymor olaf yn y coleg bron i gyd. Yno, yn yr ystafell foel ble roedd aelodau'r staff yn ymgynnull i gael paned a sgwrs yn ystod egwyl y bore a thros ginio, y newidiodd patrwm cyfforddus, clyd bywyd Sara.

Roedd wrthi'n llymeitian mygaid o goffi, ac yn diolch bod ei dwyawr gyntaf yn y dosbarth wedi pasio heb unrhyw anawsterau anorchfygol, pan ddaeth gŵr golygus i mewn i'r stafell. Fel sy'n gyffredin ymysg unrhyw griw sy'n adnabod ei gilydd yn lled dda, roedd yr athrawon wedi eu rhannu'n grwpiau ac yn mwynhau sgwrs, felly anelodd yr hwyrddyfodiad yn syth at y gadair wag yn ymyl Sara gyda'i baned.

A dyna sut y bu iddi, ymhen cwta fis, ddechrau gwneud esgusodion dros beidio â dychwelyd i Nant-yr-Onnen mor rheolaidd; dros beidio ag ymuno â'i chyfeillion am swper a sgwrs mor aml; yn wir, dros beidio ag ymrwymo i'r un dim a fyddai'n amharu ar ei safle a'i statws fel cywely Mr Dyfan Owen, dirprwy brifathro.

Ychydig fisoedd yn unig a fu er penodiad Dyfan i'w swydd. Cyn hynny roedd yn athro mewn ysgol yng nghefn gwlad, ond roedd yn ŵr uchelgeisiol, bron â thorri ei fol eisiau dyrchafiad, a chan nad oedd yna fawr o obaith y deuai hynny i'w ran yn yr ysgol honno, roedd wedi ymgeisio am y swydd yn y ddinas unwaith y gwelodd yr hysbyseb yn y papur. Ac wedi ei benodiad, dyma adael ei wraig Julia a'r plant ar ôl yng nghefn gwlad i wneud y trefniadau angenrheidiol ar gyfer gwerthu'r tŷ, a'i heglu hi oddi yno nerth ei draed.

Ac yno, wrth ddôr ei fflat ddigysur yn y ddinas, y gadawodd Sara'i gwerthoedd a'i daliadau, ei moesau a'i moryndod. Taflodd bob chwa o gallineb a gwyliadwriaeth i'r pedwar gwynt. Roedd hi dros ei phen a'i chlustiau mewn cariad. Yn glaf o gariad. Cariad cyntaf. Cariad llo bach. Cariad cris-croes, tân poeth, torri 'mhen a thorri 'nghoes.

Ac felly y treuliodd ei misoedd o ymarfer dysgu – yn gwneud ymdrech oruwchnaturiol i gadw ei meddwl ar ei gwaith yn y dosbarth; yn sgriblo gweddill ei gwaith cwrs rywsut rywsut, ben set; yn gwneud esgusodion dros beidio â dychwelyd adref i weld ei rhieni a threulio amser gyda'i chydfyfyrwyr, ac, yn fwy na dim, yn ymroi, â phob defnyn o'i chorff a'i chalon, i'w pherthynas â Dyfan. Llyncai yntau bob diferyn a roddai hi iddo, yn chwannog a chwantus. A pharai blys Sara'i hun amdano yntau iddi wrido a chywilyddio gefn golau dydd.

Yna, rai wythnosau cyn diwedd y tymor, derbyniodd lythyr gan ei thad yn dweud wrthi fod yna swydd wag yn

Ysgol Gynradd Nant-yr-Onnen ar gyfer tymor yr hydref. Os oedd ganddi ddiddordeb. Doedd hi ddim i feddwl ei fod ef yn busnesa, nac yn pwyso arni i ddychwelyd i'r pentref. Dim ond rhoi gwybod iddi, dyna i gyd.

Gwyddai Sara fod cyflwr ei mam yn gwaethygu, ac y dylai fod yn falch o'r cyfle i gael gweithio yn Nant-yr-Onnen a chynnig cefnogaeth i'w rhieni, ond pan glywodd yr wythnos ganlynol y byddai swydd ar gael yn yr ysgol ble roedd yn gwneud ei hymarfer dysgu, ni feddyliodd ddwywaith cyn cyflwyno cais am y swydd honno – a'i derbyn pan y'i cynigiwyd iddi.

Bu'r haf hwnnw'n gyfnod affwysol o rwystredig. Roedd bywyd llesteiriol ei rhieni yn y tŷ capel yn ei llethu; roedd brath ei gwayw yn ei gwneud yn ddiamynedd a didostur; roedd ei dyhead i ddychwelyd i'r ddinas ac i gôl ei chariad yn gwneud pob dydd a dreuliai yn ei chartref yn ddi-werth a di-liw.

Ni chwestiynodd ei thad ei phenderfyniad i droi'n ôl am y ddinas. Cofleidiodd hi'n dyner wrth iddi adael, gan godi i'r wyneb donnau o euogrwydd ac edifeirwch a fu'n bygwth ei boddi gydol ei siwrne'n ôl.

Roedd Sara wedi rhentu fflat iddi ei hun ar gyfer y tymor newydd, gan fwriadu ei haddurno'n gelfydd a gofalus nes y byddai'n nythle diddos iddi hi a'i chariad. Ond bellach roedd Dyfan a Julia a'r plant wedi ymgartrefu mewn tŷ sylweddol ar gyrion y ddinas, a'r peth olaf ar feddwl Dyfan oedd ailafael mewn cogio caru efo Sara. Yn wir, prin y byddai'n ei chydnabod yn yr ysgol; gwnâi ddim mwy na'r hyn oedd yn angenrheidiol i ymddangos yn sifil. Treuliodd Sara'r tymor trallodus hwnnw yn tagu o dorcalon; ei chywilydd yn ei llethu a'i llosgi am yn ail, a'i hymwybyddiaeth o'i ffolineb a'i phenwendid yn peri iddi ei chystwyo a'i cheryddu ei hun yn ddidrugaredd. Gofynnodd i'w thad gadw ei lygaid ar agor am swyddi yn ardal Nant-yr-Onnen iddi. Dywedodd wrtho ei bod wedi ailfeddwl am fywyd dinesig, a'i bod am ddychwelyd i'w chynefin.

Pan gysylltodd ei thad â hi gyda'r newydd nad oedd yr athrawes a benodwyd i Ysgol Nant-yr-Onnen wedi gallu ymgartrefu yng nghefn gwlad, a'i bod am ddychwelyd i ardal fwy trefol cyn gynted ag y byddai modd penodi athrawes yn ei lle, golchodd pwl o ryddhad a gollyngdod dros Sara fel y seithfed don. Ac wedi iddi gael ei bychanu yn y parti Nadolig, pan fu'n rhaid iddi ddygymod â Dyfan yn gafael yn dynn dynn yn Julia wrth ddawnsio gyda hi o dan ei thrwyn, tyngodd Sara lw na fyddai byth eto'n cael ei thwyllo yn y fath fodd, nac yn syrthio i fagl hunan-dwyll. Fyddai hi byth eto'n ymgreinio o flaen allor cariad celwyddog. Na blys bradychus. Na thrachwant twyllodrus.

Croesawyd Sara adref yn gariadus ac yn addfwyn gan ei rhieni, ac os oeddynt yn chwilfrydig ynghylch ei newid meddwl, ei hwyliau chwim-chwam a'i gwelwder, ni ofynnwyd yr un cwestiwn. Wedi Nadolig distaw a phrudd a ddynododd ddechrau'r diwedd i'w mam, cychwynnodd Sara ar ei gwaith yn Ysgol Nant-yr-Onnen ym mis Ionawr yn berson fesur mwy dwys a difrifol na'r ferch ddyfal ond diniwed a adawodd am y ddinas yn llawn hwyl a hyder rai misoedd ynghynt. Ac yno y bu Sara, trwy wythnosau olaf cystudd ei mam; trwy flynyddoedd olaf ei thad pan welodd, wedi cau'r capel, y gŵr tirion, triw fu'n bresenoldeb mor gadarnhaol yn ei bywyd yn digalonni nes ei fod fel cysgod o'r hyn a fu; a thrwy fisoedd olaf ansicr yr ysgol, pan leihaodd nifer y plant i'r fath raddau fel mai hi oedd yr unig athrawes ar ôl.

Ond roedd hi wedi dod i ben, wedi dygymod â'r profedigaethau a ddaethai i'w rhan, wedi ymdopi â'i chaledi a'i cholled. Yn wir, bellach roedd yn mwynhau'r hamdden oedd ganddi i ymhel â phwyllgorau niferus mudiadau'r pentref, ynghyd â photsian o amgylch y tŷ capel dan ei phwysau. Gwyddai y deuai'r dydd pan fyddai'n rhaid iddi feddwl o ddifrif am werthu'r hen gartref a bodloni ar dŷ bach neu fyngalo a fyddai'n fwy addas ar ei chyfer hi a'i hadnoddau ariannol. Rywbryd byddai'n rhaid cydnabod bod yr holl orchwylion a ddeuai yn sgil bod yn berchen clamp o dŷ ac arno doreth o waith cynnal a chadw yn drech na gwraig sengl ganol oed. Ond pethau i boeni amdanynt ryw dro eto oedd y rheini. Rhyw dro yn dyfodol. Y dyfodol pell.

Serch hynny, yn ogystal â rhoi halen ar hen friw nad oedd erioed wedi llwyr fendio, gorfodwyd Sara gan yr hysbysiad yn y papur i wynebu ei meidroldeb. Ond doedd hi ddim yn teimlo'n hanner parod i gyfarfod â'r ddrychiolaeth honno wyneb yn wyneb. Ddim eto. Ddim o bell ffordd.

```
. . . penderfynwyd cyfarfod eto ymhen wythnos i
geisio dod i gytundeb ynglŷn â'r ffordd orau i
symud ymlaen wrth wynebu'r argyfwng yma.
Yn y cyfamser, diolchwyd i'r Is-gadeirydd am ei
gynnig a mynegwyd gwerthfawrogiad y Cyngor.
Dim materion eraill.
Dyddiad y cyfarfod nesaf - 31 Mai 2003 am 7.30.
```

Rhoddodd Sara lyfr cofnodion y Cyngor Cymuned yn ôl yn ei briod le yn nrôr chwith y seidbord.

Roedd llythyr y Cyngor Sir wedi cynhyrfu'r dyfroedd yn o arw, ac roedd yn rhaid rhoi clod ble roedd yn ddyledus a diolch i Albert Parry am fod mor barod i helpu. Er bod y swm a grybwyllwyd yn uchel, roedd yn rhesymol iawn o ystyried y gwaith y byddai'n ymrwymo i'w wneud. Go brin fod ganddo gymhelliad cudd dros wneud y cynnig; wyddai Albert yr un dim am y peth tan heno, fel y gweddill ohonynt – heblaw Cledwyn a oedd wedi derbyn y llythyr, ac a alwodd y cyfarfod brys. Na, hi oedd yn ddrwgdybus o Albert oherwydd ei fod wedi treulio cyfnod maith ar ei sodlau yn ceisio'i pherswadio i werthu'r tŷ capel iddo.

Gwnaeth Sara baned iddi ei hun, ac er iddi gynnau'r teledu ar gyfer y newyddion, thalai fawr o sylw i ddigwyddiadau'r dydd gan fod yr wltimatwm dirybudd, a'r ymgais am ymateb derbyniol ac ymarferol iddo, yn hawlio'i holl sylw. Yn wir, roedd yn canolbwyntio cymaint ar broblem y neuadd nes bu ond y dim iddi fethu stori fach ddifyr ar gynffon y rhaglen am griw o aelodau Sefydliad y

Merched yn creu calendar bronnoeth er mwyn codi arian at achos da.

Horlicks 'ta Ovaltine? Bovril 'ta bîff tî? Gwenodd Sara wrthi ei hun wrth astudio'r arlwy'n un rheng o'i blaen yn y cwpwrdd. Byddai'n rhaid iddi wahodd rhai o drigolion y pentref draw am barti coctels a hithau'n berchen ar y fath ddarpariaeth. Pob copa walltog oedd dros ei ddeg a phedwar ugain, efallai; byddent yn gwirioni'n lân dan ddylanwad y fath ormodedd.

Horlicks amdani, 'ta.

Tra oedd yn disgwyl i'r llefrith gynhesu edrychodd Sara ar gloc y gegin. Pum munud wedi dau, a deng munud ers iddi edrych ar ei chloc bach ar y cwpwrdd wrth erchwyn y gwely am y canfed tro a sylweddoli nad oedd ganddi ddim tamaid mwy o siawns syrthio i gysgu'r funud honno nag oedd ganddi o hopian i ben yr Wyddfa a wardrob wedi ei chlymu rownd ei gwddw, a phiano ar ei chefn. Cymysgodd y dracht cysgu, ymlwybro yn ei hôl i'w llofft ac eistedd yn ei gwely i'w lymeitian.

Roedd yn amau erbyn hyn mai breuddwyd fu holl ddigwyddiadau'r noson honno. Do, yn sicr ddigon, cynhaliwyd cyfarfod misol cangen Merched y Wawr Nant-yr-Onnen. Ond y gweddill? Effaith y caws ar dost a gafodd i'w the oedd hynny, siŵr iawn.

Doedd hi ddim wedi bwriadu am eiliad roi'r fath gynnig gerbron. Prin ei bod yn cofio manylion y pwt o stori a welodd yn ystod y rhaglen newyddion a dweud y gwir. Ac yn ddi-os, pan agorodd ei cheg yn y cyfarfod a chlywed ei llais ei hun yn gwneud awgrym mor ynfyd, credodd ei bod wedi ei meddiannu gan ryw andras anweledig.

Ac er iddi lwyddo i gadw wyneb yn ystod y cyfarfod, a hyd yn oed fwynhau ymateb taer a dadleuon huawdl y merched, ar yr un pryd teimlai fel petai am gael cyffylsiwn cyn ffrwydro'n gwbl ddigymell yn y fan a'r lle.

Erbyn diwedd y cyfarfod roedd pob un o'r criw'n byrlymu â brwdfrydedd. Neu'n ymddangos felly, o leiaf. Ond sut byddent yn teimlo yng ngolau dydd? Sut byddai hi ei hun yn teimlo?

Rhoddodd Sara'i chwpan wag ar y cwpwrdd wrth erchwyn y gwely a diffodd y golau.

". . . ac mi fydd Sioned yma erbyn tua hanner awr wedi wyth, medda hi."

"Iawn, Marian, mi ddo inna tua'r un pryd."

"Diolch, Sara. Mae'n ddrwg gen i fod hyn mor ddirybudd, ond dywedodd Mererid mai gora po gynta y bydden ni'n trefnu dyddiad efo Harri."

"Dim o gwbl. Dwi'n dallt yn iawn. A dwi'n cymryd bod pawb arall yn fodlon i ni bennu dyddiad drostyn nhw?"

"Ydyn, am wn i. Os awn ni o gwmpas pawb yn holi pa ddyddiad sy'n gyfleus ddown ni byth i ben. Ac wrth gwrs mae popeth yn dibynnu ar pryd y bydd Harri'n rhydd."

"Ac o leia fe fydd 'na ddigon o amser i bawb wneud unrhyw drefniadau angenrheidiol."

"Bydd. Reit, 'ta, mi fydda i'n eich disgwyl chi a Sioned nos Sul felly."

"Hanner awr wedi wyth nos Sul. Diolch yn fawr, Marian."

"Popeth yn iawn. Hwyl am y tro."

"Hwyl fawr."

". . . ac mi ro i wybod i Mererid cyn gynted ag y bydd y profluniau'n barod."

"Diolch o galon i chi, Harri."

"Diolch i *chi* – mae wedi bod yn bleser ac yn fraint cael cydweithio â chriw mor lew."

"Wel, wn i ddim pa mor lew oedden ni i gyd erbyn heddiw, a'r awr dyngedfennol ar ein gwarthaf, ond o leia mi ddalion ni'n tir."

"Do, wir, ac mi fyswn i'n tynnu fy het i bob un wan jac ohonoch – tasa gen i un!"

"Diolch."

"Reit 'ta, Sara, dwi am ei throi hi er mwyn danfon y ffilmiau 'ma i'r stiwdio a chodi rhai gwag ar gyfer mynd i ryw siop siarad o swper sych heno. Mi fydd y lle'n llawn o ddynion busnes boliog, hunanbwysig yn brolio'r naill a'r llall bob yn ail â'u brolio nhw'u hunain. Ddim hanner mor ddymunol â'r gwaith fues i'n ei wneud yn ystod y dydd."

"Gwell chi na fi, Harri."

"Ia, debyg. Wel, diolch am y baned, ac mi fydda i mewn cysylltiad cyn bo hir. Hwyl fawr i chi rŵan."

"Hwyl fawr, Harri. A diolch eto."

Wrth yrru'n ôl i gyfeiriad Tresarn, a'r ffilmiau hollbwysig yn ddiogel yn eu blychau, ni allai Harri lai na rhyfeddu at ddigwyddiadau'r diwrnod. Pan gyflwynwyd cynllun y calendar iddo gyntaf, roedd wedi amau mai criw o wragedd busneslyd llawn gweithredoedd da a gafodd ddiferyn yn ormod o sieri oedd y tu ôl i'r syniad. Neu nythaid o hen ieir a fyddai'n fflapian eu hadenydd ond yn cyflawni dim heblaw gwastraffu ei amser. Neu fintai o orgoniaid oedd am hawlio grym gwragedd, doed a ddelo. Ond roedd wedi cydnabod ei gamgymeriad o fewn ychydig funudau o'r cyfarfod cyntaf hwnnw â Mererid. Ac yn sicr roedd wedi syrthio ar ei fai ar ôl cwrdd â gweddill y merched.

Doedd aelodau Merched y Wawr Nant-yr-Onnen ddim yn ferched anghyffredin (heblaw am Mererid, ond doedd ei deimladau tuag ati hi ddim yn gwbl ddiduedd), ond ar y llaw arall, doedden nhw ddim yn ferched cyffredin chwaith, roedd hynny'n amlwg. Ac am Sara . . . m-m-m . . . aderyn y nos oedd hi, ym marn Harri. Er iddo gwrdd â hi nifer o weithiau yn ystod y misoedd diwethaf, amhosibl oedd dirnad ei theimladau a'i meddyliau. Ond heddiw, wrth gael tynnu ei llun, a'i gwallt arianwyn wedi ei ryddhau o'i dorch arferol, gwyddai Harri iddi fod yn syfrdanol o hardd yn ei hieuenctid. Ac eto, yn ôl y sôn roedd hi wedi byw yn y tŷ capel gydol ei hoes bron iawn, ac yn dal ei gafael yn yr hen gartref fel gefel, er bod ei thad wedi marw ers blynyddoedd bellach. Ac yno'r eisteddai heddiw, yn gefnsyth a balch, i gael tynnu ei llun, yn feistres ar y tlodi bonheddig o'i hamgylch, ond ar yr un pryd roedd yna rywbeth pruddglwyfus yn ei chylch.

Wedi i Harri gyrraedd y stiwdio aeth Sara'n angof ganddo pan faglodd ar draws ei gyfrifiadur a chofio bod Mererid yn dod draw dros y penwythnos i'w helpu i'w osod yn ei le.

Ond ble byddai hi'n dechrau? Dyna'r cwestiwn.

Cerddodd Sara drwodd o'r gegin i'r stafell fyw ac yna i'r parlwr. Ceisiodd ystyried cynnwys y stafell yn wrthrychol. Oedd, mi roedd hon yn hen stafell drymaidd. Yn wir, fyddai hi byth bron yn rhoi ei thrwyn o amgylch y drws erbyn hyn. Yn y stafell hon y byddai ei mam yn gorffwys ar y soffa flodeuog cyn i'w salwch ei chaethiwo i wely cystudd, ac yn y stafell hon y treuliodd ei thad oriau lawer wedi cau'r capel, yn ceisio amgyffred cynllun yr Hollalluog.

Na, doedd yna fawr o atgofion melys yn perthyn i'r mawsolëwm mwll hwn o barlwr.

Camodd Sara at y ffenest ble roedd pelydryn o haul gwantan yn gwneud ei orau glas i dreiddio i fwrndra'r stafell. Rhwbiodd ymyl y llenni rhwng ei bys a'i bawd; roedd y melfed trwchus yn llychlyd ac wedi colli ei liw yn y canol lle byddai'r haul yn ei daro yn y bore, a'r llenni net hwythau wedi troi'r un lliw â chroen pwdin reis dan lygad yr haul. Pwdin reis go iawn, hynny ydi, fel y byddai hi'n ei gael i ginio dydd Sul pan oedd yn ferch fach. Ddim y llymru llawn lympiau a gâi ei werthu mewn tuniau erbyn hyn.

Wrth droi'n ôl at ganol y stafell, gwelai Sara fod y dodrefn tywyll trymllyd yn llethu'r lle, ond wedi dogn helaeth o eli penelin, meddyliodd, byddent yn sicr o fod at ddant rhyw brynwr cefnog yn rhywle a oedd yn gwirioni ar hen gelfi. Teimlodd ias yn cripian i lawr ei hasgwrn cefn. Doedd hon ddim yn stafell i fagu gwaed ynddi.

Wrth gau'r drws ar ei hôl a dychwelyd i'r gegin, doedd Sara ddim tamaid nes at ddatrys ei phenbleth.

". . . a fedra i ond ymddiheuro unwaith eto, Gwyneth, am fod mor esgeulus."

"Does dim angen, Sara, wir i chi."

"Diolch o galon, Gwyneth . . . mi wela i chi ymhen rhyw awr."

"Yr awr fawr, Sara."

"Ia wir, yr awr fawr."

"Wel, hwyl fawr tan toc."

"Hwyl, Gwyneth . . . a diolch."

Aeth Sara i fyny i'r llofft i newid, gan ddweud y drefn wrthi ei hun wrth fynd. Beth ddaeth dros ei phen hi'n mynd i fyny i'r daflod ar ôl cinio? Ac wedyn treulio'r prynhawn ar ei hyd yno'n stwna a hel meddyliau yng nghanol trugareddau'r gorffennol. A hithau wedi addo mynd draw i'r Llew Coch i helpu Gwyneth gyda'r bwyd. Er ei bod wedi bod yn glên iawn ar y ffôn, roedd Gwyneth yn siŵr o feddwl bod Sara'n hollol ddi-hid. Neu'n waeth fyth, yn colli arni ei hun. Ac mae'n debyg mai yn y lofft y byddai hi byth heblaw bod rhyw ohebydd o'r wasg leol, nad oedd wedi derbyn fawr o groeso yn y siop na'r Llew Coch, wedi cyrraedd y tŷ capel a chanu'r gloch yn ffrwst i gyd yn ei eiddgarwch i gael y blaen ar bawb a fyddai'n rhan o'r lansiad swyddogol drannoeth.

Agorodd Sara'i wardrob. Byddai'n rhaid iddi wneud penderfyniad. Ac nid ar gownt beth i'w wisgo yn unig. Byddai'n rhaid iddi benderfynu a oedd hi'n bosibl, ac yn llesol, iddi ddal ei gafael yn y tŷ capel; roedd yn bryd iddi roi'r gorau i ogor-droi.

Ond nid heno. Noson i'r genod oedd hi heno. A gyda lwc, yn ystod y misoedd nesaf byddai dyfodol neuadd y pentref yn cael ei ddiogelu, a'i osod ar sylfeini cadarn. Yna efallai y byddai hi'n gallu rhoi ei gorffennol – ei fuddugoliaethau a'i brofedigaethau – y tu ôl iddi, a symud ymlaen.

# PIN-YPS

Roedd lolfa'r Llew Coch yn llawn mwstwr gorfoledd, a phob un o'r merched wedi derbyn copi personol o'r calendar. Roeddent yn rhyfeddu at y lluniau ac yn canmol Harri, unig ŵr gwadd y noson, i'r cymylau. Aethai'r ymateb i'r fenter y tu hwnt i holl obeithion y criw, a gwerthwyd bron pob copi o'r calendar ymlaen llaw. Byddai'r arian yn y banc ymhen ychydig wythnosau ac yna byddai'n bosib cychwyn ar y gwaith o ddymchwel y neuadd. Wedyn byddai'r caban yn cael ei osod ar y safle, a chyda'r cais am arian Loteri yn nwylo'r pwyllgor perthnasol ers peth amser, roedd dyfodol man cyfarfod y pentre'n edrych yn llawer mwy diogel nag yr oedd rai misoedd ynghynt.

Daeth Tina draw at y merched a photel o siampên ym mhob llaw.

"Gan Cledwyn," meddai, gan amneidio at y bar lle safai ambell un o gefnogwyr y merched, ynghyd â selogion y tŷ.

Ymddangosodd Gwyneth gyda hambwrdd yn llawn gwydrau, ac agorwyd y siampên yng nghanol môr o ewyn a chwerthin . . .

Am unwaith roedd Albert wedi gorfod gadael ei ymerodraeth i ofalu amdano'i hun am awr neu ddwy, ac roedd yn y tŷ yn gwarchod Delyth ac Aled. Wedi setlo'r plant yn eu gwlâu estynnodd gan o gwrw ac eistedd yn wyneblaes wrth fwrdd y gegin i ailddarllen y llythyr swta a gawsai gan Prys Harris ar ran y Cyngor Sir y bore hwnnw. Wel, dyna un cynllun i droi grôt yn chwech a fyddai'n gadael twll go ddwfn yn ei boced.

Ar ben hynny roedd wedi ffraeo efo Anwen eto heno cyn iddi adael i ymuno â gweddill y criw yn y Llew Coch. Roedd llythyr Prys wedi ei daro oddi ar ei echel, heb sôn am y lansiad di-fudd 'ma. Doedd gan Anwen ddim syniad sut y dylai gwraig o'i statws hi ymddwyn. Roedd hi wedi mynd yn erbyn ei ewyllys – yn wir ei orchymyn – fel ag yr oedd hi, drwy gymryd rhan yn y cynllun cywilyddus. Byddai ei gysylltiadau busnes yn saff o wneud sbort am ei ben wedi gweld ei wraig yn gwneud y fath sioe ohoni'i hun, ac yn ei ystyried yn bric pwdin o'r radd flaenaf. Ac i goroni'r cwbl roedd Anwen wedi canfod rywsut sut roedd y gwynt yn chwythu ynglŷn â'r tai newydd ar derfyn yr ardd, ac wedi mynd yn horlics llwyr. Fyddai'n synnu dim petai Prys Harris wedi gadael y gath o'r cwd, yr hen gadiffan sbeitlyd iddo fo.

Wedi llowcio sawl llymaid o'r hylif gwinau, eisteddodd Albert yn ôl a syllu ar weddill cynnwys ei wydr yn synfyfyriol. Y funud hon, edrychai ei wydr yn hanner gwag yn hytrach nag yn hanner llawn.

Roedd pawb wedi gwirioni ar y sampler. Roedd enw pob un ohonynt yno, a'r gwaith wedi ei addurno â brodwaith cywrain.

"Mae o'n gampwaith, Mererid," canmolodd Sara.

"Ydi wir," cytunodd Marian. "Gwylia di dy hun, rŵan, neu mi fyddwn ni i gyd yn mynd ar d'ofyn di pan fydd gennym achlysur arall fel hwn i'w ddathlu."

"Achlysur arall fel hwn? Bobol bach, Marian, peidiwch â sôn am y fath beth." Roedd Phyllis wedi cynhyrfu drwyddi.

"Peidiwch â phoeni, Phyllis," cysurodd Sara hi, "go brin y byddwn ni'n gwneud dim byd fel hyn eto. Cynllun unwaith ac am byth a grewyd mewn cyfyngder oedd hwn."

"A beth bynnag," rhoddodd Brenda ei phig i mewn, wrth i ragor o haelioni edmygwyr y merched gyrraedd ar hambwrdd, "petai'r angen yn codi am ymgyrch o'r fath eto, siawns gen i mai tro'r dynion fyddai hi."

"Wel, ar ôl ein haberth ni, fyswn i'n disgwyl dim llai na'r *full monty* ganddyn nhw," meddai Tina, wrth orffen gwagio'r hambwrdd.

Rhoddodd Harri ei ddwylo dros ei glustiau. "Dwi isio mynd adra," ffug gwynodd. "Hogyn bach diniwed fel fi yng nghanol criw o ferched drwg fel chi – mi rybuddiodd Mam fi amdanoch chi a'ch tebyg . . ."

Yn y swyddfa ddosbarthu yn Nhresarn roedd dynion a merched y Post Brenhinol yn dechrau ar y shifft nos. A phetai rhywun wedi chwilio'n ddyfal ymysg y miloedd llythyrau oedd yno, byddent wedi dod ar draws un wedi ei gyfeirio at Cledwyn Tomos, Cadeirydd Cyngor Cymuned Nant-yr-Onnen. A logo'r Loteri ar yr amlen.

"Dewch, Sara," anogodd Sioned, "dywedwch air bach."

Cododd Sara ar ei thraed.

"Wel, genod," cychwynnodd.

"A Harri."

"Ia wir, a Harri. Mae gennym ni, aelodau Merched y Wawr Nant-yr-Onnen, le i ymfalchïo. Ymfalchïo yn ein safiad a'n hymroddiad. Mi wn i fod rhai ohonoch chi'n credu 'mod i'n hollol wallgo pan awgrymais i gynhyrchu'r calendar 'ma – yn wir, roeddwn i'n amau hynny fy hun ar brydiau, ond bellach aethpwyd â'r maen i'r wal, a diolch i Albert Parry fe fydd gennym ganolfan ar gyfer ein cyfarfodydd yn ystod y gaeaf. Diolch i chi, bob un ohonoch."

Eisteddodd Sara i gyfeiliant bonllef o gymeradwyaeth.

Ymgodymodd Anwen â'i hunanreolaeth. Nid dyma'r amser i ddweud wrth bawb gythraul mor dan din, diegwyddor ac awdurdodol oedd ei gŵr, ond roedd y penderfyniad cadarn a wnaethai'r bore hwnnw'n prysur fynd i'r gwynt. Ac wedi'r ffrae danbaid honno cyn iddi adael y tŷ, a'r ffaith bod ffrydiau o hylif adfywiol, bywiocaol yn llifo o'i blaen, doedd ryfedd na fu'n anodd iddi ddwyn perswâd arni'i hun na fyddai un bach yn gwneud fawr o ddrwg.

Cododd Nia ei gwydr. "I Sara," meddai, "am ein llusgo oddi ar y llwybr cul, a'n perswadio i luchio'n synnwyr a'n swildod i ebargofiant."

"I Sara," atseiniodd pawb.

Stryffagliodd Sioned ar ei thraed. "Wel," meddai'n araf, "fyswn i byth wedi credu chwe mis yn ôl y byddwn i'n eistedd yn fan'ma heno'n dathlu'r fath ddigwyddiad. Roeddwn i'n ymfalchïo fy mod i'n berson a fyddai'n fodlon gwarchod fy nheulu a'm cymdogaeth, ond sylweddolais mai egwyddorion cadair freichiau oedd gen i. Heb gefnogaeth y gweddill ohonoch chi, mi fyswn i wedi wfftio'r cynllun yma ganwaith, à dwi'n dal i amau ar adegau a oedd cymryd rhan ynddo yn beth doeth. Ond bid a fo am hynny, yr hyn dwi'n credu rydyn ni wedi ei brofi yw – gyda chyfeillgarwch a chefnogaeth ein gilydd, does dim pen draw i'r hyn y gallwn ei gyflawni."

Clywyd sawl "amen" wedi'r araith yma, ac edrychodd y merched ar ei gilydd yn synfyfyriol.

Ond cyn i'r awyrgylch droi'n rhy ddwys neidiodd Harri ar ei draed. "Fel yr unig ddyn yma heno," meddai, "a diolch yn fawr am y gwahoddiad," ychwanegodd, "er, roedd gen i amheuon ynglŷn â'i dderbyn, rhag ofn i chi 'mwyta i'n fyw . . ."

" Mae 'na ddigon o amser yn weddill," galwodd Tina o'r bar.

"Fel y dywedais," aeth Harri yn ei flaen yng nghanol y rhialtwch. "Fel yr unig ddyn yma heno, mi fyswn i'n lecio deud gair bach. Dwi wedi gweithio efo llawer o bobol ers imi ddechra ar fy ngyrfa fel ffotograffydd, ond yn fy marn i, fel dyn di-ddallt a di-asgwrn-cefn, rydach chi i gyd yn deilwng o bob canmoliaeth. Ac yn fwy na hynny, rydach chi i gyd yn llawn haeddu eich statws fel pishys y pentre." Cododd ei wydr. "I bin-yps Nant-yr-Onnen."

"Pin-yps Nant-yr-Onnen!"

# CYNFFON CI CADNO

Ystafell 8b
Cysgodfan
Rhyd-y-Gro
Mwynfil

26 Tachwedd 2003

Annwyl Nia,

Diolch i ti am dy gerdyn. Dwi'n ymddiheuro am beidio sgwennu ynghynt, ond wir i ti, tydw i erioed wedi teimlo mor uffernol yn fy mywyd. A dweud y gwir, dwi'n dal i deimlo'n uffernol. Heblaw am Delyth ac Aled, tydw i ddim yn credu y byddwn i'n gallu magu'r plwc na'r nerth i drio ffeindio fy ffordd allan o'r pwll diwaelod 'ma - dwi'n teimlo fel taswn i'n trochi mewn triog.

Tydw i ddim yn siŵr faint o'r stori rwyt ti'n ei wybod - mae'n debyg mai rhyw fersiwn wedi ei hidlo mae Albert wedi ei weini i bawb - ond ar ôl i mi golapsio wedi'r lansiad swyddogol, a chael fy rhuthro i'r ysbyty, bu'n rhaid i mi gydnabod maint fy mhroblem yfed. Wrth gwrs, roedd Albert yn gandryll - ei wraig barchus, drwsiadus ef yn gaeth i'r botel? Nag oedd, siŵr; celwydd noeth. Ac wedi bytheirio a chwarae'r diawl efo hynny o staff oedd ar ddyletswydd, mi drefnodd i mi gael fy nghludo i fan'ma'r bore wedyn. Ac yma ydw i byth.

Diolch o galon i ti am ofalu am Delyth ac Aled nes i Mam a Dad gyrraedd. Tydw i ddim yn gwybod beth fyddai wedi digwydd iddyn nhw heblaw amdanat ti. Dwi'n cael eu hanes gan Mam bob dydd ar y ffôn, ond wnaiff Albert ddim dod â nhw i'm gweld i, na gadael i Mam a Dad ddod â nhw. Rhyw unwaith yr wythnos y daw Albert yma, ac a dweud y gwir does gennym ni fawr i'w ddweud wrth ein gilydd. Daeth y ffrae honno gawson ni noson y lansiad answyddogol, pan ddeallais fel yr oedd wedi palu celwyddau wrtha i ar gownt y tai newydd yng ngwaelod yr ardd, yn agos i orffen pethau. A bod yn onest, mi fydda i'n reit falch ei weld o'n gadael ar ôl ei ymweliadau.

Dwi'n dechrau cryfhau'n gorfforol erbyn hyn, ac yn benderfynol o fynd adre cyn Dolig. Dwi'n swp sâl eisiau gweld Delyth ac Aled, ond ddim yn edrych ymlaen at fod dan yr un to ag Albert eto. Tydw i ddim yn credu y bydd o byth yn gallu maddau imi am ddwyn y fath anfri arno. Wn i ddim beth ddaw ohonon ni. Ac mi fydd hi'n anodd iawn wynebu pobl Nant-yr-Onnen hefyd. Wedi'r cwbl, beth bynnag yw'r stori mae Albert wedi'i lledaenu, mae'n siŵr y bydd sibrydion a sisial amdana i o gwmpas y lle.

Ond adennill fy nerth yw'r peth pwysig rŵan, a chael dod adre at Delyth ac Aled. Yn y flwyddyn newydd bydd rhaid i mi bwyso a mesur fy sefyllfa, a thrio cael trefn ar fy mywyd.

Rhaid i mi gloi am y tro; mae'n amser y seiat ddyddiol – 'Pam y deuthum yn alcoholig' – fel taswn i angen f'atgoffa!

Diolch o galon i ti am bob dim – mi fydda i'n edrych ymlaen at dy weld ar ôl dod adre (os byddi di am fy ngweld i, hynny ydi). Sgwenna os bydd amser gen ti. Sut mae pethau efo criw'r calendar? Ydi o'n dal i werthu? Ydi Elsi wedi pwdu am byth?

Cofion at Ifan, a sws bob un i Dafydd a Gruffudd a Glesni.

Diolch am fod yn ffrind mor driw.

Hwyl am y tro, a chariad mawr,

Anwen

Tyddyn Taflod
Nant-yr-Onnen
Cedog

28 Tachwedd 2003

Annwyl Anwen,

Fedra i ddim dweud wrthat ti mor falch yr oeddwn o dderbyn dy lythyr. Rydan ni i gyd wedi bod yn poeni amdanat, a dwi'n difaru nad oeddwn i'n ddigon meddylgar i fod wedi helpu mwy arnat. A finna i fod yn ffrind gorau i ti.

Yn ystod y dyddiau cyntaf ar ôl i ti ddiflannu, wydden ni ddim beth oedd wedi digwydd yn iawn. Doedd wiw gofyn i Albert, ac roedd o wedi dwyn perswâd ar dy rieni i gadw'n ddistaw hefyd, ond llwyddais i gael dy gyfeiriad a rhywfaint o'th hanes gan dy fam yr wythnos diwetha.

Bellach mae Albert wedi'i berswadio'i hun mai arnom ni griw'r calendar mae'r bai am yr hyn y mae'n cyfeirio ato fel dy "anhwylder". Mae wedi lledaenu stori o amgylch y pentref mai mewn fferm iechyd wyt ti yn cael iachâd trwy orffwys llwyr ar ôl cael d'arwain ar gyfeiliorn gennym ni. Mae'r genod yn amau'n gryf nad yw'r stori'n wir bob gair, ond tydw i heb ddweud dim mwy wrthynt - dy ddewis di fydd hynny. Dwi'n falch dy fod ti'n dechrau teimlo'n fwy cadarn, ac yn edrych ymlaen yn fawr at dy weld cyn y Nadolig.

Mae'r pentref wedi bod yn ferw gwyllt drwyddo ers y lansiad. Mae newyddiadurwyr a gohebwyr o bob lliw a llun wedi bod yma'n tynnu lluniau ac yn cyfweld nifer ohonom. Mae pob copi o'r calendar wedi ei werthu, a'r argraffwyr wrthi fel lladd nadroedd yn trio cael rhagor allan cyn y Nadolig. Mae Cledwyn wedi derbyn llythyr sy'n dweud bod y cais am arian Loteri wedi bod yn llwyddiannus, felly rhwng hynny ac arian y calendar mae dyfodol y neuadd yn ddiogel. Ac i goroni'r cwbl, tydi Elsi heb dorri gair efo fi ers wythnosau. Nefoedd ar y ddaear!

Ond mae gen i ofn na tydi pethau ddim yn fêl i gyd. Mae Dewi wedi diflannu ers tua tair wythnos, wedi gadael Sioned heb ddweud gair, a heb unrhyw eglurhad. Ŵyr hi ddim ai mynd ynteu dod y mae hi ar hyn o bryd, y greadures. Mae arni ofn y bydd yr efeilliaid yn gweld bai arni hi - wedi'r cwbl, roeddynt yn gwpwl mor fodlon, mor rhadlon, tan y misoedd diwethaf 'ma. Beryg fod cynllun y calendar, a'r ffaith bod y plant wedi gadael am y brifysgol, wedi peri iddo gael rhyw greisis canol-oed, ac yntau'n un dwfn, ddim yn hawdd siarad â fo.

Mae gweddill y criw yn iawn yng nghanol yr holl brysurdeb, ac yn gofyn amdanat ti'n arw. Mae Mererid yn canlyn yn selog efo Harri erbyn hyn; mae Marian wrth ei bodd oherwydd bod Griff wedi derbyn swydd yn Nhresarn, ac mae Phyllis wrthi fel slecs gyda'r peiriant gwau. Dwi'n amau bod rhyw newid ar droed ym mywyd Brenda, er tydw i ddim yn siŵr sut fath. Mae hi wedi prynu tipyn o ddillad newydd eitha ffasiynol, ac wedi cael torri ei gwallt mewn steil mwy modern – ac mae hi'n sôn am gael *contact lenses*! Mae Tina'n rhedeg yr ysgol feithrin fwy neu lai ar ei phen ei hun ar hyn o bryd, ac yn dal ati gyda'r cwrs mae wedi cychwyn arno yng Ngholeg Tresarn. Mae Gwyneth ac Alison wedi dod yn ffrindiau mawr, gydag Alison yn treulio tipyn o'i hamser sbâr yn y Llew Coch yn helpu allan, sy'n gyfleus iawn gan nad yw'n bosibl i Tina ei dal hi 'mhob man, ac mae'r dafarn a'r siop wedi elwa o'r cyhoeddusrwydd sydd wedi dod i bentre bach Nant-yr-Onnen yn sgil y calendar. Mae Sara wedi penderfynu rhoi'r tŷ capel ar y farchnad a symud i dŷ llai, ac mae Caryl a Morris wedi gwahanu'n swyddogol, felly mae tŷ Caryl ar werth hefyd. Mae hi'n dweud – er mawr syndod i ni i gyd – ei bod hi am aros yn Nant-yr-Onnen oherwydd mai yma mae ei ffrindiau. Dipyn o aderyn y nos ydi hithau hefyd! Ond efallai bod rhai o'r helyntion 'ma wedi digwydd cyn i ti fynd yn sâl.

Tydan ni ddim wedi trefnu parti Nadolig ar gyfer y gangen eto – doedd Sioned ddim yn y cyfarfod mis diwetha, ac roedden ni am ddisgwyl nes cael gwybod sut oeddet tithau hefyd. Wedi'r cwbl, fedrwn ni ddim cael dathliad heb gael y criw i gyd at ei gilydd. Sut bynnag, hwyrach ein bod ni wedi dathlu digon am eleni, ac efallai mai disgwyl tan y flwyddyn newydd fyddai orau, i bawb gael cyfle i gael eu gwynt atynt. A chofia, mi *fyddi* di'n dod, ac mi *fydda* i'n eistedd wrth d'ymyl di – ac mi yfa i lemonêd efo chdi hefyd.

Canolbwyntio ar wella a chryfhau sydd ei angen arnat ti rŵan; rhaid i ti fod yn dy lawn iechyd cyn gwneud unrhyw benderfyniadau. Paid â phoeni gormod am Delyth ac Aled (er y gwnei, mi wn). Er eu bod nhw'n colli Mam, maen nhw'n cael eu difetha gan Nain a Taid, ac mi fyddi di'n ôl efo nhw mewn dim o dro.

Mae'r genod i gyd yn cofio atat ti, ac yn edrych ymlaen at dy weld di wedi i ti fendio'n iawn.

Cofion oddi wrth Ifan a'r plant.

Ta-ta tan toc.

Cariad,

Nia

xxx